STP 1358

Field Instrumentation for Soil and Rock

Gary N. Durham and W. Allen Marr, editors

ASTM Stock Number: STP 1358

ASTM
100 Barr Harbor Drive
West Conshohocken, PA 19428-2959

Field instrumentation for soil and rock / Gary N. Durham and W. Allen Marr, editors.
(STP ; 1358)
Papers from a symposium held in Atlanta, Ga., June 18-19, 1998.
Includes bibliographical references and index.
ISBN 0-8031-2604-2
1. Engineering geology--Instruments. 2. Soil-structure interaction. 3. Geophysical instruments. I. Durham, Gary N., 1942- II. Marr, W. Allen. III. ASTM special technical publication ; 1358.

TA705 .F48 1999
624.1'51'028--dc21

99-046576

Peer Review Policy

Each paper published in this volume was evaluated by two peer reviewers and at least one editor. The authors addressed all of the reviewers' comments to the satisfaction of both the technical editor(s) and the ASTM Committee on Publications.

To make technical information available as quickly as possible, the peer-reviewed papers in this publication were prepared "camera-ready" as submitted by the authors.

The quality of the papers in this publication reflects not only the obvious efforts of the authors and the technical editor(s), but also the work of the peer reviewers. In keeping with long standing publication practices, ASTM maintains the anonymity of the peer reviewers. The ASTM Committee on Publications acknowledges with appreciation their dedication and contribution of time and effort on behalf of ASTM.

Printed in Chelsea, MI
October 1999

Foreword

This publication, *Field Instrumentation for Soil and Rock*, contains papers presented at the symposium of the same name held in Atlanta, Georgia, on June 18–19, 1998. The symposium was held in conjunction with the June 14–17, 1998 standards development meetings of Committee D-18 on Soil and Rock and its Subcommittee D18.23 on Field Instrumentation, the symposium sponsor. The symposium was chaired by Gary N. Durham, Durham Geo Enterprises Inc., GA; W. Allen Marr, Geocomp Corporation, MA, was the co-chairman. They also both served as STP editors of this publication.

Contents

Instrumentation for Measuring Physical Properties in the Field

Overview

This overview summarizes the results of the Symposium on Field Instrumentation for Soil and Rock held in Atlanta, Georgia June 18–19, 1998 as part of Committee D18 scheduled meeting week. ASTM's subcommittee D 18.23 on Field Instrumentation sponsored the symposium. Major session topics include:

- Field Data Acquisition and Data Management
- Geotechnical Instrumentation of Landfills
- Instrumentation for Project Cost Management

The following is an overview of the information presented in this volume.

Instrumentation Associated with Soil Structure Interaction and Construction Monitoring

Three papers address field instrumentation associated with culverts and buried pipe. McGrath et al. presented the results of backfilling on pipe of various material and diameter. Unique methods were presented for instrumenting corrugated pipe. Webb et al. discussed instrumentation of concrete and corrugated steel arch culverts with spans of about 9 m. Various instruments were installed to monitor structural strain and deformation and culvert-soil interface pressure. The arches were backfilled and tested under live loads. Deformation measurements of the walls of the culverts were made with a custom-built laser device developed for this research. Yang et al. presented convincing evidence on earth pressure instrumentation of concrete box culverts beneath highway backfill that current ASSHTO guidelines underestimate earth pressures perhaps as must as 30%.

Barley et al. described instrumentation of strain gauged steel bars used in a soil nailed slope some 10–m high. Montannelli describes instrumentation of geogrid used in a reinforced slope. In situ pullout tests that were performed were in good agreement with scaled laboratory pullout tests. The study found good agreement between observed stress and deformations with those predicted from FEM modeling.

The paper by Chen et al. discussed the effectiveness of in situ instrumentation of diagnosing the pavement layer conditions under full scale accelerated traffic loading. Multi-depth deflectometers were used to measure both permanent deformations and transient deflections resulting from traffic loading and falling weight deflectometer tests.

Russel et al. presented a case history of monitoring preload performance for a highway alignment traversing deposits of organic silt and peat. Flentje and Chowdhury used a case history to point out the unique problems associated with instrumentation programs installed in urban areas.

Boston's Central Artery/Tunnel Project

Two papers were presented concerning the geotechnical instrumentation aspects of the Central Artery/Tunnel Project, Boston, Massachusetts. Bobrow and Vaghar discussed the Geographic Information System (GIS) that were developed for rapid analysis and reporting of instrumentation and survey data on the Central Artery/Tunnel Project. The paper by Hawkes and Marr discussed the role of automation and reviewed the data management system developed for processing large quantities of field data on a daily basis.

Instrumentation to Monitor Landfills

Four papers were specific to geotechnical instrumentation of landfills. The paper by Taylor et al. discussed landfill liner lead detection systems. Their work indicated that a below liner monitoring grid combined with above line surveys to pinpoint leaks accurately offer a successful approach to leak detection of lined waste sites. The paper by Thomann et al. presented a case history of the unique problems connected with installing geotechnical instruments within refuse fill and the underlying foundation to monitor pore pressures, lateral and vertical movements and temperature. They reported that the attrition rate of instruments in a refuse environment is about 10 to 15% per year; therefore, redundancy of instruments is mandatory to insure a successful program. Byle et al. discussed instrumentation of a landfill liner constructed of dredge spoil. Moo-Young et al. presented a method using thermistor and conductivity probes for measuring frost penetration that might deteriorate the permeability of the landfill's liner or cap. Another paper by Benson and Bosscher discussed results of their study indicating frost depth can be more reliably measured using electrical resistivity or dielectric constant rather than with thermistors.

Data Acquisition and Data Management

A paper by Marr described the application of recent developments in electronics and instrumentation to three geotechnical cases to illustrate some of the benefits of good automated field instrumentation systems. He also reviewed the considerations involved in deciding whether to automate a geotechnical instrumentation sytem. Welch and Fields described the automation of the field instrumentation associated with a large dam. Fiber optic cable was used extensively from the various sensors to eliminate voltage surges. The automation included satellite as well as land line remote transmission of the data. A related paper by Brokaw discussed the use of geographic information systems (GIS) coupled with global positioning satellite technology for field data acquisition and data management. Hansen furthered the overall review with a discussion of relating field instrumentation data that has been incorporated into GIS framework and the implications to other geospatial data.

Acknowledgments

The editors express grateful appreciation to the authors and the many engineers and scientists who provided peer review for all papers. The ASTM editorial staff deserves special credit for their support and encouragement throughout.

Gary N. Durham
Durham Geo-Enterprises Inc.,
Stone Mountain, GA 30087;
Symposium chairman and editor

W. Allen Marr
Geocomp Corporation, Boxborough,
MA 01719;
Symposium cochairman and editor

Instrumentation Associated with Soil Structure Interaction

Paulo L. Pinto[1], Brian Anderson[2], and Frank C. Townsend[3]

Comparison of Horizontal Load Transfer Curves for Laterally Loaded Piles from Strain Gages and Slope Inclinometer: A Case Study

---

REFERENCE: Pinto, P. L., Anderson, B., and Townsend, F. C., "**Comparison of Horizontal Load Transfer Curves From Strain Gages and Slope Inclinometer: A Case Study,**" *Field Instrumentation for Soil and Rock, ASTM STP 1358*, G.N. Durham and W.A. Marr, Eds., American Society for Testing and Materials, West Conshohocken, PA, 1999.

ABSTRACT: Laterally loaded deep foundations are commonly analyzed using the Winkler model with the soil-pile interaction modeled through nonlinear springs in the form of *p-y* curves. Computer programs such as FloridaPier and COM624P use default *p-y* curves when performing lateral analyses. These curves are based on input soil properties such as subgrade modulus, friction angle, undrained shear strength, etc. Soil properties must be deduced by laboratory testing or correlation to in situ test results. This is a source of uncertainty. In a few cases, lateral load tests are performed on instrumented piles, and the validity of such assumptions can be assessed. Test piles are commonly instrumented with strain gages and/or inclinometers. *P-y* curves can be back computed from these data, and the curves obtained with the two methods should agree closely. Results from a field test on a concrete pile are presented and the critical factors for the analysis are discussed. One added difficulty with concrete is its nonlinear behavior particularly near structural failure. The curves obtained are also compared with those developed from the Dilatometer/Cone Pressuremeter Test and the Standard Penetration Test.

KEYWORDS: piles, deep foundations, instrumentation, data reduction, lateral load test, P-y curves, strain gages, slope inclinometers

---

[1]Assistant Professor, Department Engenharia Civil, Universidade de Coimbra, 3004 Coimbra, Portugal.

[2]PhD candidate, Department of Civil Engineering, University of Florida, Gainesville, FL 32611-6580.

[3]Professor, Department of Civil Engineering, University of Florida, Gainesville, FL 32611-6580.

Introduction

The analysis of laterally loaded piles usually utilizes the concept of a beam on an elastic foundation represented by nonlinear curves of soil resistance versus horizontal deflection (*p-y* curves). It is a problem of soil-structure interaction, such that deflection of the pile depends on the soil response and the soil response depends on the pile deflection. A *p-y* curve represents the total resistance at a particular depth to the lateral displacement of a horizontally loaded pile.

In current practice, the load transfer curves are generated based on soil type, strength and deformability properties, and geometry of the pile as well as loading conditions. Several authors (Reese et al. 1974, 1975; O'Neill and Murchison 1983) have proposed curves, for different types of soil and loading conditions, based on a limited number of field tests. In these tests, the piles have been instrumented with strain gages or with inclinometers. It is of great interest to compare the *p-y* curves from lateral load tests on piles with both strain gages and inclinometers. The inclinometer has become the instrumentation of choice for concrete piles, for reasons that will be presented later.

Theory

P-y Curves From Strain Gage Data

A laterally loaded pile can be analyzed as a beam under composed bending (pure bending plus axial load). The governing equation is,

$$EI\frac{d^4y}{dx^4} + P_x\frac{d^2y}{dx^2} + E_s y = 0 \qquad (1)$$

where EI is the flexural rigidity of the pile, P_x is the axial load, E_s is the secant modulus of soil reaction (slope of *p-y* curve), y is the lateral deflection and x the depth. On most lateral load tests, the effects of the axial load may be neglected and the second term is dropped.

During a lateral load test, strains are measured at given depths, for each load increment. With the strains, the curvature, Φ, and the bending moment at the section are determined.

$$M = EI\frac{d^2y}{dx^2} = EI\Phi \qquad (2)$$

The horizontal displacements are obtained by double integration of the moment diagram, and the lateral earth pressures are computed by double differentiation. In order to obtain the horizontal displacements some boundary conditions, such as the deflection and slope at the top of the pile, have to be known. For a pile with linear elastic behavior the solution is straightforward. The integration and derivation of the moment distribution can be performed using several numerical techniques. Due to the scarcity and spacing of the data points, they are commonly fit using a polynomial or a spline. This continuous function is differentiated and integrated. The procedure is repeated for each load increment. At specific depths, the pressure from each load increment is plotted versus the corresponding deflection resulting in a *p-y* curve.

For a nonlinear analysis, the flexural rigidity of the pile has to be adjusted, depending on the strain level. The procedure is more elaborated and a correct characterization of the nonlinear behavior of the pile is crucial. Table 1 and Figure 1 show the nonlinear variation of stiffness with curvature for a prestressed concrete pile. It was computed by applying increasing loads at the top of a pile fixed at the tip and determining the corresponding curvature.

TABLE 1--*Nonlinear variation of flexural rigidity with curvature for test pile T1*

Moment (kN-m)	ε_{smax} (tract)	ε_{smin} (comp)	**Curvature** (1/m x 10^{-6})	**EI** (kN-m2 x10^3)
0			0.000	912.6
113	3.70E-05	-3.70E-05	0.124	911.6
226	7.40E-05	-7.40E-05	0.248	911.6
339	1.11E-04	-1.11E-04	0.372	910.1
452	1.50E-04	-1.53E-04	0.506	893.2
565	1.89E-04	-1.97E-04	0.647	873.8
678	2.28E-04	-2.42E-04	0.787	861.3
791	2.70E-04	-2.90E-04	0.940	841.8
904	3.34E-04	-3.50E-04	1.145	789.2
1017	5.40E-04	-4.61E-04	1.678	606.1
1073	8.78E-04	-5.62E-04	2.412	444.9
1130	1.03E-03	-6.24E-04	2.777	406.8
1186	1.20E-03	-6.90E-04	3.158	375.7
1243	1.36E-03	-7.56E-04	3.538	351.3
1356	4.23E-03	-1.22E-03	9.122	148.6

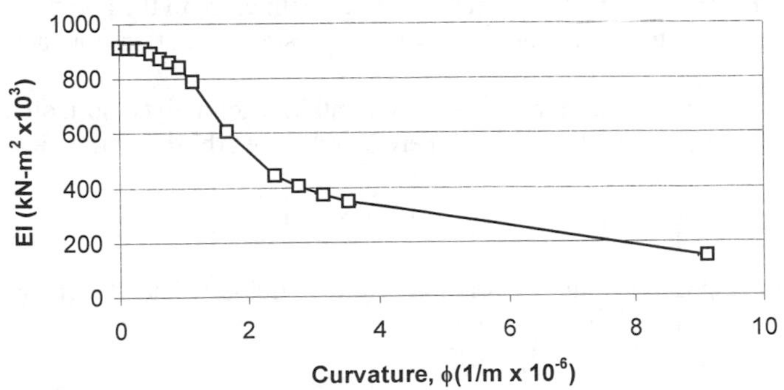

FIG. 1—*Graphical depiction of nonlinear variation of flexural rigidity with curvature for test pile T1*

The main advantage of using strain gages is that stresses and bending moments can be computed directly from the measured data. The *p-y* curves can be obtained without estimating soil properties. The major drawback is that the gages are often damaged during the installation of the piles and drilled shafts. Strain gages are sensitive equipment that can easily be damaged by improper handling.

P-y Curves From Slope Inclinometer Data

The use of slope inclinometers to measure lateral deflection of the pile has substantial advantages compared to the traditional strain gages. A grooved PVC pipe is inserted in the pile or drilled shaft and the variation from the original vertical position is measured using a slope inclinometer. It is quite economical since the only material not recovered is the PVC pipe. When used in concrete piles/shafts it avoids the difficult and expensive task of installing strain gages. The main disadvantage is that data reduction requires more judgement. A direct computation of soil pressure is hindered by numerical errors, because the slope distribution would have to be differentiated three times.

A method based on the assumption of a predefined *p-y* curve and on the optimization of the soil parameters that match the deflected shape has been proposed (Brown et al 1994), with good results. The disadvantage of such an approach on the calibration of *p-y* curves is the assumption of the shape of the curve. Additional effort is also required for the iteration over two or more soil parameters (angle of internal friction and modulus of lateral soil reaction for sands, or undrained shear strength and strain at 50% of strength for clays).

The authors have used a simpler approach, by iterating on a single parameter, the modulus of lateral soil reaction, k (F/L^3). For uniform sands, the initial slope of the *p-y* curves increases linearly with depth, z, and is equal to

$$Es(z) = k\ z \tag{3}$$

Once the pile is loaded and deformation occurs, the secant modulus of the *p-y* curve will no longer have a linear variation as shown in figure 2. It is reasonable to consider that if the pile is flexible, the horizontal displacements at a depth close to the pile tip will be very small. In this case the secant modulus will have a small variation, for each load increment.

The variation of the secant modulus with depth (z), for a given horizontal load, is approximated to a parabola with equation where a and b are arbitrary coefficients:

$$Es(z) = az^2 + bz \tag{4}$$

Note that this corresponds to a linear variation of the modulus of lateral soil reaction, k, with depth given by:

$$k = \frac{dEs}{dz} = 2az + b \tag{5}$$

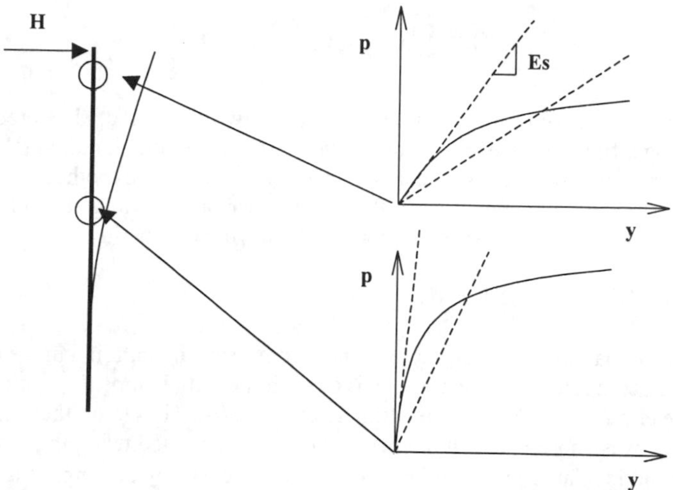

FIG. 2—*Evolution of secant modulus with depth and deformation*

If the secant modulus at the pile tip Es^{max}, is assumed constant for all loads, the parameter a is defined and k will depend solely on the value of the modulus of lateral soil reaction at the surface, b (figure 3).

$$a = \frac{\left(E_s^{max} - bz_{max}\right)}{z_{max}^2} \tag{6}$$

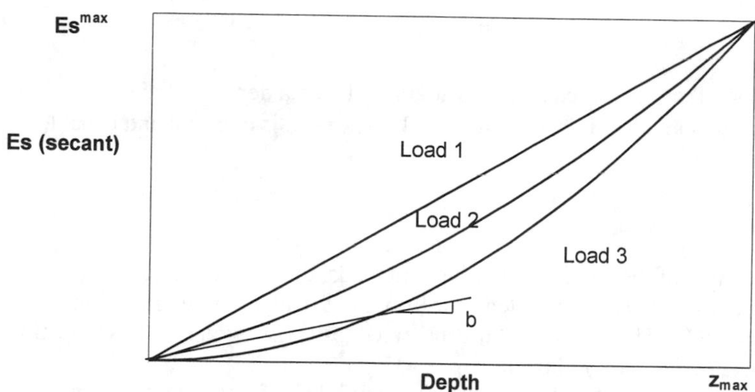

FIG. 3—*Assumed evolution of secant modulus with depth and load level*

The parameter b is iterated upon, until good agreement of the deflected shapes is found. The measure of the error is given by

$$\varepsilon_d = \sum_{i=1}^{n} \left| y_{predicted}(i) - y_{measured}(i) \right| \tag{7}$$

where n is the number of points where the deflection was evaluated. The prediction is obtained using a finite-difference technique where the governing equation (Eq 1) is solved for particular points along the depth. This is similar to the method used in the program COM624 (Wang and Reese 1991) but was performed using a simple spreadsheet. The same spreadsheet is capable of carrying out the iterative analysis.

Curves from DMT and Cone Pressuremeter

As a comparison to the curves obtained from instrumentation, curves based on in situ tests were computed. A composed curved with the initial slope, K_1, estimated from the dilatometer modulus (E_d) and the ultimate pressure (p_{lim}) based on the cone pressuremeter was chosen. The dilatometer modulus is obtained using the Marchetti dilatometer and is related to the initial slope of the *p-y* curve by an empirical factor F_ϕ, varying from 2 to 3.

$$K_1 = F_\phi \; E_d \tag{8}$$

The combined curve is linear up to a horizontal displacement of 1.1-mm (DMT B reading) and hyperbolic from then on. The ultimate *p* is estimated using the cone pressuremeter limit pressure multiplied by the pile width, D, and an empirical factor α (Robertson et al, 1985) and is given by:

$$p_{ult} = \alpha \; p_{\lim} \; D \tag{9}$$

Robertson and co-authors proposed a linear variation of α to a depth of 4 diameters. At the surface α is zero and at 4D it is equal to 1.5. Thereafter it is constant and equal to 1.5.

Results

As an example of the application of both methods, the authors selected a pile tested as part of a large (4x4) group tested near Roosevelt Bridge, in Stuart, Florida (Townsend et al. 1997). This was a well instrumented test, with extensive in situ testing performed to characterize the soil layers.

The prestressed concrete pile was 16.2 m long (2.4 m of free length, above the mud line) and had a square section of 0.76 x 0.76-m. (f_c'=41 MPa, Ec = 30.4 GPa). It was located at the right corner of the trail row (pile 1 in figure 4). The pile was instrumented with a series of nine full bridges of strain gages. A slope inclinometer guide pipe was installed in the center of the pile, to a depth of 9.0 m from the ground surface. The reduced data from both the strain gages and slope inclinometer are shown in figures 5and 6, respectively.

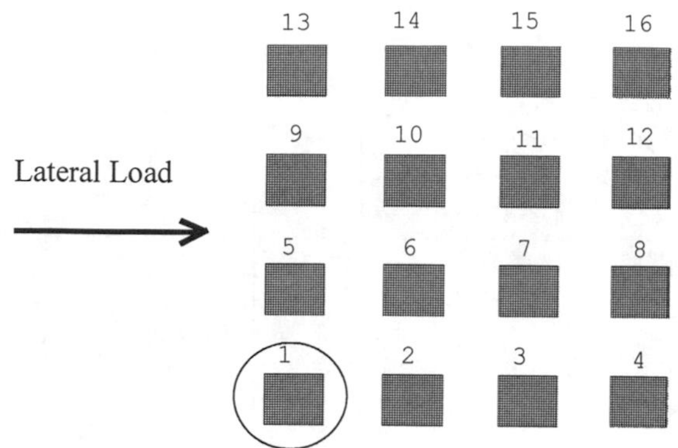

FIG. 4—*Location of piles in test group*

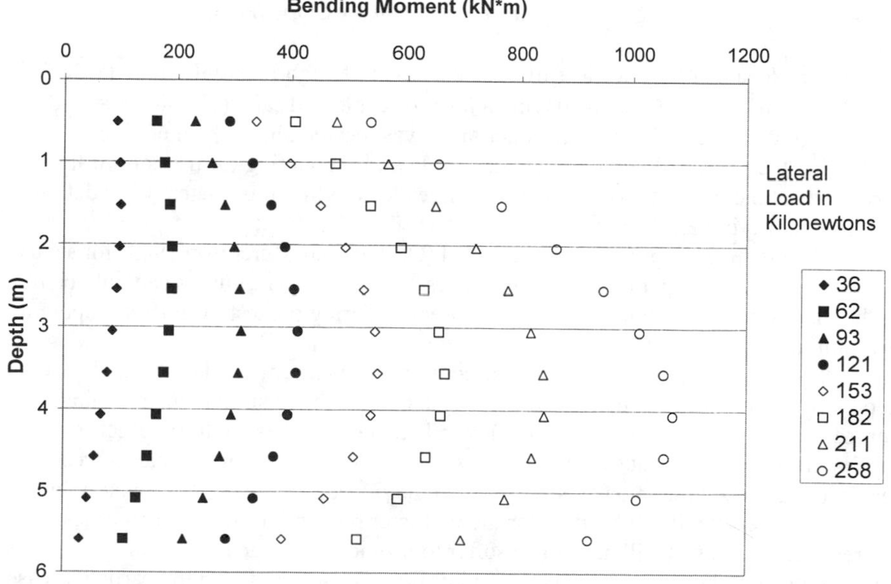

FIG. 5—*Bending Moments from Strain Gage Readings*

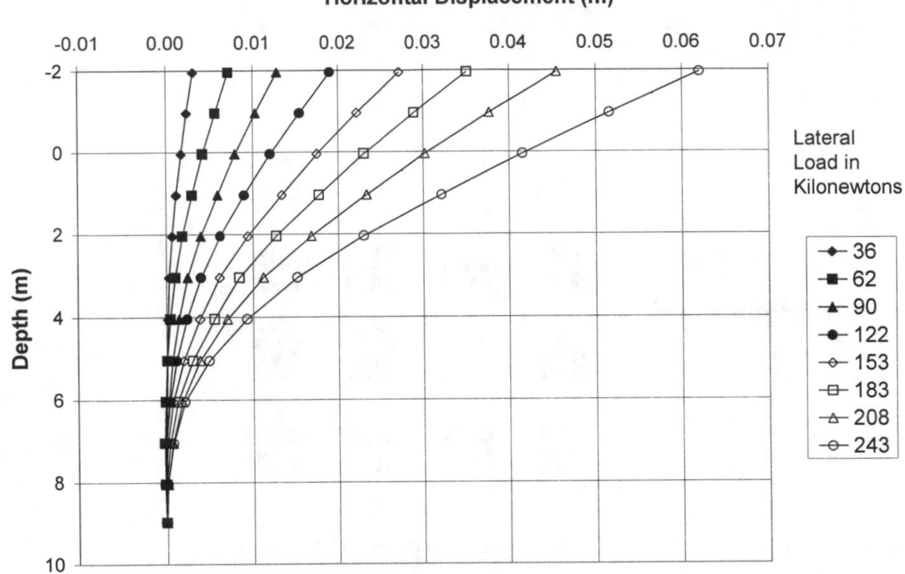

FIG. 6—*Displacements from the Slope Inclinometer*

The local geology consists of a surface layer of slightly silty fine sand (ϕ' =35° γ'=10.4 kN/m^3), 4.0 m thick overlaying a layer of cemented sand (ϕ' = 44° γ'=12.0 kN/m^3), 10.0 m thick. The ground water table was located above the mud line. Following the steps outlined for strain gage data and accounting for the nonlinear properties of the concrete section, the *p-y* curves for the first four meters were determined as those shown in Fig. 7.

The shape is hyperbolic, as suggested (O'Neill and Murchison 1983) for sands, and the limiting force per unit length increases with depth. The curves can only be traced to the maximum horizontal displacement observed during the test. Proper extrapolation is required for larger displacements.

As shown in Fig. 8, the method used with the inclinometer data yielded curves in good agreement with the strain gages. A key factor to obtain similar and meaningful curves is the correct assessment of the flexural stiffness. It has contrary effects on the results, depending on which data are analyzed. A pile stiffness higher than correct will result in higher bending moments, shifting up the *p-y* curves obtained from strain gage data. The inverse will be observed for inclinometer results. If the pile stiffness is overestimated, the soil will have to be softer to provide the same horizontal displacements. Therefore, the curves will be shifted downward. On the particular case of concrete piles, loaded past the moment that causes the concrete to crack, the proper modeling of the non-linear moment curvature relation is critical. The inclusion of nonlinear pile behavior marginally increases the complexity in the data reduction from the strain gages. That is not the case for the inclinometer, where the iterative loop would have to consider the curvature of each section and the bending stiffness would be

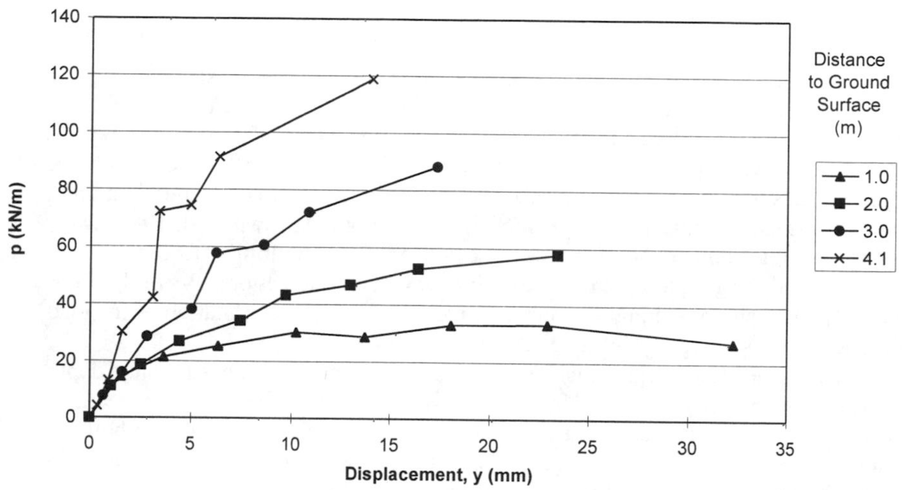

FIG. 7--*Curves from Strain Gages*

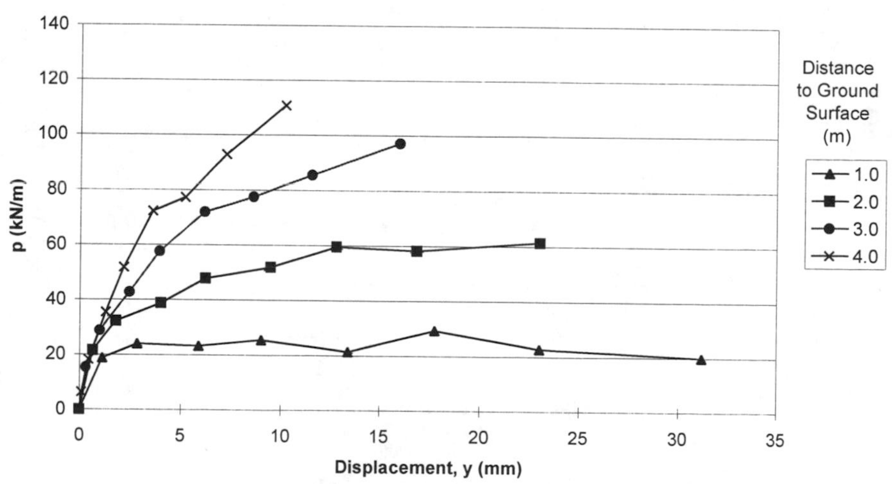

FIG. 8--*Curves from Inclinometer*

computed for each iteration. Although not extraordinarily difficult, that would limit the benefit of the simple procedure outlined and implemented on a spreadsheet.

Curves Based on In situ Testing

The fact that the pile was included in a group complicates the scope of the problem. Ideally, a single pile would have been chosen, but the data available for concrete piles is very scarce and few tests have been conducted with both strain gages and inclinometers. For the purpose of comparison of the curves with the curves based on the in situ tests the situation is not ideal. The fact that the pile was included in a group greatly influences the pile-soil interaction. A *p-y* multiplier (Brown et al. 1988) was required to allow comparison of curves. The concept of the multiplier forces the curve to shift down, reducing capacity, to account for the pile-soil interaction. *P-y* multipliers are dependent on position within the group, relative to the point of application of the load, pile spacing and soil type. (McVay et al. 1995)

The curves calculated using DMT and pressuremeter data measured at a depth of 2 meters are displayed in Fig. 9. The DMT curves seem to have slightly lower soil resistance but are still in reasonable agreement with the curves obtained from both inclinometers and strain gages. A multiplier of 0.3 (70% of reduction in the curve) had been back computed based on the overall group performance.

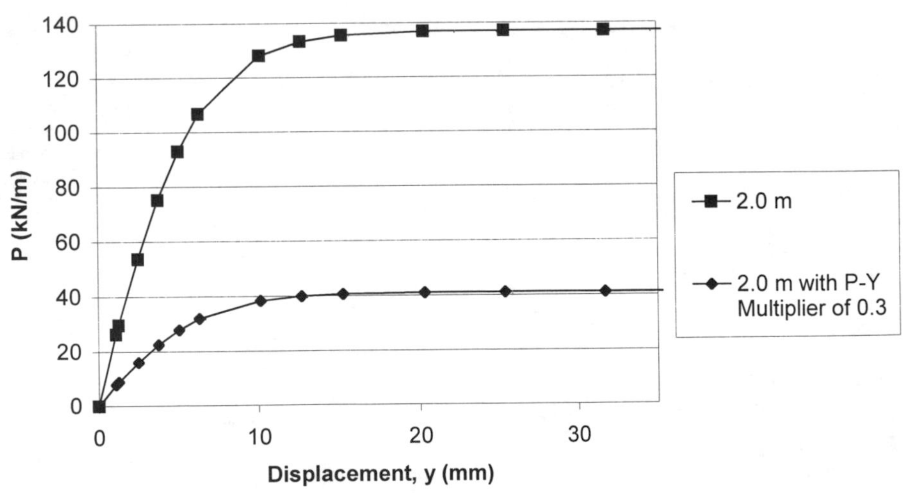

FIG. 9--*Curves from Dilatometer and Pressuremeter*

If Oneill Sand P-y Curves (O'Neill, 1983) are used with ϕ'=35° and γ'=10.2 kN/m^3 estimated from SPT tests and a multiplier of 0.3, the resulting curves are plotted in

Fig. 10. The resulting p-y relationship is stiffer than the prior and would have resulted in smaller displacements than observed in the test.

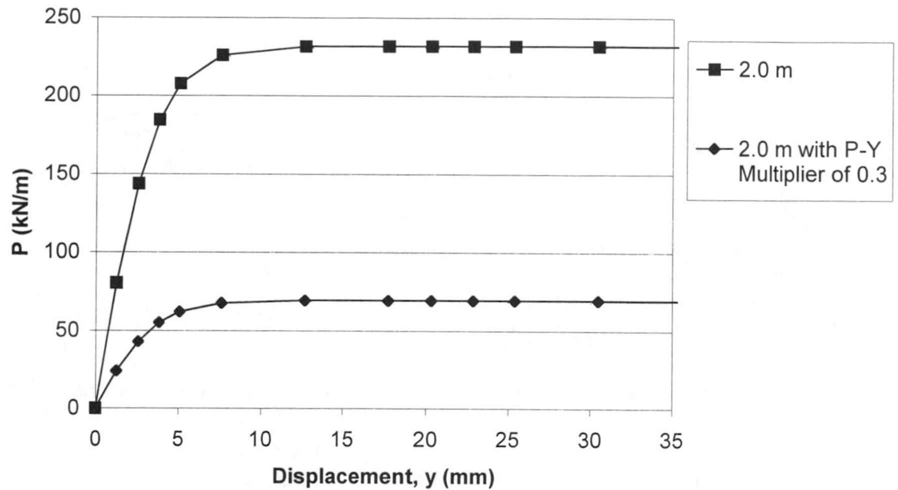

FIG. 10—*Theoretical curves based on SPT*

In order to evaluate the *p-y* curves obtained from instrumentation, a program with full nonlinear analysis capability was required. The program FloridaPier (McVay et al. 1998) was selected for that purpose. User defined curves, based on strain gage and inclinometer data, was used to a depth of 4.5 m. From that point to the pile tip, the soil springs were modeled using O'Neill curve for sands with ϕ'=15°, k=16,290 kN/m^3. These low values correspond to the application of a multiplier of 0.3. The predicted pile head displacements match well with the values observed during the load test (Fig. 11). The pile behaves almost linear until the moment caused by a horizontal load of 200 kN cracks the concrete section.

Conclusions

Lateral load tests in deep foundations are important to access the validity of *p-y* curves used in design. Two methods used to determine the *p-y* curves using strain gage and inclinometer test data were presented and discussed. A simple method to reduce the inclinometer data, for cohesionless uniform soils, was proposed and the resulting curves compared well with the curves deduced from strain gage data. The scarcity of lateral load tests on concrete piles/shafts, instrumented with strain gages and slope inclinometers prevented the extension of the current comparison to a larger database.

Reinforced concrete piles loaded close to structural failure exhibit nonlinear behavior. The consideration of this nonlinear behavior is important to interpret the test

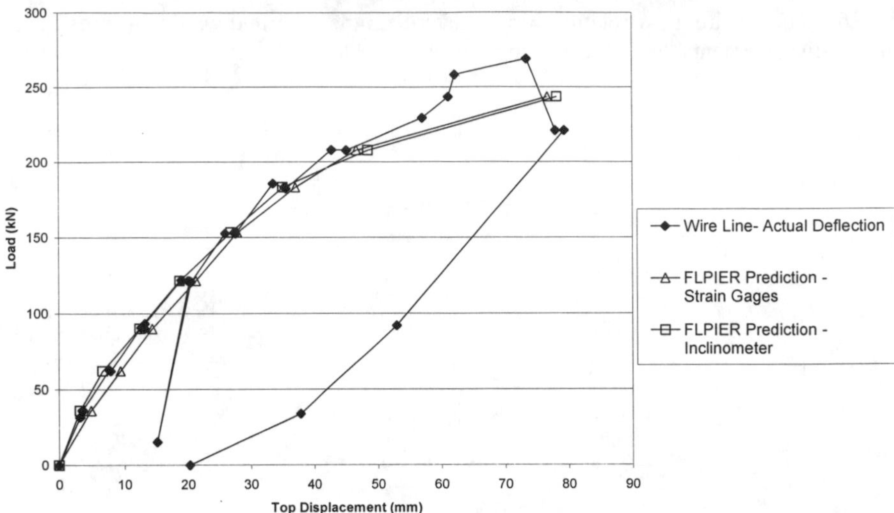

FIG. 11 -- *Pile Head Deflection, comparison between test results and predictions*

results correctly. The poor evaluation of the bending stiffness will cause the p-y curves from strain gages and inclinometer to move in opposite directions.

For the current case, where the pile was included in a group, comparison of the curves based on instrumentation with curves based on in situ testing required the use of *p-y* multipliers. If a value of 0.3 is used, the curves agree reasonably. When the instrumentation curves are used in a full nonlinear analysis, Figure 11 shows there is an excellent match of pile head deflection with the behavior observed in the load test.

References

Brown, D. A., Morrison, C. and Reese, L., 1988, "Lateral Load Behavior of Pile Groups in Sand," ASCE, *Journal of Geotechnical Engineering*, Vol. 114, No. 11, pp. 1261-1276.

Brown, D. A., Hidden, S. A. and Zhang, S., June 1994, "Determination of P-Y Curves Using Inclinometer Data," *Geotechnical Testing Journal*, Vol. 17, No. 2, pp. 150 –158.

McVay, M. C., Casper, R. and Shang, T., 1995, "Lateral Response of Three- Row Groups in Loose to Dense Sands at 3D and 5D Pile Spacing," ASCE, *Journal of Geotechnical Engineering*, Vol. 121, No. 5, pp. 436-441.

McVay, M., Hays, C. and Hoit, H., 1998, "User's Manual for FloridaPier for NT Version 1.22," Department of Civil Engineering, University of Florida, Gainesville.

O'Neill, M. W. and Murchison, J. M., 1983, "An Evaluation of P-y Relationships in Sands," Research Report No. GT-DF02-83, Department of Civil Engineering, University of Houston, Houston, Texas.

Reese, L.C., Cox, W.R. and Koop, F.D., 1974, "Analysis of Laterally Loaded Piles in Sand," Paper No. OTC 2080, *Proceedings*, Sixth Annual Offshore Technology Conference, Houston, TX.

Reese, L.C., Cox, W.R. and Koop, F.D., 1975, "Field Testing and Analysis of Laterally Loaded Piles in Stiff Clay," Paper No. OTC 2312, *Proceedings*, Seventh Annual Offshore Technology Conference, Houston, TX.

Robertson, P. K., Campanella, R. G., Brown, P. T., Grof, I. and Hughes, J. M., 1985, "Design of Axially and Laterally Loaded Piles Using In situ Tests: A Case History," *Canadian Geotechnical Journal*, Vol. 22, No. 4, pp. 518-527.

Townsend, F. C., McVay, M. C., Ruesta, P. and Hoyt, L., 1997, "Prediction and Evaluation of a Laterally Loaded Pile Group at Roosevelt Bridge," Report No. WPI 0510663, Department of Civil Engineering, University of Florida, Gainesville, FL.

Wang, S. T. and Reese, L. C., 1991, "Analysis of Piles Under Lateral Load – Computer Program COM624P for the Microcomputer," Report No. FHWA-SA-91-002, U.S. Department of Transportation, Federal Highway Administration, Washington, DC.

Filippo Montanelli,[1] Piergiorgio Recalcati,[1] and Pietro Rimoldi[1]

An instrumented geogrid reinforced slope in central Italy: field measurements and FEM analysis results

---

Reference: Montanelli, F., Recalcati, P., Rimoldi, P., "**An instrumented geogrid reinforced slope in central Italy: field measurements and FEM analysis results,**" *Field Instrumentation for Soil and Rock, ASTM STP 1358*, G.N. Durham and W.A. Marr, Eds., American Society for Testing and Materials, West Conshohocken, PA, 1999.

Abstract: A 15 m high green faced reinforced slope was built in 1996 to stabilise a landslide situated on the Montone hill in the province of Perugia (Italy). The reinforced slope was built using locally available soils as fill material and HDPE mono-oriented extruded geogrids as reinforcement. The reinforced slope has been instrumented with strain gages connected to the reinforcing geogrids and with total pressure cells; full scale in situ pull-out tests have been performed as well.

The preliminary results demonstrate the good performance of the geogrids and define possible failure mechanisms for slopes with stepped geometry. The in situ pull-out test results validate laboratory tests performed in the past. To evaluate the field stress and deformation behavior of the reinforced slope a finite element analysis has been carried out. The paper describes the model technique developed to evaluate the field stress and deformation of steep reinforced slope using special interface elements. The results of finite element analysis are in agreement with the field measured results.

Keywords: slope reinforcement, field instrumentation, geogrid, FEM analysis

---

Introduction

The Montone hill, located in the province of Perugia (Central Italy), shows important instability phenomena along its slopes, resulting from a complex evolutive situation. Many factors have an important role, including the lithology, the structural

[1]TENAX SpA, Viganò, LC, Italy.

arrangement of the geological units, the still active neotectonic phenomena. The hills also suffered from hydrological and hydrogeological conditions, and the anthropic modification suffered by the territory during the last decades.

During the last decades few landslides concerned the North-East slope of Montone hill, particularly along the Montone-Pietralunga main road. These phenomena involved an area about 15 m high and 200 m long; in the proximity of the Fosso Fornaci a road failure occurred (Fig.1).

In one of the sub-projects for the stabilisation of the hill, along the Montone-Perugia main road, the use of geosynthetics was foreseen for different functions, and more precisely:

- soil reinforcement to build reinforced slopes (using mono-oriented HDPE geogrids);
- erosion control to protect the slopes affected by erosive phenomena (using geocells, polypropylene geomats and biomats composed of coconut and straw fibres);
- separation and filtration for the construction of drainage trenches (using nonwoven geotextiles).

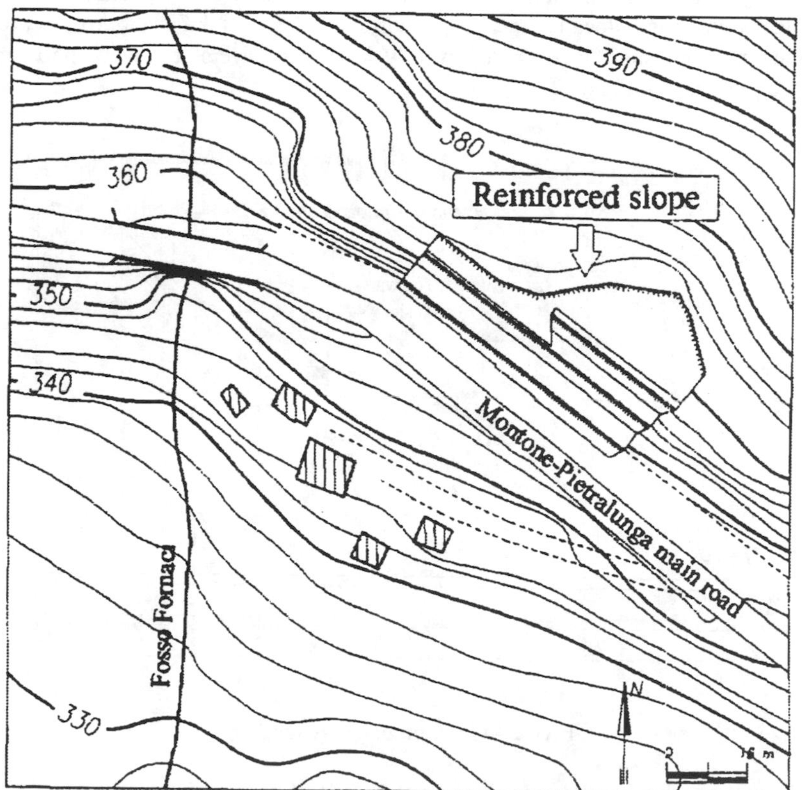

FIG.1 - *Reinforced slope location.*

The reinforced slope presents a maximum height of 15.5 m and a horizontal length of about 53 m (see Figure 1). The geogrid reinforced steep slope was divided in 3 steps of about 5 m each. The instrumented section was equipped with 40 strain gages (electrical wire resistance type) bonded on the geogrid ribs of the first and the second reinforced block; 3 total vertical pressure cells positioned at the bottom of the first reinforced block; 2 vertical inclinometers installed on the first block and on the top of the third block; 1 pore pressure cell and 1 acquisition system installed on top of the first block.

Field pull-out tests have been performed up to failure on dummy geogrid layers purposely embedded into the bottom step of the slope.

Geotechnical characteristics

Due to the extension of the area and the complexity of the problems, a large scale geophysical and hydrological test campaign has been carried out, with boreholes, piezometric and penetration tests, plus extensive geotechnical laboratory testing (Coluzzi et al., 1997). The area involved in the works is characterised by the presence of a silty soil of lacustrine origin, locally emerging in correspondence of the instability phenomena. Along the main landslide surfaces, yellow silty sands with layers of gravel, up to 1 m thick, are present (see Figure 2).

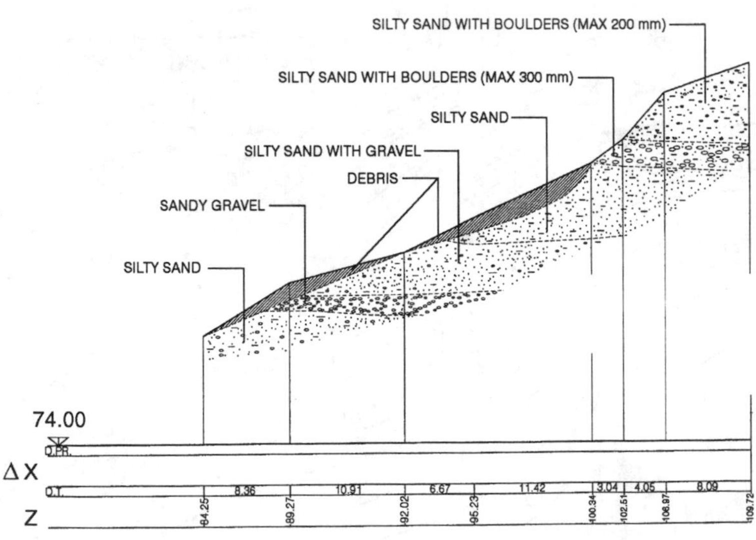

FIG. 2 - *Geological cross section.*

The water table in this area is located at 2 m depth from the ground level. A geological cross section has been prepared for the main landslide as shown in Figure 2.

Through traditional slope stability analysis, a sliding-rotational failure mechanism was identified in this section very close to the ground surface (approximately at 2.5 m depth). This failure was due both to the slope cutting operations for the construction of the road and to the loss of mechanical characteristics in the above mentioned soils due to the saturation under the water table level. Old landslide bodies have been found in the area by site investigation. The geotechnical parameters for the surface colluvial layer of soil and for the deeper subgrade are reported in Table 1.

TABLE 1 - *Geotechnical characteristics of the soils involved in the landslide.*

Soil	Surface colluvial layer	Deeper subgrade
Unit weight [kN/m^3]	21.2	20.3 - 21.7
Friction angle [°]	19.7	22.4 - 27.5
Undrained cohesion [kPa]	0	8.3 - 95
Cohesion [kPa]	-	37 - 69
Water content [%]	-	15.7 - 19.9

A back-analysis of the global stability using Janbu method has been performed on the failed slope: considering that Montone is in a highly seismic area and therefore applying a seismic load corresponding to an acceleration ratio a/g equal to 0.07, and considering the in-situ soil with a unit weight of 19.5 kN/m³ and a cohesion of 0 kPa, an overall friction angle of 26° was found. This value was in very good agreement with the geotechnical data of the tested soils. For both economical and practical reason (access to the site, easiness and speed of construction), it was decided to stabilise the slope by rebuilding it with the local soil reinforced with HDPE geogrids.

Design and construction of the reinforced slope

The problems to be faced were the presence of a soil with poor geotechnical characteristics and the presence of a very high water table.

Therefore, for a running length of 50 m, the body of the landslide has been completely removed (Figure 3). The existing ground surface has been modelled to a maximum height of about 15 m, in three 5 m high blocks, at 60° slope, separated by two berms, 3.7 and 4.5 m wide (Figure 4). After excavation and before reconstructing the slope, the exposed soil was protected from rainfall with a light polyethylene membrane.

A drainage layer made up of about 0.5 m gravel, separated from the existing surface by a 800 g/m^2 nonwoven geotextile, has been placed beneath the reinforced blocks. This drainage layer was connected to a water removal system consisting in steel pipes embedded in a draining trench (as shown in Figure 4).

FIG.3 - *The landslide body is removed, and the ground surface modelled.*

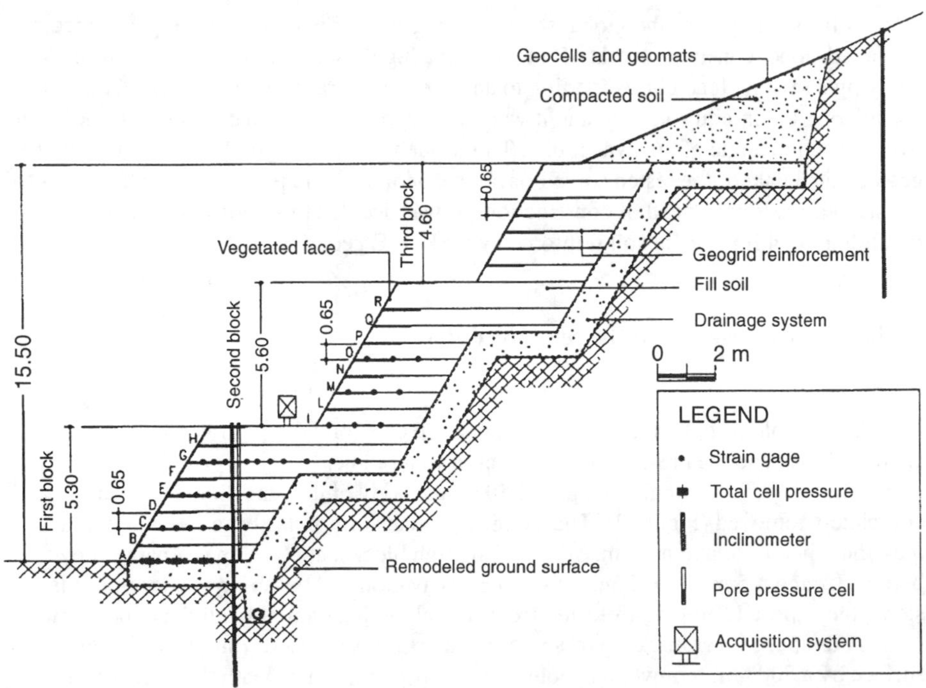

FIG.4 - *Cross section of the instrumented geogrid reinforced slope.*

The steep reinforced slope has been designed according to the Jewell method; (Jewell, 1991) from calculations, Tenax TT060 SAMP HDPE geogrids, having 60 kN/m ultimate (peak) tensile strength and 25 kN/m Long Term Design Strength have been used. (ICITE N. 508/98, 1998) The main geogrid properties are reported in Table 2.

A global stability analysis on the reinforced section (with the geotechnical data previously reported), using the STABGM code, has been performed. This software, based on Bishop's modified method (Duncan et al. 1985), determines the circular surfaces, passing through a specified point or tangent to horizontal surfaces, whose Factor of Safety (FS) is minimum.

The program takes into account the presence of reinforcement layers in the determination of the resisting moment, by multiplying the tensile strengths of the reinforcements by their arms. The complexity of the geometry verifies not only the circular surfaces passing through the toe of the first step, but also through the toe of the upper steps and through other points along the face of the first step of the slope.

The same analysis was performed with surfaces tangent to deeper horizontal planes.

TABLE 2 - *Main properties of the geogrids used in the project.*

Properties	Value	Unit	Test method
Polymer type	HDPE		
Structure	Extruded mono-oriented		
Unit weight	400	g/m^2	ISO 9684
Peak tensile strength	60.0	kN/m	GRI-GG1
Strain at peak	13.0	%	GRI-GG1
Strength @ 2% strain	17.0	kN/m	GRI-GG1
Strength @ 5% strain	32.0	kN/m	GRI-GG1
Long Term Design Strength	25.0	kN/m	ASTM D5262

The results of the analysis are presented in Figure 5: the failure mechanism with the lowest Factor of Safety seems to encompass several failure surfaces; in fact, low Factors of Safety are found both for the circular surface tangent to a horizontal layer (FS=1.289), and for circular surfaces passing through the toe (FS=1.264) and through the external end of the third geogrid (FS=1.355). These considerations are supported by the instrumentation results, as shown later.

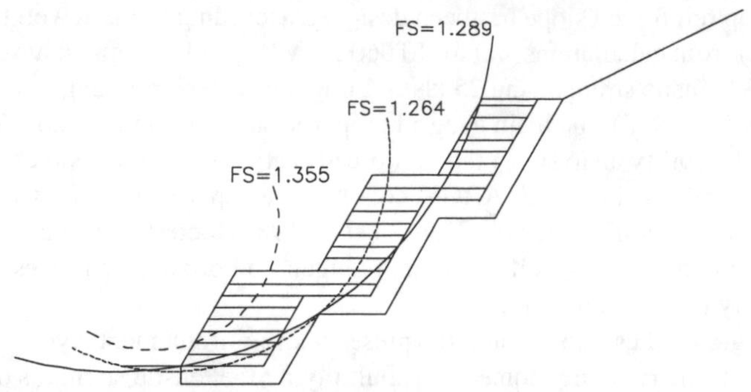

FIG. 5 - *Global stability analysis*

The steep reinforced slope has been built using the Tenax RIVEL System (Rimoldi and Jaecklin, 1996) (Figure 6). It consists in the use of sacrificial steel mesh formworks that help in the construction of the face slope and obtaining an uniform geometry of the slope; moreover the time necessary for the construction and hence the costs of it are quite low. It is possible to create steep slopes, completely vegetated thanks to the use of biomats, which provided a very good medium for preventing the washout of the soil and for the support of the growing plants. The vegetation of the face was further enhanced by hydroseeding the face at the end of the construction of the reinforced soil structure.

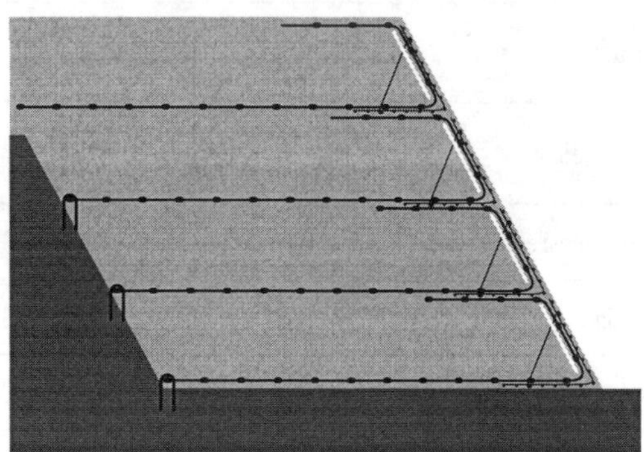

FIG. 6 - *Tenax RIVEL System*

To build the reinforced slope it was necessary first of all to prepare the geogrid layers cut at the required length, and to bend a steel mesh sheet at the required angle. These sacrificial "formworks" were lined at the face of the slope, overlapped for about 50 mm and jointed by steel wires.

The reinforcing geogrids were placed and anchored by means of "U" shaped staples with an overlapping of two ribs. A biomat was placed on the internal side of the geogrid, in the wrapping-around of the slope face. Hooked bars were fixed to the formworks (one every 500 mm) to avoid outward bending during soil compaction (Fig.7).

Following the common practice for geogrid reinforced slopes, the soil was laid and spread on the geogrid in 300 mm lifts and compacted by using a hand roller near the face of the slope (within 1 m), while the rest of the soil (more than 1 m away from the face) was compacted with a heavy roller compactor (Fig.8). Finally the geogrid was wrapped around the face, stretched and fixed with "U" shaped steel bars.

When the work was completed, the face was hydroseeded to encourage fast vegetation growth.

The Tenax RIVEL System (Fig.9) allowed more than 50 m² of face area per working day to be built, with a typical crew of 4 people equipped with 1 excavator, 1 roller and 1 hand compactor.

FIG. 7 - *Steel mesh formworks, hooked bars and biomats.*

FIG. 8 - *Compaction with a heavy duty roller.*

FIG. 9 - *View of the reinforced slope before hydroseeding.*

Instrumentation

Due to the importance of the project, the slope has been instrumented, in order to verify the long term behaviour of the structure. In particular, strain gages and total pressure cells data are hereby reported.

Strain gages

The dimension of the geogrid ribs required the use of strain gages with very small dimensions (3.18 mm x 2.54 mm). Figure 10 shows a typical Tenax TT 060 SAMP geogrid rib with a glued strain gage. Similar specimens have been used for calibration of the stress strain curve in the laboratory by either static or dynamic tensile testing.

The strain gages were completely encapsulated in polyamide resin for protecting them and to allow for their use in soils. The strain gages are self-temperature compensated within a range of -20°C up to 60°C and with a strain limit up to 5% of the gage length.

The installation of this kind of gages on the geogrid ribs was particularly difficult due to their dimension and their fragility. The strain gages have been electrically connected with a 3 wires quarter bridge system to an automated data acquisition system.

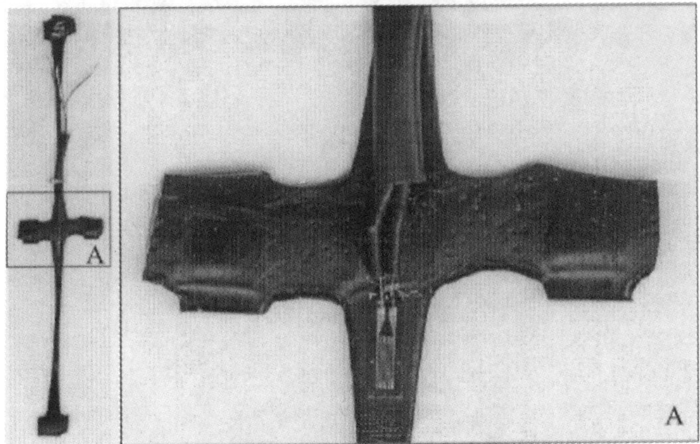

FIG. 10 - *Tenax TT 060 SAMP geogrid rib with a glued strain gage.*

First of all, the geogrid to be instrumented was prepared in the laboratory by making it perfectly flat; it was cleaned with alcohol, then the gage area was abraded with 180 and 400 silicon-carbide paper to increase the roughness. A light flame was quickly passed over the rib to eliminate any residual powder from the abrasion and to increase the number of radicals. A neutraliser was applied to the gage area, keeping the surface wet by scrubbing it with a cotton tipped applicator not to allow evaporation. Immediately after applying the neutraliser, the residual was removed with a gauze sponge. Then, the gages were glued using a cyanoacrylate adhesive to the ribs, and cured in an oven at 40° for at least 8 hours. The strain gages have been coated with a layer of silicon rubber to protect and waterproof them.

In particular, it was very difficult to apply a thin and uniform adhesive layer (due to the non planarity of the surface); then, taking into account the fact that the strain gages had to be glued on a 10 m long geogrid layer, it was difficult to handle it in all phases (from preparation to transportation on site and installation) with the necessary care. The number of "dead" strain gages (9 out of 37) can therefore be considered a success.

The instrumented geogrids were (from the bottom) the first, third and fifth layers of the first step, with gages at 0.5, 1.0, 1.5, 2.0, 2.5, 3.0 and 3.5 m from the face; the 7th geogrid with gages at 0.5, 1.0, 1.5, 2.0, 2.5, 3.0, 3.5, 4.5, 5.5 and 6.5 m from the face; the first and fifth geogrids of the second step, with gages at 0.5, 1.0 and 1.5 m from face (Fig.11).

The geogrid layers have been installed in tension and the strain gages and cables covered with 150 mm of fine sand.

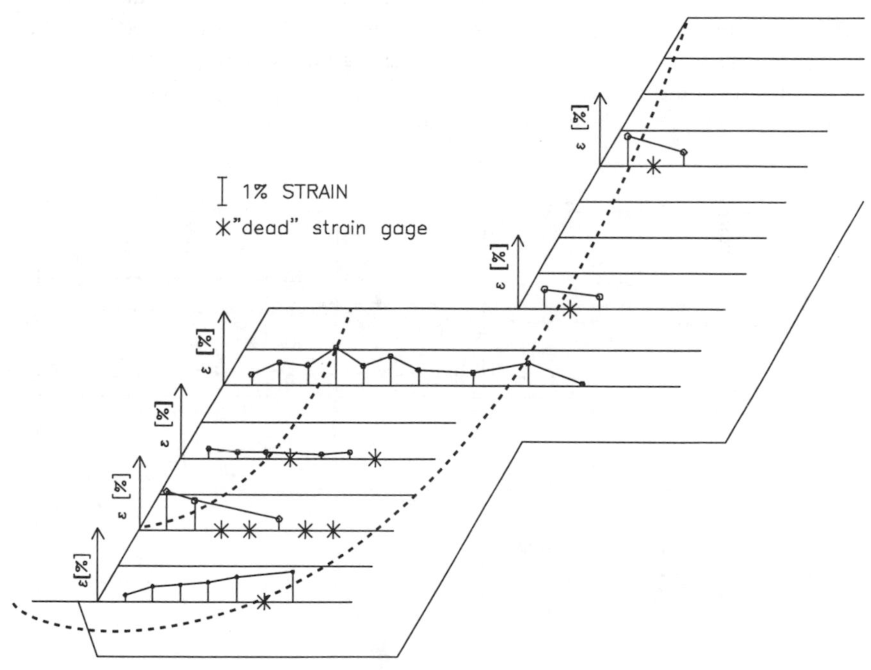

FIG. 11 - *Gages outputs, 90 days after the installation.*

The outputs of the strain gages are shown in the same Figure 11, 90 days after the installation.

From the analysis of these preliminary data, it is possible to see the presence of strain peaks in the strain envelopes, corresponding to a stress concentration in the geogrids. Analysing the 7th geogrid, it is possible to see an important peak (about 1.7% strain) at 2.0 m distance from the face which, together with the peak at the face point of the third geogrid, seems to be justified by the circular surface found during the global stability analysis and passing through the toe of the third geogrid layer (see Figure 5). The second peak of the 7th geogrid, at 4.5 m from the face, together with the peak of the first geogrid (at 3.5 m from the face) could belong to the deep seated surface (interesting also the upper step), which provides the lowest FS in Figure 5. The loss of 9 strain gages unfortunately does not allow a more accurate interpretation, but the result of this instrumentation is considered much more than satisfying. Some of the relatively high strains recorded near the face may be a result of the contractor using the heavy duty roller too close to the slope face and the consequent face bulging and geogrid tension.

Total pressure cells

Three total pressure cells where placed at the bottom of the first reinforced step, respectively at 1.0, 2.0 and 3.0 m from the face (Fig.4), in order to verify the actual distribution of the vertical load on the base. The pressure cells made available were the vibrating wire type with 300 mm diameter, with a standard working range from 0 to 2070 kPa.

Installation of the cells consist simply in laying down horizontally the cell, and covering it with a thin sand layer before spreading the fill soil.

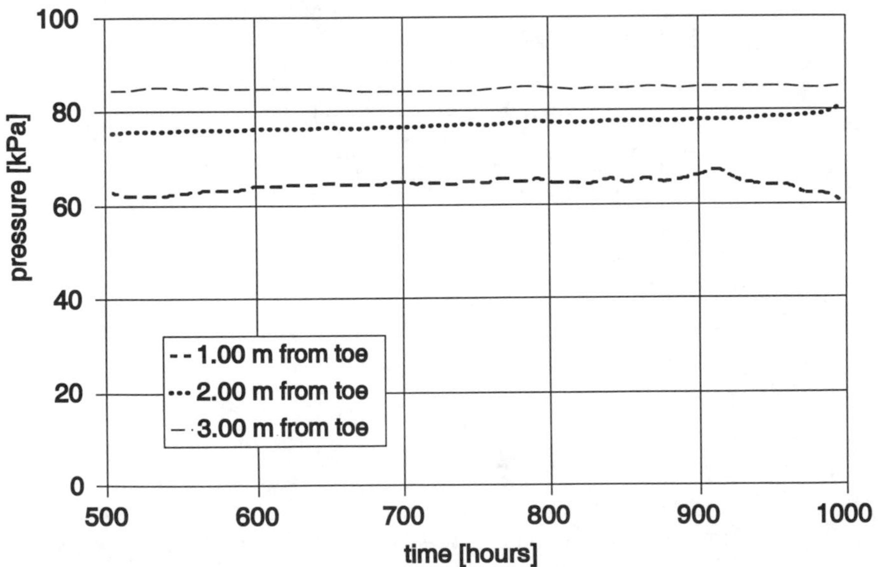

FIG. 12 - *Total pressure cells outputs (after end of construction).*

The total pressure cells output signal, after an initial phase with a lot of discontinuity due to the construction of the whole slope, stabilised itself as shown in Figure 12.

In Figure 13 the base pressure envelope has been compared with the "static" pressure due the weight of the embankment. The larger output of the cell 1, closest to the face, compared to the static pressure, seems to be due to the presence, very close to the face itself, of the road embankment (constructed while the reinforced slope was built) and to the eccentricity of the vertical load on the base, due to the thrust of the backfill.

The values of the inner cell C3, instead, are not so far from the expected static pressure.

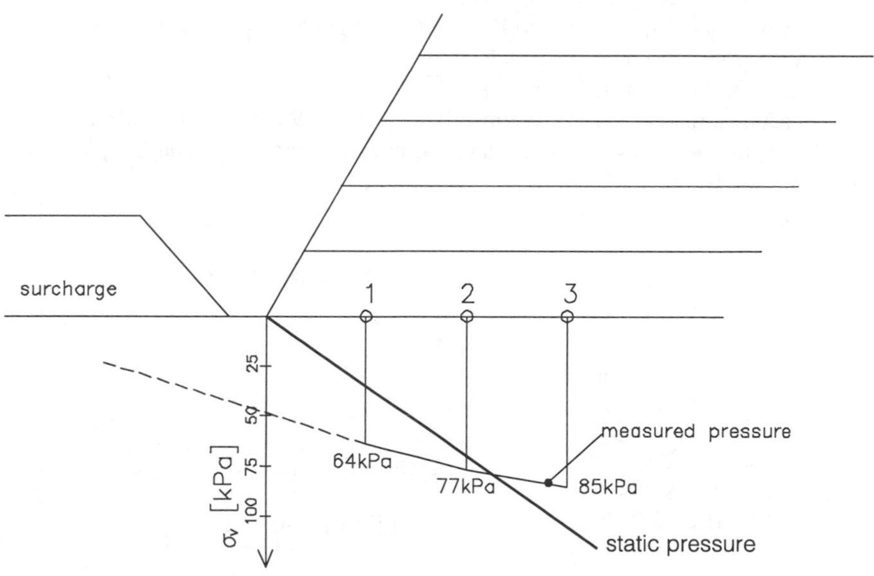

FIG.13 - *Envelope of pressure at the base, from pressure cells readings. after 500 hours*

In situ pull-out tests

In situ pull-out tests have been performed on sacrificial geogrid layers, about 500 mm wide, with different anchorage lengths and positions along the slope. In particular, three tests, with an anchorage length of 1000, 1500 and 2000 mm have been performed. About 1000 mm length of geogrid has been left outside the slope, in order to allow for the connection of the geogrid with the tensile apparatus.

The installation procedure for the slope area interested by these tests has been modified as follows:

1. lay down the steel formwork, the geogrid and the biomat for the layer inside which the geogrid specimens for pull-out test will be installed;
2. cut a rectangular slot, 550 mm wide x 50 mm high, in the steel mesh at half height of the steel formwork itself. Then both the geogrid and the biomat at the face have been cut;
3. lay down and compact a 300 mm lift of soil;
4. lay down, in correspondence to the previously prepared slot, the required sacrificial geogrid layer. As said before, a 1000 mm length of geogrid has been left outside the steel formwork; this geogrid specimen was placed exactly in the middle of the slot, in order to avoid any friction between the geogrid and the steel mesh during the pull-out tests;
5. lay down and compact the soil on the special geogrid specimen;
6. wrap around the main geogrid layer;
7. continue the construction as usual.

A reaction structure, built with timber boards and steel pipes, was placed against the slope face in order to avoid any damage to the whole steep slopes during pull-out tests (see Figure 14).

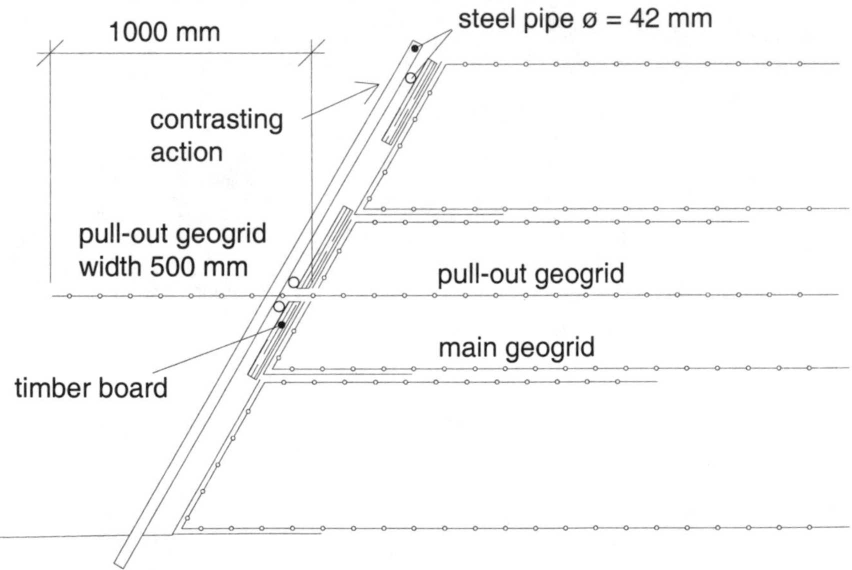

FIG. 14- *Set-up for in situ pull-out tests.*

The operation was repeated in three different locations (one for each pull-out tests), keeping the geogrid specimens always at the same depth from the top of the berm. It is important to notice that these pull-out tests will not affect the behaviour of the whole slope; in fact, the geogrid specimens were not calculated in the internal stability analysis (so, no loss of performance is suffered due to the extraction of these geogrids). Furthermore, the dimension of the slot allowed only a very small loss of soil during the pull-out tests.

Finally, the total pull-out was limited to about 560 mm, hence the disturbance to the soil in the slope was minimal.
In fact no further movements occurred in the adjacent soil or at the face.
Figure 15 provides an overall front view of the geogrids during the pull-out tests.

Tests were performed using a hand tensioning apparatus, connected to a special clamp attached to the geogrid. A load cell was inserted between the clamp and the tensioning apparatus. Reaction was made with the backhoe of the excavator. The pull-out rate was about 10 mm/min. Load (measured at the load cell) and position of the clamp were measured every 15 seconds.

The results were converted in kN/m by dividing the value of the measured load by the number of geogrid ribs tested and multiplying by the number of ribs in one metre width. The use of the hand tensioning apparatus allowed a nearly constant test speed.

From the analysis of results, it appears that the anchorage length of 1.5 m and 2.0 m were far larger than required to allow the pull-out of the geogrid sample. In both cases, in fact, the failure was due to tensile failure of the geogrid, and not to pull-out. In other words, the minimum required anchorage length was lower than 1.5 m. The results for the three tests are presented in Figure 16.

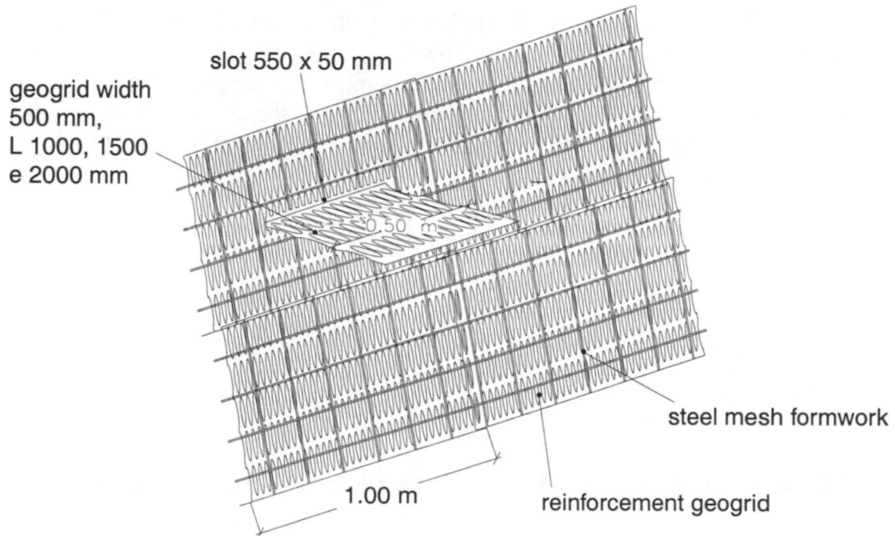

FIG.15 - *Overall front view for the pull-out tests.*

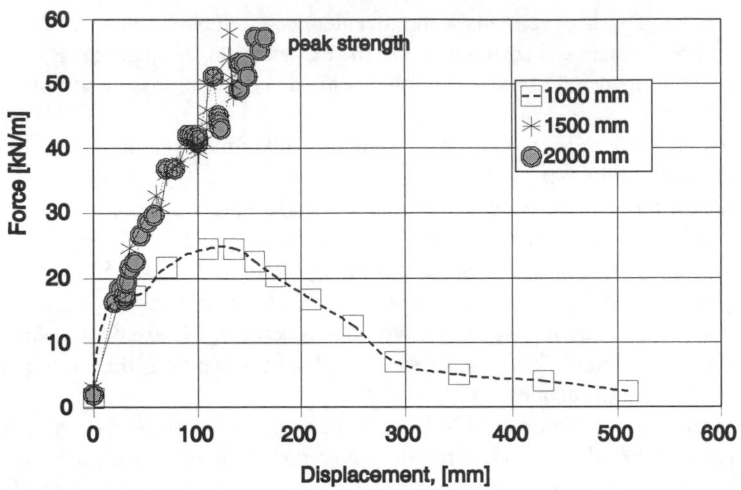

FIG. 16 - *Results of the pull-out tests.*

The two upper curves can be considered as a sort of wide width confined tensile tests, and not pull-out tests. The failure, in fact, was due to tensile failure of the geogrid, and not to pull-out. The failure load was very close to the peak tensile strength of the geogrid used (60 kN/m); the difference observed (a few kN/m) is due to the different test conditions in respect to laboratory wide width tests, both in term of speed (not strictly constant) and of application of the force (not perfectly distributed).

The only valid curve for the pull-out evaluation is the lower one in Figure 16. From this curve, considering the vertical load transmitted by the total weight of the soil on the geogrid specimen, it is possible to find a pull-out coefficient for Tenax TT geogrids with this soil, as follows.

The resistant shear stress to pull-out of the geogrid is given by

$$\tau_b = \sigma_n' \, f_{po} \tan \phi' \tag{1}$$

where

f_{po} = factor of pull-out;
σ'_n = normal stress = 21.16 kPa;
ϕ' = soil friction angle = 26°.

The maximum pull-out resistant force that the reinforcement can develop is given by

$$T_b = 2\, L\, \tau_b \tag{2}$$

where

L = length of the reinforcement in the anchorage zone.

Since it is

$L = 1.00$ m,
$T_b = 24.53$ kN/m

we have

$\tau_b = T_b/2\,L = 12.26$ kPa

and, finally,

$f_{po} = \tau_b/\sigma_n' \tan \phi' = 1.16$

This value seems to be in good agreement with the results obtained in laboratory pull-out tests on the same kind of soil and geogrid (Cancelli et al., 1992). The value obtained in situ is higher in respect to the laboratory one (1.00); this is probably due to soil arching effect caused by the boundary conditions provided by the reaction structure during the test.

Finite element analysis

To evaluate the field stress and deformation behavior of the Montone reinforced slope, a finite element model of the entire structure has been developed (Ghinelli and Sacchetti, 1998). The numerical simulation has been performed using the CRISP90 computer code (Britto et Al., 1990).

The model uses the following elements:

- "BAR" elements, with linear-elastic or linear elasto-plastic behavior, for reinforcements.
- "LSQ" (Linear Strain Quadrilateral) and "LTS" (Linear Strain Triangle) elements, with linear elastic-plastic behavior, for fill soil.
- "SLIP" elements, with behavior based on the Goodman & Taylor (1968), for soil reinforcement interface.
- "LSQ" and "LST" elements, with linear-elastic behavior, for foundation soil.

Material properties

Field tests on the reinforced slope and laboratory tests on materials taken from the reinforced block were carried out for assessing the most important mechanical parameters of the numerical model.

The tests undertaken are:

- Field density tests, carried out on all three reinforced blocks, that supplied the unit weight of the fill soil.
- Field pull-out tests, previously explained.

• Rigid plate load tests, carried out on the first reinforced block using circular plates with 300 mm and 600 mm diameter, that supplied the principal deformation parameters of the fill soil.
• Classification tests, performed on several samples of fill soil taken from the reinforced slope, that supplied the limit indexes and the particle size distributions.
• Direct shear tests with 60x60 mm standard box, performed on samples of fill taken from the reinforced slope, that provided the shear resistance of fill with particle size finer than ASTM number 10 sieve.

The most important test results achieved are summarized in Table 2.

TABLE 2 - *Geotechnical properties of fill soil and mechanical properties of reinforcement geogrids adopted for finite element analysis.*

	Unit	Design Value
FILL SOIL		
Unit weight	[kN/m^3]	18.4
Friction angle	[°]	33
Effective cohesion	[kPa]	0
Initial elastic modulus	[MPa]	35.2
Poisson ratio	[-]	0.2
FOUNDATION SOIL		
Unit weight	[kN/m^3]	20.0
Friction angle	[°]	23
Effective cohesion	[kPa]	100
Elastic modulus	[MPa]	10
Poisson ratio	[-]	0.4
INTERFACE		
Pull-out coefficient	[-]	1.00

Figure 17 shows the comparison between the distribution of reinforcement strains obtained with the numerical analysis (light dots and dashes) and from field measurements (dark dots) at two months after the end of construction, relating to the first reinforced block (layers A to H in Fig.4). Figure 18 shows the same comparison relating to the second reinforced block (layers I to R in Fig.4).

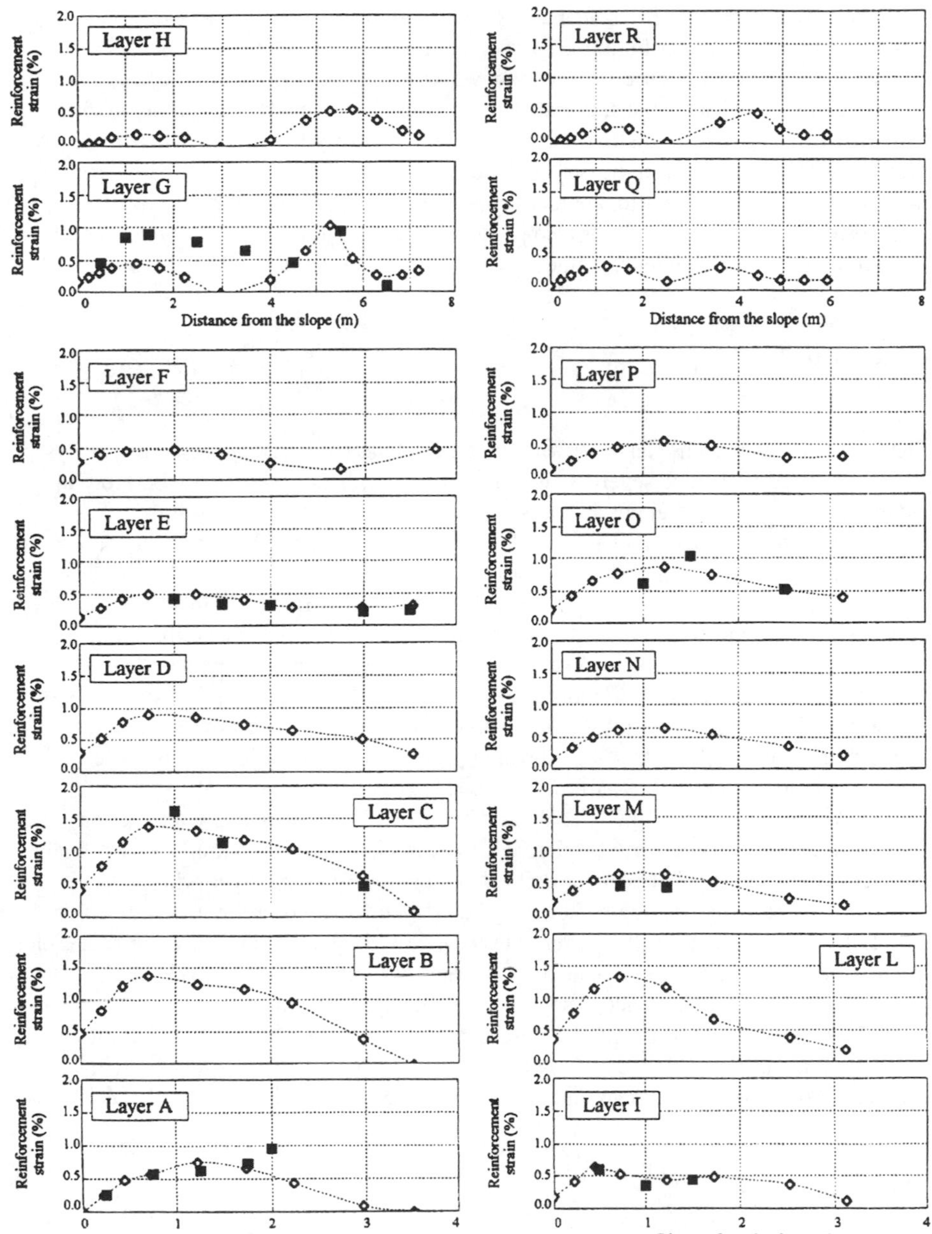

FIG. 17 - *Strain distributions on the geogrids of the first reinforced block (dark dots: field data; light dots: FEM analysis).*

FIG. 18 - *Strain distributions on the geogrids of the second reinforced block (dark dots: field data; light dots: FEM analysis).*

With reference to numerical analysis results, all layers present more or less the same pattern of strain distribution. In particular, the second layer of each block (respectively B and L) has the largest strain whereas the layers closer to the top (respectively F and P) has the smallest. The peak reinforcement strains in the bottom layers occurred closer to the slope surface than in the top layers. The very particular strain profiles of layers G and H in the bottom block and Q and R in the second block are mainly related to their specific boundary conditions (stepped embankment).

From Figures 17 and 18 it is also possible to highlight a good affinity between measured values and calculated values.

Conclusions

The following conclusions can be drawn from the field measurements and the FEM analysis:

1. In a stepped slope several potential failure mechanisms exist and therefore the envelope of the tensile strength developed along the geogrid layers is complex;

2. The strain gages readings provide, so far, qualitative results which can be well explained with the global stability analsyis;

3. The pull-out behaviour of the geogrids, observed in full scale in situ tests, are in very good agreement with the results of laboratory pull-out tests for a similar type of soil;

4. For a complex geometry like a 3 steps slope the pressure at the base is the result of a rather complex superposition of the effect of the self weight of the fill soil and the thrust from the backfill. It is anyway the Authors' opinion that in a simpler geometrical situation the self weight of the fill soil would prevail and the envelope of the base pressure would be very close to the static pressure one.

5. When performing a FEM analysis the use of "SLIP" elements with frictional and adhesive behavior, which permits relative displacements, allows to simulate the complex interaction mechanisms between the soil fill and the reinforcement geogrids.

6. The strain distribution results of finite element analysis are in good agreement with field measurement results. Nevertheless the boundary conditions have a remarkable influence on the base reinforcement layer and on the geogrid layers between each reinforced block. This influence is negligible in the other reinforcement layers.

7. To the Authors' opinion, the good agreement achieved confirm the validity of modeling technique developed and highlights the possibility to use FEM to obtain detailed information about stress and deformations under field condition.

Aknowledgments

The field measurements here presented were part of a project in association with the Department of Civil Engineering of Florence University (Florence, Italy), TENAX SpA (Viganò, Italy) and R.P.A. SpA consultants (Perugia, Italy). The contributions of these organizations is gratefully aknowledged.

References

Britto, A., Gunn, M., 1990, *Critical State Soil Mechanics via Finite Elements,* Ellis Horwood Limited, Chichester, England.

Cancelli, A., Rimoldi, P., Togni, A., 1992, "Frictional characteristic of geogrids by means of direct shear and pull-out tests" *Proceedings of IS Kyushu '92 International Symposium on Earth Reinforcement Practice,* Fukuoka, Japan.

Coluzzi, E., Montanelli, F., Recalcati, P., Rimoldi, P., Zinesi, M., 1997, "Preliminary results from an instrumented Geogrid Reinforced Slope for the Stabilization of the Montone Hill in Central Italy" *International Symposium on Mechanically Stabilized Backfill (MSB),* Colorado, USA.

Duncan, M., et Al., 1985, *STABGM: a computer Program for slope stability analysis of reinforced embankments and slopes.* Virginia Polytechnic Institute and State University.

Ghinelli, A., and Sacchetti, M., 1998, "Finite Element Analysis of instrumented geogrid reinforced slope" *Proceedings of the 6th International Conference on Geosynthetics*, Atlanta, USA.

ICITE, 1998, *Sistema di rinforzo e stabilizzazione di terreni mediante geogriglie Tenax TT SAMP.* Certificato di idoneità tecnica n. 508/96, ICITE - CNR, Milano, Italy.

Rimoldi, P., Jaecklin, F., 1996, "Green faced reinforced soil walls and steep slopes: the state-of-the-art" *Geosynthetics: Applications Design and Construction*, De Groot, Den Hoedt & Termaat Editors, Balkema, Rotterdam, pp.361-380.

W. Allen Marr[1]

Uses of Automated Geotechnical Instrumentation Systems

REFERENCE: Marr, W.A., "**Uses of Automated Geotechnical Instrumentation Systems**," *Field Instrumentation for Soil and Rock, ASTM STP 1358*, G. N. Durham and W.A. Marr, Eds., American Society for Testing and Materials, West Conshohocken, PA 1999.

ABSTRACT: Developments occurring in electronics and instrumentation promise to lower the cost and improve the reliability of electronic systems to monitor geotechnical instrumentation. The miniaturization of electronics, reduction of power consumption, reduced component cost, and improved component reliability all help to make new instruments possible and geotechnical instrumentation more cost effective.

This paper describes the application of some of these developments to three problems and summarizes the potential benefits to the engineer from their use. The increased capabilities of instrumentation and data acquisition equipment combined with their improved reliability and lower cost will make future applications of geotechnical instrumentation more cost effective.

KEYWORDS: field instrumentation, tilt meters, piezometers, automated data acquisition

Introduction

The geotechnical practitioner faces many unknowns in design. In many cases, the effort required to remove these unknowns during design is too costly. Construction must proceed anyway. As a result, geotechnical engineers rely on field instrumentation to monitor the constructed facility and forewarn them of adverse performance resulting from these unknowns. Where required, the design or construction method is adjusted to avoid unacceptable performance. On most major geotechnical projects, field instrumentation constitutes an important element of the engineer's overall design. The results from field instrumentation programs have contributed in a major way to many of the advances in our profession.

---

[1]Chief Executive Officer, GEOCOMP Corporation, 1145 Massachusetts Ave., Boxborough, MA, 01719, USA, e-mail: wam@geocomp.com.

The potential benefits of good field instrumentation programs are many and varied. Evaluated data from field monitoring programs can reduce risks, avoid failures, avoid litigation, reduce delays and reduce costs. While our profession learns more each day about geotechnical behavior and pushes the envelope of knowledge further out, we are also faced with increasingly complex site conditions, tighter restrictions, new construction materials and processes, pressure to cut costs, and concerns about litigation. Field instrumentation can many times help us deal with these requirements.

With all these potential benefits, field instrumentation systems are not as widely used as one might expect. Cost has always been a hindrance. The costs to procure the equipment, install the instrumentation, collect the data, and interpret the results can become substantial. On the other hand the benefits cannot always be quantified. In a strict cost-benefit picture, it becomes hard to justify many of these systems, especially the automation component. A second hindrance to more widespread use of field instrumentation has resulted from problems with reliability. Many clients have experience where they spent a lot of money only to obtain confusing and contradictory data. These clients become reluctant to spend money for field instrumentation on their next project. Thirdly, inexperience is a major hindrance to more use of instrumentation. I am speaking of the inexperience of some who try to do instrumentation projects without the knowledge and experience to do the job properly, the inexperience of those procuring instrumentation systems who do so solely on the basis of price, and the inexperience of those who don't recognize the benefits of a well conceived and executed instrumentation program. Some of the new technology discussed in this paper help us overcome some of these hindrances by lowering the cost and increasing the reliability of field instrumentation systems.

Characteristics of Typical Applications of Automated Systems

Cost considerations have previously limited the use of data acquisition systems for field instrumentation. Typically a data acquisition system can only be justified in one or more of the following circumstances:

large number of sensors ≈ more than 50
need for frequent readings ≈ more than once per day
need for simultaneous readings ≈ take all readings within seconds of each other
difficult access ≈ safety issues or long access time

The given numbers are rough guidelines from my experience.

Most automated field data acquisition systems being installed in the US employ variations on two systems. The CR10 unit from Campbell Scientific is highly adaptable to geotechnical requirements and forms the backbone for systems provided by GEO-KON, RockTest and Slope Indicator. Geomation has released a new product that reads up to 20 channels of voltage, current, or frequency input and provides a variety of ways to transmit the data to a central location. These units have been adapted to readily accept input from the commonly available geotechnical instruments.

The typical costs of an automated instrumentation system are revealing. Table 1 summarizes these in a general way for a typical installation on a per sensor basis. These costs are just those costs for automation. The costs of support services to get the sensor in place, such as drilling costs, are not included as these costs more-or-less apply equally to automated and non-automated instruments.

Table 1: Cost Components for Automating a Field Instrument

Component	Cost	Percent of Total
Sensor	\$100 - \$600	13 - 20
Wire	\$100 - \$300	10 - 13
Data acquisition, including communications	\$300 - \$500	17 - 40
Install sensor, excluding drilling	\$50 - \$500	6 - 17
Install wire	\$100 - \$500	13 - 17
Install data acquisition system	\$50 - \$200	6 - 7
Maintain system	\$50 - \$300	6 - 10
TOTAL	\$750 - 2,900	--------

The costs in Table 1 are only approximate and typical. There are plenty of exceptions. Nevertheless in a general way, Table 1 shows typical cost components and some conclusions are possible. The cost of components makes up 50-70% of the cost of automating a field instrument. Installing the system makes up about 25-35% and maintenance of the system during operation consumes 5-10% of the total cost. New technology can potentially lower each of these cost elements by lowering the cost for the electronic components, reducing the cost of installation and reducing the need for maintenance by providing more reliable systems.

The total cost in Table 1 is significant. If we consider that a typical field technician charged with manually reading sensors might cost \$50 per hour, we would have to save 15 to 60 person hours per instrument to recover the cost of automation. Clearly a large number of readings per instrument, a large amount of time to get a reading per instrument, or factors that preclude manual readings push the decision to automation.

Merely committing the money to automate an instrumentation system does not insure success of the field monitoring program. I am aware of recent situations where large sums of money were spent to automate the instrumentation in embankment test sections placed on soft ground with drains added to accelerate consolidation. While these projects used current products, they failed to produce the data required for a meaningful assessment of the test sections. Problems with installation and system operation produced too many obstacles to interpreting the data. Perhaps even more than with conventional systems, automated field monitoring systems must follow the 25 links to success set forth by Dunnicliff (1988) for implementing a field instrumentation system.

Current Developments

Recent electronics has considerably lowered the cost of data loggers, increased their capabilities, and improved reliability. Availability of compact and low-cost computers with high reliability to log and process data provide additional opportunities. New types of piezoresistive pressure transducers provide highly stable readings but cost considerably less than vibrating wire sensors. These transducers may help reduce the sensor cost for piezometers, total stress cells and settlement monitoring gages. Strain gage technology has improved considerably as has the data acquisition equipment to read strain gage output. Advances in electronics for wireless communications are dropping the cost and improving the reliability to send data through the air rather than through a wire. This makes installation easier, troubleshooting faster, and lowers costs.

Whole new classes of low-cost sensors are being developed to measure groundwater chemistry, flow rates, changes in position, pressures, and forces to name a few. I fully expect the next five years to provide the geotechnical community with an entirely new set of sensors and readout equipment that costs considerably less than today's equipment, is easier to install and has higher reliability. We have been working with some of these new approaches. The next three cases describe some of our experiences.

Case 1 - Concrete Form Carrier for Chicago Tunnel

The 73-060-2H Contract on the North Branch Chicago River of Chicago's TARP project consisted of 50,000 ft of machine bored tunnel lined with concrete. For concreting, the Contractor assembled a set of five cars which carried collapsible forms. Each car was 35 ft long and rode on four wheels running over a curved concrete invert. The wheels consisted of a 17-inch diameter steel drum covered with a 2 inch thick, 10 inch wide solid polyurethane tire.

Within the first 2,000 ft of concreting, several of the tires experienced catastrophic failure. The polyurethane tire was separating from the steel hub. Each failure completely stopped concreting operations for several hours until the wheel could be changed. This problem posed a major threat to the project schedule and cost.

Our evaluation of the problem indicated three unknowns. What loads were being exerted on the tires? What stresses did those loads cause in the tires? What was the strength of the polyurethane and the bond between the polyurethane and steel hub? Visual observations of the cars during a move cycle indicated a widely varying load distribution on the wheels. The start of the cycle involved collapsing the forms onto the car. Sometimes the form would be frozen to the cured concrete. By manipulating the hydraulic system, the operator could set the form into a dynamic oscillation that would produce more force to jar the form loose. While moving, the car would tend to weave from side to side over the curved invert. Since the wheels had no suspension, this would cause a wheel to lose contact with the invert. At times during a move, a wheel was observed to experience sudden slip sideways of as much as one inch. Substantial lateral loads were developing that were causing unknown shear stresses in the polyurethane. Clearly the load in each wheel was far different than the weight of the car divided by the number of wheels.

We decided to instrument the wheel support system to measure the force in each wheel during a complete move cycle. Due to the dynamic nature of the move cycle, the measuring system needed to be capable of collecting data quite rapidly. The Contractor was anxious that this work not disrupt his normal production. We decided to use strain gages welded onto the struts for each wheel and read them with a laptop-based data acquisition system. The entire system would sit on the car and operate while the car was moved from one setup to the next. One strain gage was welded to each side of each steel strut. We intended to use the average strain in each of the four gages on a strut to compute axial force and the difference in strain on two opposite sides to compute a moment from which we could determine the lateral force acting on the tire. This made a total of 16 strain gages. In addition each wheel was supported by a hydraulic cylinder contained within the strut that transfers the weight of the forms to the wheel. Each cylinder was instrumented with a pressure transducer.

The sensors were connected into a signal conditioning box which provided excitation to the sensor and gained the sensor output to the voltage level required by the A/D. We used a standard signal conditioning box normally used for laboratory test equipment. The strain gages were 120 ohm 1/4" single element weldable strain gages coupled with a microchip Wheatstone bridge from all from Micro Measurements. The A/D was a Computer Boards PPIO-AI08 external card with a 32 channel add-on multiplexor card. The A/D connected to the parallel printer port of a laptop. We modified some in-house data acquisition software to read the 20 channels of input and save them to a data file. The entire system was assembled, calibrated and debugged in our facilities prior to shipping them to the site. All wires running from the data acquisition system to the individual sensors were precut, fitted with connectors and labeled prior to delivery to the site.

The complete pour cycle was taking 24 hours if everything went well. It would take about eight hours to make the pour, eight hours for the pour to cure, and eight hours to move and reset the forms. We had a window of about 16 hours during which we could install the instrumentation. By preparing as much of the system in advance as possible, our main task in the tunnel was mounting the strain gages. This required grinding the struts at each gage location to fresh steel, cleaning the steel, then welding the gage. Figure 1 shows a typical installation. For scale the leg is approximately 18 inches by 10 inches (460 by 250 mm). The installation was greatly hampered by the working conditions. Water, grease and dirt prevail in a tunnel. These elements ruin strain gaging operations quickly.

FIG. 1 – *Strain gages on two sides of strut.*

We spread the work over two move cycles. During one cycle, we installed the gages and tested their output with a standard readout box. During the next cycle we added the wires from the gages to the data acquisition system and monitored the readings during the move. We were frequently barraged with water from every direction. This made it very difficult to protect the wiring connections and data acquisition components. We kept the data acquisition components boxed as long as possible, then covered them with plastic to try to keep water away.

We decided to read each channel 10 times per second. With 20 channels of data and a move time of up to 30 minutes, the data file would become unwieldily with more than 300,000 individual readings. However, the move was not one continuous operation. The operator usually halted several times to deal with unexpected problems or difficulties. We adapted the data collection to these starts and stops. Just before the operator would begin an activity, we would start the data logging. When move operations stopped we would stop the data logging. In the interim we would keep a written log of what was happening to the car so that we could later compare the data we were collecting to the operations of the car. During the stopped periods we could check that the data logging operation was working correctly and that the sensors were continuing to give output.

Figure 2 gives a sample output obtained from the strain gages. It shows the axial force in the strut determined by taking the average strain measured in the four strain gages on that strut times the cross sectional area of steel on the strut times the modulus of elasticity of steel. From examining the data for the entire move, we were able to determine two important conclusions. First, whenever the operator manipulated the

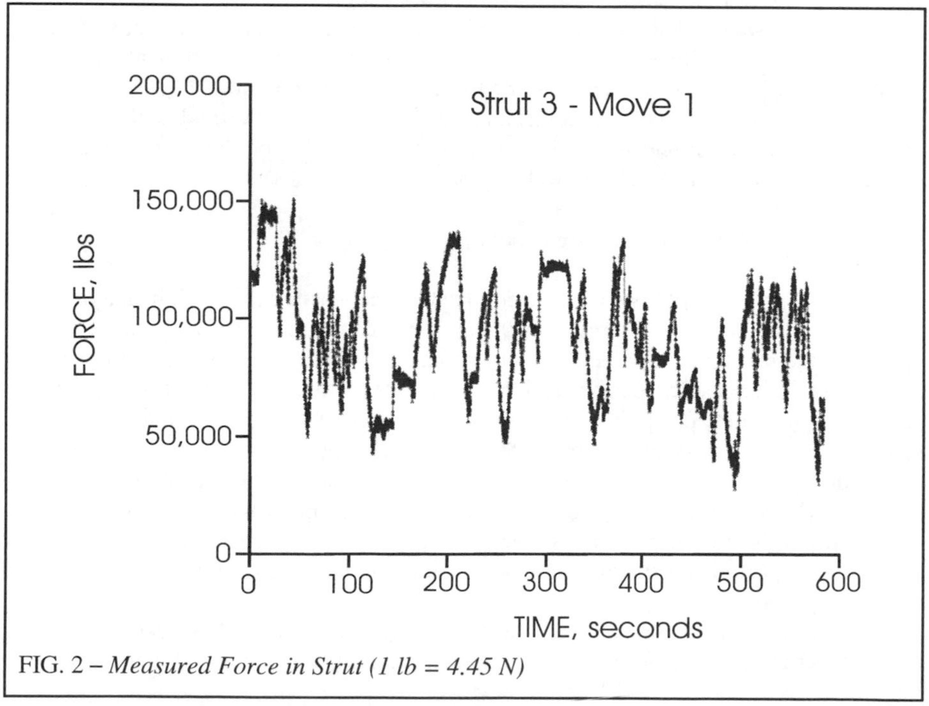

FIG. 2 – *Measured Force in Strut (1 lb = 4.45 N)*

hydraulics to try to jockey a stuck form loose, he would create a large force in one or more wheels. The hydraulic jacks were pushing the forms against the obstruction with more force than the weight of the car. Second, on several occasions during the move, the full weight of the car would be carried mostly by only two wheels. This happened when the axis of the car did not align with the axis of the curved invert. Since the car had no suspension, the entire weight would be transferred to one front tire and the rear tire on the opposite side.

Contrary to our expectations, measured lateral forces were not substantial. While lateral forces of several tons would develop, the side slip that we observed occurred when the vertical weight was suddenly reduced on the tire which then allowed it to easily slip over the invert to relieve any lateral forces.

The end result of this effort was a determination that the tires were being loaded to as much as 150,000 lb., a value which far exceeded the apparent design load of approximately 80,000 lb. The maximum force was developed during the jockeying of the car to free it from an obstruction. We recommended that the hydraulic system be altered to limit the force that could develop in the tires from the hydraulics. The measured forces were used to assess the stresses in the polyurethane with which a revise wheel geometry and polyurethane molding process was developed. New wheels were placed on each car. The remaining 46,000 ft of tunnel were completed with only a few additional tire failures. Those failures were attributed to the cars running over sharp steel objects and not to overstressing.

This case illustrates several important points. Field measurements can be essential to determining the cause of failure and what corrective actions to take. Fairly elaborate measuring systems can be deployed relatively quickly and successfully operated in severe conditions. Portable electronics now makes it possible to instrument many construction operations without hampering the Contractor's activities. These capabilities give us intelligent ways to help Contractors identify the cause and formulate the cure for unexpected behavior instead of relying solely on the traditional tool box containing a bigger hammer, a cutting torch and welding rod.

Case 2 - WDAS for tilt of building facade

This opportunity arose from a need to determine whether the brick facing of historical buildings will be deformed by underground construction activities in the vicinity of the building. Previous monitoring methods such as crack gages and visual inspections were not sufficiently sensitive and reliable to prevent construction related cracks. From a safety perspective, the damage of concern is primarily architectural in nature, i.e., minor cracks and limited differential settlement. The structural integrity of the buildings and their foundations is not expected to be impacted by construction activities. However cracks and differential settlement are things owners can see. They cause high levels of concern to the people involved. Hence the need to find some means to measure impact of construction activities on adjacent buildings before the visual damage appears.

Other instrumentation methods such as plumb lines dropped from the roof, inclinometer casings attached to the building face, and tiltmeters attached to the face of each building, were considered. Each of these had serious problems, such as: having to bother building residents to gain access to read the instrument, vandalism, overall expense for the equipment or the time to take readings, and the limited number of readings that could be obtained considering the necessity to arrange for readings in advance.

Of the options available, we concluded that tilt meters connected to data loggers using radio telemetry of the data offered a possible solution. Multiple units would be required on each side of a building to increase the chances of measuring localized movement of the building face. However, putting multiple units of commercially available data loggers onto each of the large number of buildings involved would be expensive. We located a less expensive data logger with built in wireless capability but its resolution was only one part in 256. This was no where near the required resolution for the project of one part in 10,000.

After considering the scope of the project, we decided to build a system to meet the job requirements. Our goal was to have a system with the following capabilities:

-read data to at least one part in 10,000
-read temperature at the sensor location
-transmit the data to a handheld data logger in the street upon command
-operate for several months without servicing

Figure 3 shows basic components. The design concept is an integrated CPU - A/D - RF (central processing unit - analog to digital converter - radio frequency transceiver) on a single board. All electronic components are standard, off-the-shelf items. The CPU controls all components and can operate in a sleep mode to conserve power. The A/D is a 22-bit unit with an input range of five volts. That means that an optimally configured unit can read the sensor to an accuracy of one part in 4,000,000. However it is almost impossible to remove sources of small levels of noise so we use the 22 bit A/D but ignore the lowest 6 bits of the reading which contain the noise. This gives us an accuracy of one part in 65,500, which more than meets the requirements for the project. An electronic temperature sensor mounted on the board provides temperature readings directly to the digital inputs of the CPU. An off-the-shelf radio transceiver placed directly onto the board receives incoming signals requesting readings and transmits outbound data. To avoid having to obtain an FCC broadcast license, the transmitting power is limited to 10 micro watts. This limits the broadcast range to a theoretical line-of-sight range of 1,000 ft (300 m) and a practical working range of about 300 ft (100 m). The system is referred to as the WDAS unit, which stands for Wireless Data Acquisition System.

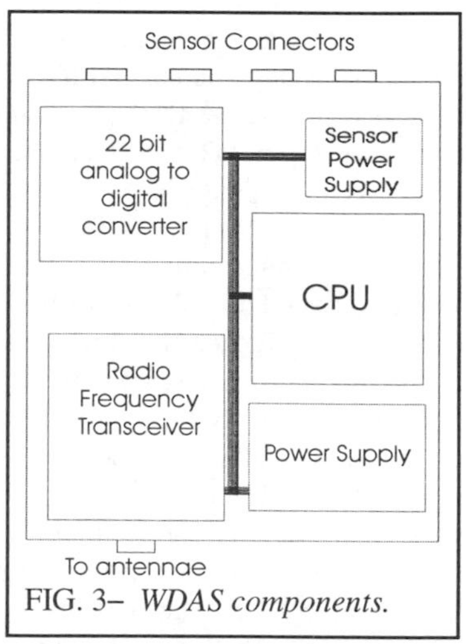

FIG. 3– *WDAS components.*

The electronics are normally shut down to conserve power. The CPU looks at the RF transceiver every few seconds to see if a request to read has been received. If a request has been received, the CPU checks to see if the request is for it. If the request is for this cell, it wakes up all components, powers the sensor and takes a reading. The reading is transmitted out the RF transceiver to be picked up by the host data logger.

FIG. 4 – *Components of WDAS.*

Figure 4 shows a typical cell. The board is mounted inside a 4 in. by 6 in. by 3 in. (100 by 150 by 75 mm) NEMA enclosure. An Applied Geomechanics Model 800 tilt meter sensor is also placed inside along with batteries to supply power. Figure 5 shows the unit installed on the side of a brick-faced building. Figure 6 shows a readout unit, in this case a hand-held Psion Workabout. This ruggedized computer has a serial port that connects to a host communications unit. The host communications unit contains the same radio transceiver as those on the individual cells. The software on the Psion sends a request to read a particular cell ID. The cell receives the request, wakes up and powers the sensor, takes the reading, and broadcasts the reading. The Psion receives the reading and adds it to its data base.

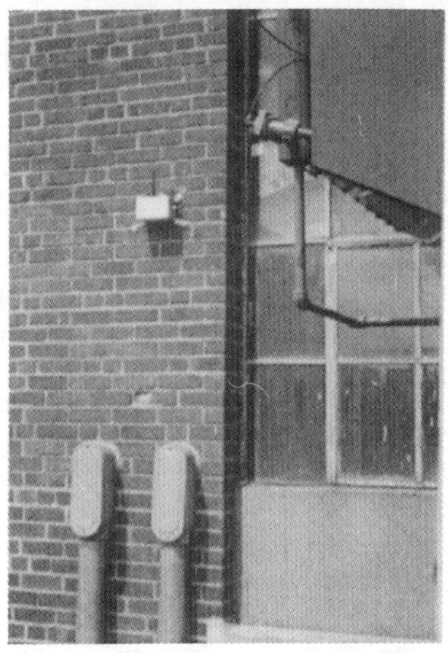

FIG. 5 – *WDAS installed on brick wall.*

The original concept was that a technician would drive by the instrumented buildings every few days and read each cell. Figure 7 shows a sample of the type of data received with this approach from two units mounted on the sides of a brick faced building. The data show considerable variation with a range of over 300 arc seconds. Data on a similar unit located on a concrete slab and kept at constant temperature varied less than five arc seconds of tilt. We suspected that the sensors were picking up movements of the building face caused by temperature changes and wind blowing against the building. These effects were probably being amplified by the type of mounting brackets used for the trial installation.

FIG. 6 – *Handheld readout unit for WDAS.*

Cracking of panel walls shows up

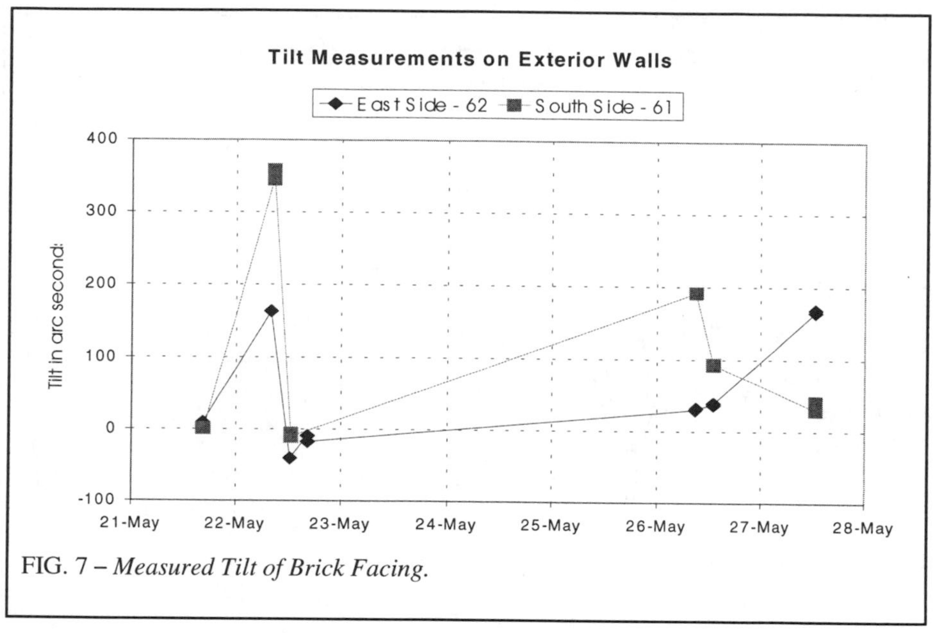

FIG. 7 – *Measured Tilt of Brick Facing.*

at tilts of 1/300 which equates to 700 arc seconds. Standard practice to avoid cracking of panel walls is to keep tilt below 1/500, or 400 arc seconds. Since the trial installation showed changes of over 300 arc seconds. from environmental effects, i.e., temperature and wind, the measurement system and approach of taking a reading every few days would not work. It would create a situation where separating construction effects from normal environmental effects at tilt levels below those which cause cracking would be impossible. Construction effects should be less time dependent than temperature and wind effects. We also expected that temperature and wind effects should be recoverable, i.e., there should be a response signature for the building that reflects temperature and wind effects. Any measured response outside this signature, excepting that from an unusual condition, would more likely be related to construction. These considerations lead us to conclude that the tilt readings had to be taken at close enough intervals to measure the environmental response envelope for the building. Since the building goes through a thermal cycle each day, we concluded that we needed to take readings hourly every day. These readings should be taken for some time prior to the start of construction in the vicinity of the building to establish the response signature for the building to normal environmental effects.

With readings taken each hour of every day it was no longer cost effective to have a technician collect data with a hand-held data logger. Fortunately the design of the system permitted this new requirement to be easily met. By making use of a rebroadcast feature of each cell, we can use the cells to transmit reading requests and data over considerable distances. As long as each cell is within 300 ft of another cell and at least one cell is within 300 ft of the host unit, each cell in a group can be automatically read by

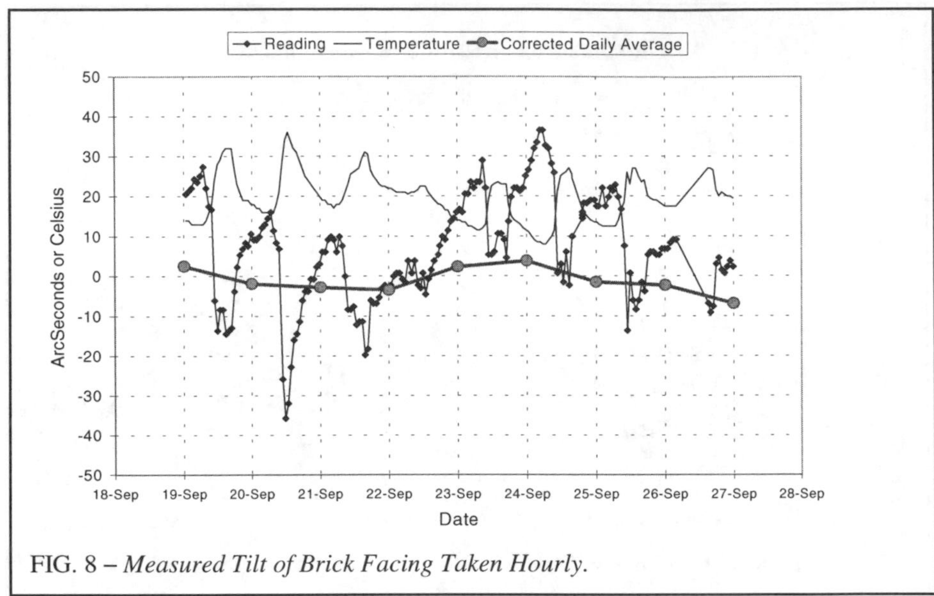

FIG. 8 – *Measured Tilt of Brick Facing Taken Hourly.*

the group host as frequently as we desire. Figure 8 shows a typical set of data from the revised system. These data come from a unit rigidly mounted to the face of a brick wall using a three point mounting system with anchor bolts inserted into the mortar. The data clearly show a cyclic response of up to 50 arc seconds that correlates with temperature. Analysis of the data shows another 10-20 arc seconds of variation that does not correlate with temperature. We think it results from wind effects and extraneous vibrations of the building. By correlating the tilt reading with temperature to obtain a temperature correction, removing the temperature effect and averaging the resulting data for each 24 hour period, we obtain one reading per day which is called the corrected daily average. For the case shown in Figure 8, the corrected daily average fluctuates less than 10 arc seconds over the eight day reading period. This is a major improvement over the 300-400 arc second variation obtained on the original system and the 70 arc second range in the raw data show in Figure 8. It is important to recall that these variations are coming from the building itself and its response to environmental conditions, and not from the instrumentation. In separate tests, we measured the variations due to instrumentation to be less than 10 arc seconds.

This project illustrates several important points. First, electronics are available today which permit us to inexpensively collect data from inaccessible locations. I expect wireless technology to permit us to greatly expand the situations where monitoring is cost effective. Second, we face increasingly strict limits on how construction activities can affect adjacent facilities. Some of these restrictions approach levels of normal environmental change. Monitoring these situations requires round-the-clock readings to define the response signature to normal environmental events. Taking a reading every few days does not provide sufficient data to make meaningful evaluations.

Case 3 - Groundwater monitoring

We frequently encounter situations which involve tens of observation wells and/or piezometers to monitor effects of construction on groundwater levels. For extensive dewatering applications, each sensor may have to be read daily. This is a potential application for automation. However, time and again we have found that the cost of automating the job is not competitive with dedicating a person full-time to data collection.

Some recent developments have changed the cost equation. One development is the wireless system described in the last case. The wireless cells remove the cost for wire and multiplexers to get the signal or readings back to the host unit. These are significant costs. Wire connections and multiplexers are prone to fail with time. Problem resolution requires a site visit by a trained technician, further increasing the cost. The wires always seem to "attract" construction equipment and become damaged. The wireless units can be prefabricated and tested. Field installation consists of installing the sensor, mounting the cell and plugging the two together. These tasks do not require a data acquisition expert. A defective unit can be quickly replaced by site personnel. By using a hand-held readout unit, each cell can be checked when it is installed. This greatly simplifies problems and system troubleshooting.

The other development is the availability of new low cost piezoresistive pressure sensors made for industrial applications. These sensors use silicon circuitry etched directly onto the stainless steel sensing diaphragm. The result is a sensor with exceptional accuracy, stability, repeatability and temperature response. Several manufacturers are offering sensors with combined linearity and hysteresis less than 0.1%, long term stability of span and offset over one year less than 0.1% and repeatability less than 0.01%. These values are an error in the reading as a percent of the full scale reading. Table 2 provides the approximate error in measured water level using various ranges of these transducers. We have used some of these sensors in laboratory applications for about one year and can confirm their excellent stability. We are just beginning to deploy them in field applications.

Table 2: Measurement Error from Modern Piezoresistive Pressure Transducers

Nominal Pressure Range psi (kPA)	Range in Water Head ft (m)	Expected Maximum Error ft (mm)
10 (70)	23 (7)	0.02 (7)
25 (170)	57.7 (18)	0.06 (18)
50 (350)	115 (35)	0.12 (35)
100 (700)	230 (70)	0.23 (70)

Figure 9 shows a model that has a water proof cable. It has a maximum outside dimension of 1 inch. The wire can be fitted with a connector to plug it directly into a data acquisition system such as the WDAS described above. This combination can be calibrated as a complete system, then sent to the field for installation. The complete assembly, including pressure transducer, cable and wireless cell costs less than just the sensor for some piezometers.

A low air entry stone can be fitted to the sensor end to conveniently obtain a closed deaired system for long-term monitoring of pore water pressure. This particular model has a thin plastic tube embedded in the wire that normally serves to vent the back side of the transducer diaphragm to atmospheric pressure. Readings with the transducer are relative to atmospheric pressure. This tube can be used to check the sensor. By applying gas pressure to the tube, we can return the diaphragm to its zero condition, (i.e., increase the pressure until the reading of the transducer equals the original reading at zero pressure). The required gas pressure should equal the reading of the transducer.

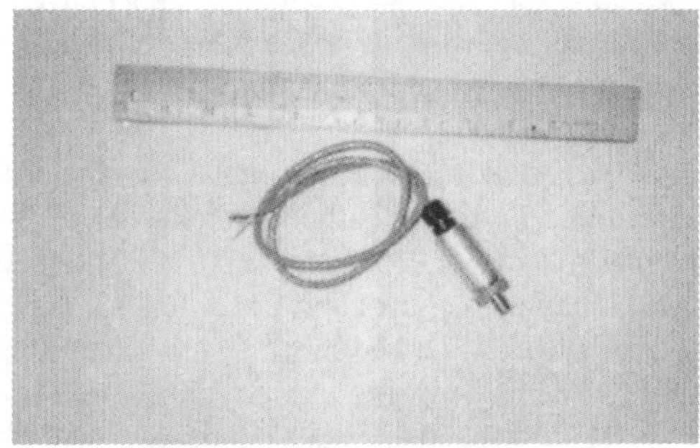

FIG. 9 – *Piezoresistive pressure transducer.*

It is useful at this point to review some historical information. Piezoresistive pressure transducers were first used in geotechnical applications about 30 years ago. That initial experience was not always successful, particularly in situations requiring measurements over several months or years. Readings from these sensors tended to drift with time. For field applications, voltage losses in cable extensions altered the sensor calibration which produced errors in the measurement. Many installations experienced serious electrical noise problems. These problems led geotechnical engineers to exploit the benefits of vibrating wire strain gage sensors. Vibrating wire sensors transmit a frequency that does not change with cable length. Vibrating wire sensors available in the 1970s gave much better stability than the piezoresistive sensors. With time the geotechnical community has come to rely heavily on vibrating wire type sensors. Essentially all of the electrical piezometers, strain gages, and load cells on the Boston Central Artery Tunnel are vibrating wire sensors.

The problem with the piezoresistive sensors lay in their method of manufacture. A strain gage was bonded to a thin metal diaphragm with epoxy resin. Pressure on one side of the diaphragm caused the diaphragm to bend. The sensitive strain gage responded to the bending with a change in resistance which could be read as a change in voltage. The problem was that the epoxies tended to creep with time. This creep led to drift in the output of the transducer. This problem has been solved by etching the strain gage circuits in silicon directly onto the back of the thin metal diaphragm. There is no longer any epoxy to creep. The new process was not only better but less costly. Manufacturing quantities have also increased dramatically to supply sensors for the automotive and industrial control markets. Consequently the cost of piezoresistive pressure sensors has dropped more than 50% over the past 30 years.

I believe we will see an increase in the use of these sensors in geotechnical field applications in the coming years. It is much easier, with off-the-shelf equipment, to excite and read a voltage-based device than it is for a frequency-based device. Additionally, the problem of line losses in voltage type devices can be solved by using current loop versions of these sensors. In this case the sensor gives a current output that is proportional to pressure. At the reading end, the current can be passed through a resistor and the voltage drop across the resistor input to off-the-shelf readout equipment. Electrical noise problems have been greatly reduced by using shielded cable and noise canceling readout techniques. By using the vent tube to pressurize the back of the diaphragm, it is possible to obtain a check on the reading of a suspect sensor. Finally, today's piezoresistive type sensors cost less than the alternatives.

Summary and Conclusions

The cases described in this paper illustrate some of the benefits of a good automated field instrumentation system. There is no substitute for the right data from a field instrumentation system to give the geotechnical engineer control over field performance. Recent developments in electronics, sensors and software provide the tools to help us implement automated field instrumentation systems that are reliable, responsive and affordable. However, automation systems have previously suffered from problems of cost, reliability and ignorance.

Many new low-cost sensors are available and more are coming. Cost of sensors, readout equipment and related materials are dropping. Off-the-shelf software, sometimes as simple as using a spreadsheet, provides low cost but powerful ways of reducing and reporting data to meet project needs. Off-the-shelf hardware components are making it easier to automate data collection from field instrumentation. These components are becoming more reliable and easier to use.

For projects with a lot of instruments and/or lot of readings, it may be cost effective to design a project specific data collection/management system tailored to the specific needs of the project (Hawkes and Marr, 1999). Such systems can save many man-hours and more importantly can get accurate data to the place where it is needed when needed. These systems involve knowledge beyond the domain of the typical geotechnical engineer. To effectively use today's technology in automated field instrumentation systems requires individuals with current knowledge and experience in software, instrumentation, electronics, signal processing as well as geotechnical engineering. A team with a weakness in any of these areas can create the opportunity for failure.

Once an instrumentation system is automated, the cost of collecting data more frequently is relatively low. Frequent data collection, i.e., several times a day, can help us establish a response signature for the facility that includes the normal fluctuations from temporal environmental effects. With this response signature, it becomes much easier to separate the true effects of our activities from the normal response of the facility. This more complete data set helps us avoid the difficult situation of what to do when we have a single reading that has suddenly changed and there is no explanation for the change.

Acknowledgments

The dedicated efforts of Martin Hawkes, Salim Werden and Chafik Hankour, engineers at GEOCOMP, helped develop the described systems and put them into practice. Mr. Louis Brais of Perini Corporation provided the opportunity to put our capabilities to work in the Chicago Tunnel case. Dr. Thom Neff of Bechtel/Parsons Brinkerhoff encouraged the development of the tilt monitoring system. The cooperation and assistance of these gentlemen are acknowledged and appreciated.

References

Dunnicliff, J. (1988), Geotechnical Instrumentation for Monitoring Field Performance, John Wiley & Sons, New York.

Hawkes, M. and Marr, W. A. (1999), "Data Acquisition and Management for Boston's Central Artery/Tunnel Project," *Field Instrumentation for Soil and Rock, ASTM STP 1358*, G. Durham and W.A. Marr, Eds., American Society for Testing and Materials, 1999.

Peck, R.B. (1969), "Advantages and Limitations of the Observational Method in Applied Soil Mechanics," *Geotechnique*, Vol. 19, No. 2, pp. 171-187.

Anthony D. Barley,[1] Michael C.R. Davies,[2] and Alun M. Jones[3]

Instrumentation and Long Term Monitoring of a Soil Nailed Slope at Madeira Walk, Exmouth, UK

REFERENCE: Barley, A.D., Davies, M.C.R., and Jones, A.M., **"Instrumentation and Long Term Monitoring of a Soil Nailed Slope at Madeira Walk, Exmouth, UK,"** *Field Instrumentation for Soil and Rock, ASTM STP 1358*, G.N. Durham and W.A. Marr, Eds., American Society for Testing and Materials, West Conshohocken, PA, 1999.

ABSTRACT: This paper describes the field instrumentation of a soil nailed slope undertaken at Madeira Walk, Exmouth, UK. The instrumentation consists of strain gauged steel bars that are employed as soil nails to maintain stability of a slope. The nature of the stabilisation works and the instrumentation used to monitor the axial load induced in the soil nails are described. Initial results are presented showing the changes in axial load induced in the nails.

KEYWORDS: field instrumentation, soil nails, slope stabilisation, strain gauges

---

Introduction

Exmouth is a small seaside resort located on the United Kingdom's South Devon coastline at the mouth of the River Exe. Madeira Walk is a footpath that stretches the length of this section of coastline, between 400 and 500 metres from the current high tide line. The walk runs along the base of a slope known as the 'Maer'. Historically the Maer was the cliff line above the beach. Subsequently the sea has regressed some 500m.

---

[1] Director of Engineering, Keller Ground Engineering, Thorp Arch Trading Estate, Wetherby, West Yorkshire, LS23 4SJ, United Kingdom.

[2] Professor of Geotechnics, Civil Engineering Department, University of Dundee, Dundee, DD1 4HN, United Kingdom.

[3] Research Student, Division of Civil Engineering, University of Wales Cardiff, P.O. Box 917, Cardiff, CF2 1XH, United Kingdom.

Tourism is one of the main industries of Exmouth. Madeira Walk provides a focal point for walks along the coastline and is also regularly used by locals. The safety aspects of the cliff and slope above the path were of major concern to the local authority when localised failures began to occur.

Background

The cliff is formed from New Red Sandstone, which has weathered heavily leaving a residual soil mantle several metres thick. Slope angles on the weathered face are between 45° and 60°, standing up to 12 metres high. Bent trees point to long term movement of the steep slope, and tension cracks are evident along the footpath running along the crest.

Generally, movements of this weathered zone have been by creep, but in the winter of 1994 a larger slip blocked Madeira Walk. Following this, East Devon District Council commissioned an assessment of the stability of the slope. The result of the ground investigation (Geotechnics Limited 1995) was a proposal to improve the stability of the slope using soil nails.

Objectives

The objective of the programme described here was to investigate the forces induced on the soil nails by further movements of the slope over the period of a number of years. The project was considered suitable for monitoring since it was a relatively small scheme where soil nailing was the main remedial works being undertaken. Access to the site was easy, and the cooperation of the consulting engineer and the contractor was significant if the instrumentation was to be successfully installed. In addition, with only a few soil nailing projects currently being undertaken in the UK, this project was within suitable distance of the research group to allow attendance during the installation of the instrumented nails and regular monitoring.

Slope Stabilisation Design

Site Investigation

Three boreholes were drilled at the crest of the slope to determine ground conditions. The boreholes were sunk until bedrock was reached between 4.5 and 5.5 metres below the crest. Drilling was continued 10.5 meters below ground level to confirm the bedrock.

The borehole near the section to be instrumented indicated made ground for the first 0.5 metres, consisting of tarmacadam over gravel sub base. This was underlain by a 1.3 metre thick layer of firm red brown clayey silt with some gravel and occasional decomposed organic matter. A head of stiff red brown silty clay with frequent gravel sized blocks of mudstone proceeded the rock head. The rock was encountered at a depth of five metres. This consisted of Exmouth Mudstone, a red brown, closely fissured, very silty weak Mudstone. No groundwater was encountered whilst cable percussive drilling.

Table 1 summarises the soil tests conducted in the field, and the laboratory tests results from samples taken are given in Table 2. These results were used for the basis of the geotechnical design.

TABLE 1 -- Soil properties derived from field tests.

Depth (m)	Standard penetration Test (blow counts)	Rock Quality Designation, %	Fracture Index
1.0	10	...	...
2.4	20	...	...
5.4	...	98	6
6.5	...	92	3

TABLE 2 -- Soil properties derived from laboratory tests.

Sample	Moisture content, %	Plastic limit, w_P	Liquid limit, w_L	Plasticity index, w_L-w_P	Bulk unit weight, kN/m^3	Undrained shear strength, kN/m^2
Firm, dense weathered red brown clayey silt with occasional clay pockets	15	23	43	20	20.9	99
Firm, dense friable red brown clayey silt, with calcareous silt pockets	21	...	...	...	20.2	111

Design of Slope Stabilisation Works

Stabilisation of the slope was required to prevent any further localised slippage and the possibility of a major slide. Access to the slope was relative easy on foot, but to mobilise plant and machinery in a built up area would be difficult. The solution proposed was a regrading of the slope together with stabilisation using soil nails.

The environmental sensitivity of the project meant that there were two major components to the stabilisation work. The design of a soil nailing system to prevent further creep of the slope, and the design of a facing to minimise localised surface movement between the nails and to facilitate the re-establishment of vegetation.

Nail Design

The design of the nail layout was undertaken according to the Department of Transport design guide for the reinforcement of highway slopes by reinforced soil and soil nailing techniques (HA68/94 1994). The nails were designed to a working load of 40kN and were inclined at an angle of 15° to assist with the grouting process.

The final design section layout is given in Figure 1 (Rust Environmental 1996). The nails were spaced at 1.5 metres centres and installed in a diamond grid pattern. The nails consisted of 20mm diameter bars grouted into 100mm diameter holes (Tables 3 and 4), varying from 4 to 9.5 metres in length depending on location. Each nail was protected from corrosion using a single 70mm diameter corrugated duct.

TABLE 3 -- *Nail properties.*

Nail	Bar diameter, mm	Young's modulus, kN/mm^2	Ultimate Tensile Strength, kN
MacAlloy 500, galvanised steel	20	205	173

TABLE 4 -- *Grout properties.*

Grout	Compressive Strength, kN/m^2	
	14 day	28 day
Ordinary portland cement	53 – 62	67 – 87

Surface Protection

Removal of the existing vegetation was required prior to nail installation. Therefore, a system of surface protection was designed, with resistance to localised failure and facilitated the growth of new vegetation.

Geogrid netting (Table 5) was used to prevent localised movement between the nails and was fixed to each nail head using a 250mm by 250mm steel plate. The growth of the vegetation was facilitated by using a biodegradable hessian mat ('geojute' impregnated with seed laid over the geogrid). In addition to the vegetation, trees and shrubs were planted to improve the root matrix with the aim of reducing surface soil erosion.

TABLE 5 – *Geogrid properties.*

Geogrid	Thickness, mm	Grid size, mm^2	Mass per square metre, kg/m^2	Strength, kN/m
Tensar GM4	1.7	62 x 62	0.34	15

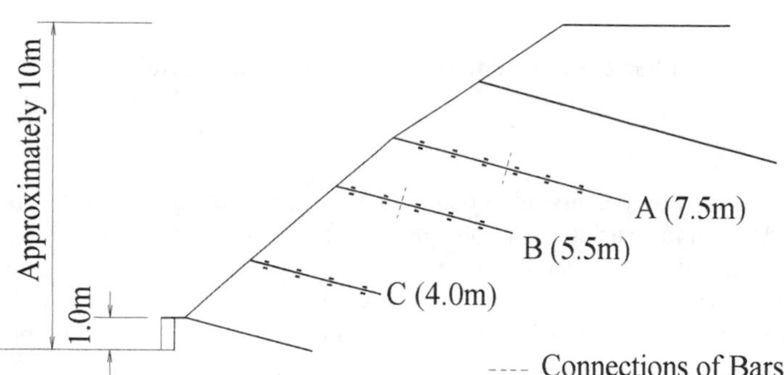

FIG. 1-- Nail layout: Section view.

Instrumentation

The objective of the instrumentation was to provide a monitoring system to assess the contribution of the nail system to the stability of the slope, within the budget limitations. This allowed for installation, and monitoring for 18 months, of three strain gauged soil nails.

Strain Gauged Nails

The strain gauges are installed to provide measurements of strains induced by the soil mass loading the nail. The gauges are bonded to the prepared surface of the bar. Protection against damage and waterproofing are made by potting with Flexonoss polyurethane (Figure 2). The lead wires from the gauges were three strand waterproof Teflon. This system was both effective and suitable for the harsh conditions, with minimal change to the surface properties of the nail.

The gauges used were of 350-ohm resistance, and temperature compensated for steel. The active gauge length was 3.2mm, with a range of ±1500 microstrain and a cyclic loading life of 10^6 cycles. Details of the gauge set-up are given in Figure 2.

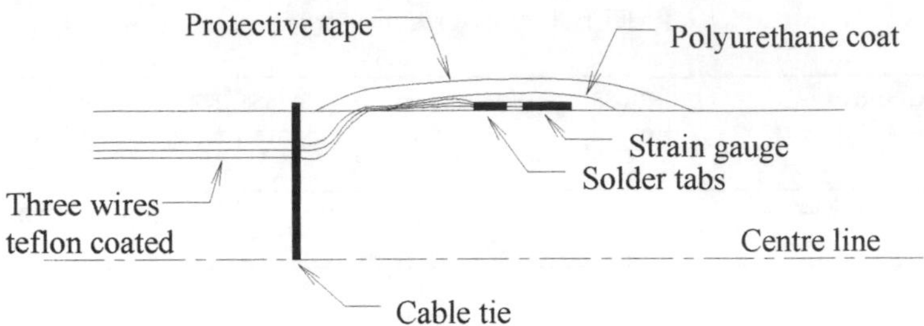

FIG. 2 -- Gauge mounting and protection detail.

The gauges were positioned in pairs at one metre spacing along the bars as shown in Figure 3. At each measuring point, one gauge was mounted on the top of the bar, the other on the underside. This accounted for any strain variation between top and the bottom of the nail to be monitored. The cables were then fed along the nail to the surface. Three bars were instrumented: 7.5m, 5.5m and 4m length with 6, 5 and 4 pairs of gauges respectively. To allow transport to site, the 7.5m and 5.5m nails were in two sections and couplers were used on site.

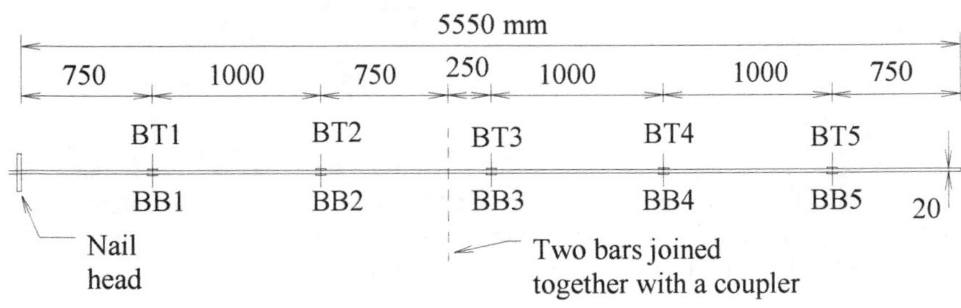

FIG. 3 -- Strain gauge layout along the nails (Example: nail B).

Wiring and Bridge Completion System

The lead wire length of the nails (up to 10 metres) meant that using a standard two wire ¼ bridge arrangement would lead to temperature fluctuations having a significant effect on the stability of the bridge measuring system. This is countered by implementing a 'three wire' configuration. The resistance change in the wires now appears on both arms of the bridge, and measurement is made at the gauge.

The wires are terminated at the head of each nail using a multi-way plug with gold plated pins. The plug is then connected to a switching box that completes half of the measuring bridge with another 350-ohm resistor, and allows up to 10 gauges to read.

Data Logger

The strain changes are measured using a HBM DMD 20A digital strain meter. The meter completes the second half of the bridge with two 120-ohm resistors, and has a gauge factor setting that allows the direct digital readout in microstrain. The supply voltage to the completed bridge is one volt and has a measuring range of ±19 999 microstrain.

The HBM allowed 10 strain gauges to be monitored in sequence using a switching mechanism. Measured values are recorded manually during each visit and input into a spreadsheet for processing.

Construction Works

The nails were delivered to site at Exmouth prior to installation for testing to ensure all gauges were working correctly prior to installation.

Location in the Slope

The section the nails were installed in the slope was chosen for a number of reasons:

- the section was fairly constant in height and slope inclination
- from the borehole logs, rock head appeared to be furthest from the surface
- the section was adjacent to a borehole
- good access from both the toe and the crest of the slope in this area
- minimal trees to influence the soil nailing system.

The instrumented nails were to be in the second, third and fourth rows of the five rows of nails installed in this section (Figure 1). Due to financial constraints, not all the nails in the section could be instrumented. The three selected were chosen since they were in the central body of the slope.

Slope Preparation

The slope surface before nailing began comprised of large areas of brambles and small trees. A number of these had to be removed to provide access to the slope surface. In areas, a certain amount of regrading was required to maintain the stability of the slope.

To allow operation of the drilling equipment to be undertaken safely, a scaffold arrangement of temporary works was erected before nail installation could begin. The temporary work was erected to provide a platform to give access for drilling for the top row first. As nailing was completed, a level of scaffolding was removed and the next row installed.

Nail Installation Procedure

Once the platform had been erected the nail positions were marked out and the drilling rigs moved into placed. A Twin Tech 4 rig was used to drill 100mm diameter holes, using 1.5 metre extension rods to reach the required lengths. The method of drilling involved a short length of casing then open hole using air flush.

The tensile element of the nails were 20mm diameter MacAlloy 500 hot dipped galvanised ribbed steel bar (ultimate tensile strength of 173kN). These were provided to site in up to four metre lengths. Where necessary lengths were connected together using threaded couplers.

The design of the nail included a 70mm diameter Drossbach plastic corrugated duct, encapsulating the nail for corrosion protection. The duct was installed in the drilled hole first, using 'Chinese lantern' centralisers to position them in the hole. A grout hose was then coupled to the head of the duct and grout pumped down to the base until it returned up the outside of the duct to the surface. The grout hose was then removed and the steel bar with centralisers at 1.5m spacing pushed down the centre of the duct.

The instrumented nails were carefully installed in this way, ensuring the strain gauges were in the correct orientation and that the lead wires were safely fed along the nail to the surface.

As nailing proceeded, drainage pipes were also installed in the slope at ten metre intervals between the nails and of similar length. These were intended reduce pore water pressure within the slope.

Surface Replacement

Once a row of nails had been installed, the scaffolding platform was lowered and the next row of nails installed. In parallel with this activity, the surface of the previous levels was protected. The 'soft, flexible' facing applied was a geogrid. Lengths of the geogrid were rolled down the slope and fixed to the surface at the nail heads using facing plates bolted to the nail heads, and in between the nails using 0.5 metre long, 10mm diameter pins.

The face plates providing connection between the facing and the nails consisted of a 250mm by 250mm galvanised steel plate embedded on a concrete base around the nail protruding from the slope. A nut was tightened down so the plate was firmly bedded onto the concrete. The remaining length of nail was then cut off at the nut. A grease filled cap was then applied to the nail head to protect it from corrosion.

Once the nailing had been completed, hessian 'geojute' was rolled down the slope, and fixed to the geogrid. Small shrubs and trees were then planted through the surface protection.

Monitoring and Results to Date

Once the nailing work was complete, a base reading was taken from the strain gauges. The initial monitoring period was from August 1996 to January 1998. This involved a number of visits to the slope, approximately every eight weeks to take a measurement from the gauges (Jones 1998).

In conjunction with the strain measurements, a visual survey was made of the slope during this period to observe whether any further cracks had developed, or if there were any other signs of localised failure.

The strain gauge results presented here are for the period October 1996 to January 1998. There was initially a difficulty in acquiring a stabilised zero reading for the strain gauges. This was diagnosed as a capacitance problem that required an earth of the system once the nails were installed. Once this had been resolved, a base reading was taken. All changes calculated were relative to this.

Pullout Tests

Pullout tests were undertaken during nail installation to verify the design pullout capacity of 10kN per metre length. The tests were undertaken using a hollow ram jack with hand pump, a pressure gauge, and a 50mm dial gauge to measure displacement. The dial gauge was mounted remotely from the stressing frame.

Nine tests were carried out in accordance with the contract specification. Incremental loading and unloading took place over two cycles with movement being recorded at 1,2,3,4,5,6 and 10 minutes for each load increment/decrement. During this 10 minute period, the nails showed a small creep movement. This is shown by an increased displacement at a constant load. Figures 4 and 5 present the loading and

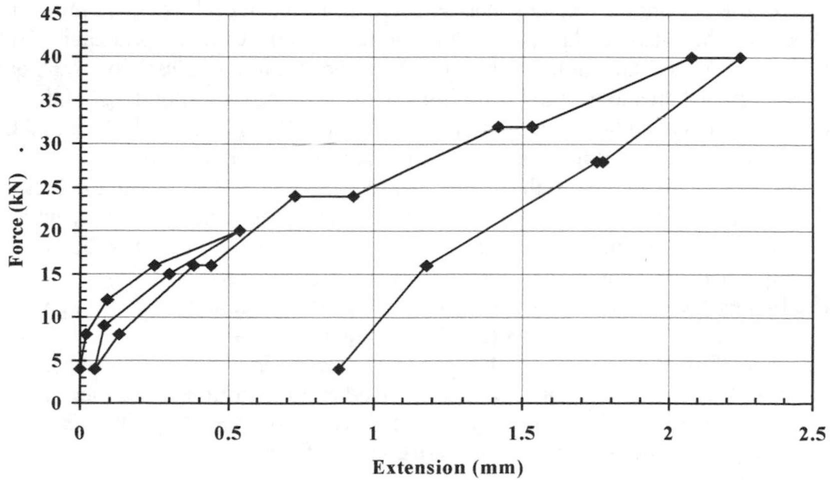

FIG. 4 -- Force versus extension plot for pullout test PT6.

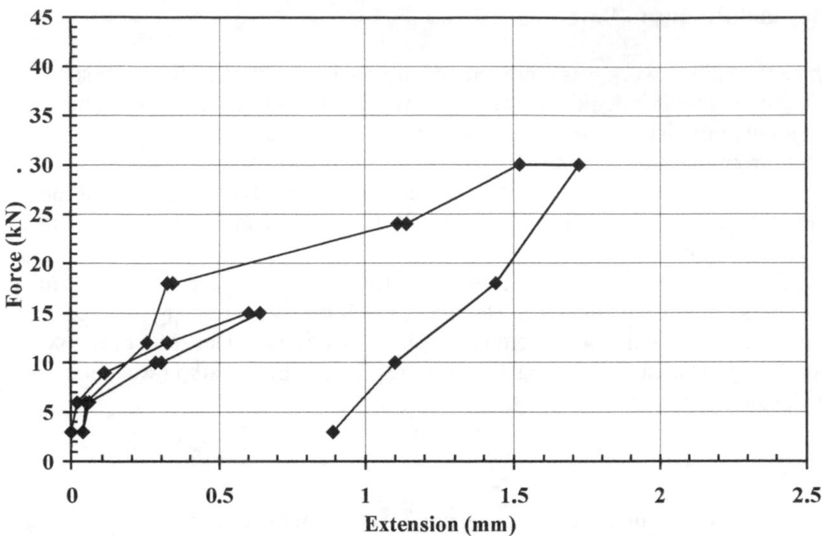

FIG. 5 -- Force versus extension plot for pullout test PT9.

unloading curves for pullout test nails PT6 and PT9, each installed at 3m and 9m below the crest level respectively. The bonded length of test nail PT6 was 4.0m and was loaded to 40kN. The bonded length of test nail PT9 was 2.5m and was loaded to 30kN.

Nail Loading

Figures 6-8 present the changes in axial force measured at each gauge point over the initial period. This was calculated by averaging the strain measurements at each point shown in Figure 3. This axial strain was multiplied by the modulus of elasticity of the bar and the cross sectional area to give the resulting axial force. The corresponding axial force profiles developed along the nails are given in Figures 9 to 11, for nails A, B and C, respectively. The axial loads induced are quite small; hence they are currently contributing only a small amount to the stability of the slope. The peak load in the bottom level nail is about 6kN, but loading is typically 3kN. The loading pattern along the nail, consistent with expectation, increases towards the middle and tails off towards the distal end of the nail, albeit at such low loads the trend is not strongly marked.

The figures of axial force versus time for each gauge position does show a general trend of increased loading. This appears to oscillate slightly, probably due to the varying moisture content of the slope. This is highlighted more clearly in Figures 12-14 where gauge point results are compared for each instrumented nail for particular distances from the slope face. Though generally they show an increase in load, some drops in load (shrinkage in slope) occurring in the summer are consistent.

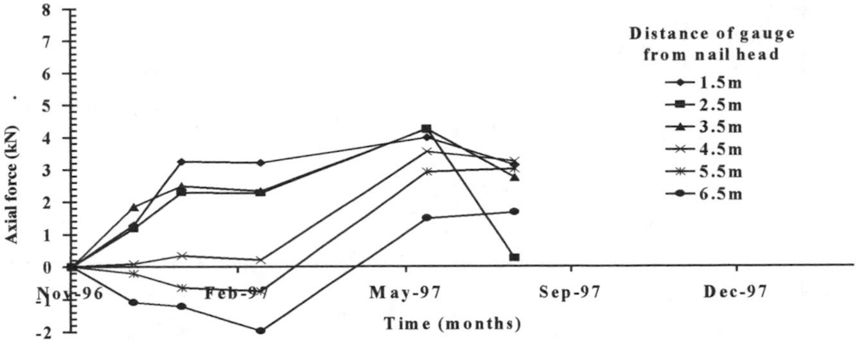

FIG. 6 -- Nail A, Axial force at each gauge point versus time.

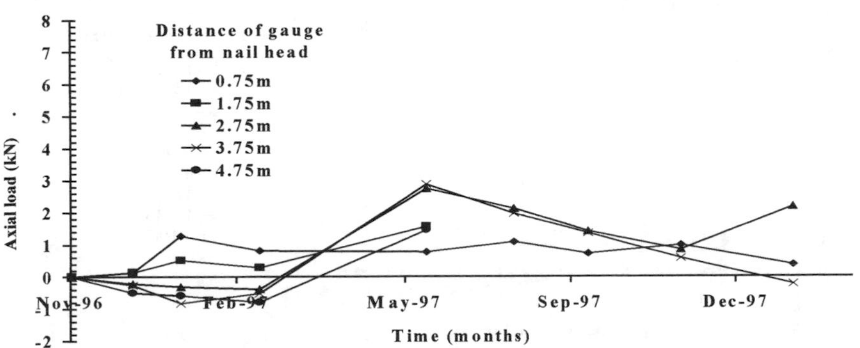

FIG. 7 -- Nail B, Axial force at each gauge point versus time.

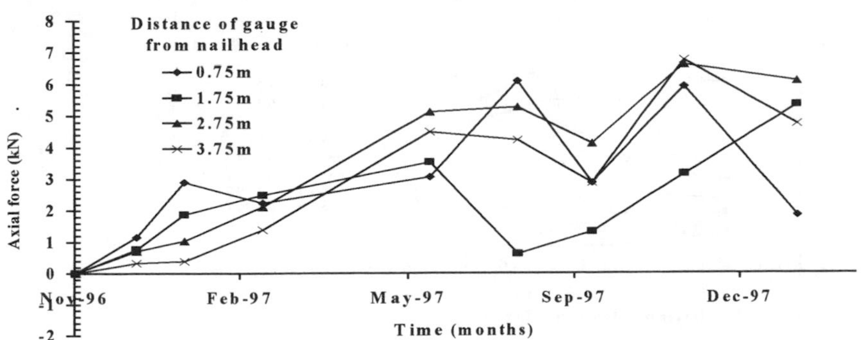

FIG. 8 -- Nail C, Axial force at each gauge point versus time.

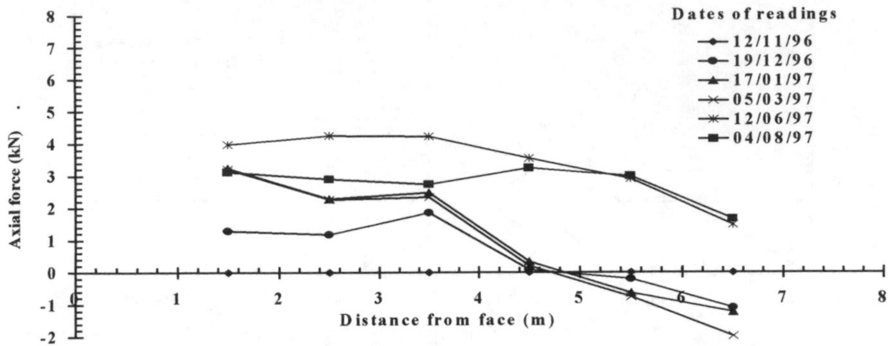

FIG. 9 -- Nail A, Axial force developed along the nail.

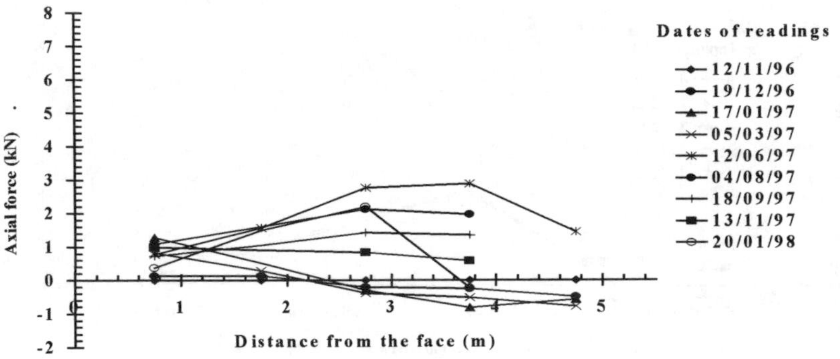

FIG. 10 -- Nail B, Axial force developed along the nail.

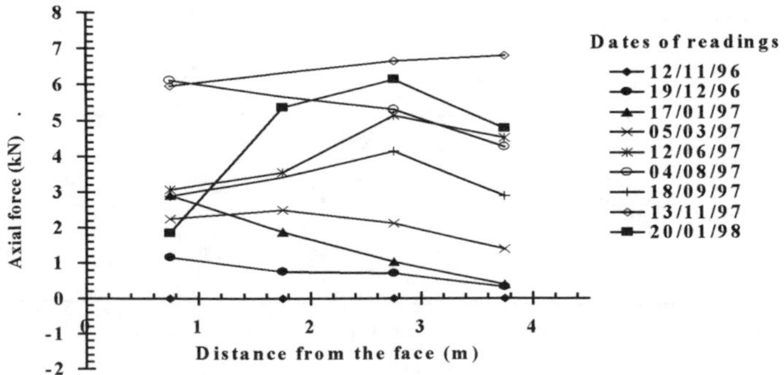

FIG. 11 -- Nail C, Axial force developed along the nail.

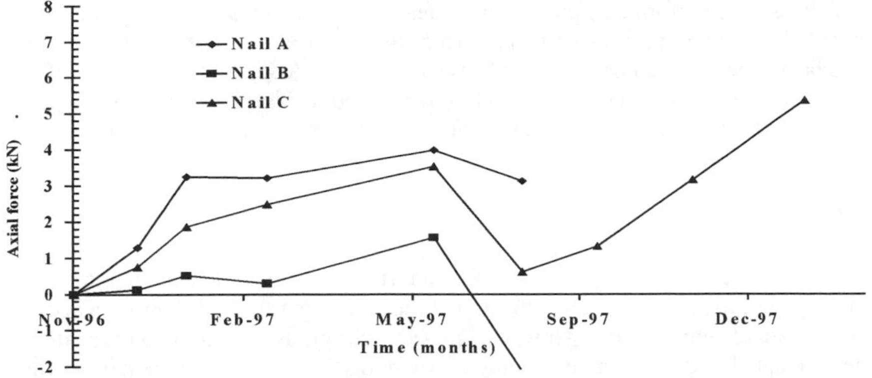

FIG. 12 -- Axial force in each nail at 1.75m from the slope face.

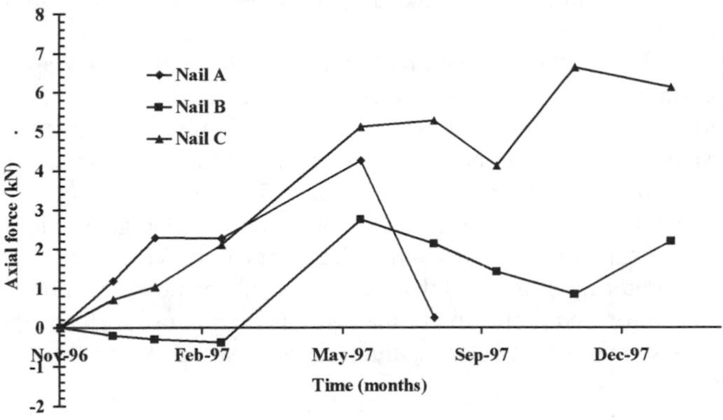

FIG. 13 -- Axial force in each nail at 2.75m from the slope face.

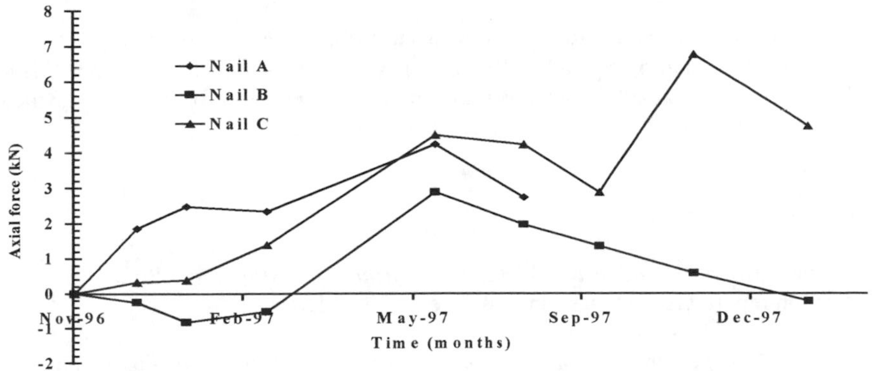

FIG. 14 -- Axial force in each nail at 3.75m from the slope face.

Most gauges have performed well and show steady changes in results, though there appears to be some problems with nail A (hence the absence of recent data points). The gauges have recently been providing rather erratic results. Also, contraction and expansion of the soil due to climatic conditions may induce the negative axial forces measured at some points along the nails, but these loads tend to be very small.

Conclusions

Over the eighteen month period of monitoring no visually detectable movement of the slope occurred which is an indication of the soil nails fulfilling their intended purpose of stabilising the slope. However, slope movement is necessary to cause loading of the soil nail. The developments of only nominal loads (3 to 6 kN) in the nails during the monitoring period is consistent with the occurrence of only small undetectable amounts of movement. It is noted that these changes are measured from post construction in an attempt to distinguish the load induced by the soil mass as opposed to the effects of the curing of the grout.

It may well be that the installation of drainage holes within the slope itself made a major contribution in reducing loading and enhancing soil strength. However, the monitoring period of eighteen months is a very short spell of the intended life span of the soil nailing system. Over a longer period of say 15 to 20 years the monitoring would detect greater nail forces and a more significant contribution.

The project has shown the successful application of strain gauges to monitor nail loading in geotechnical works with a small budget that can provide valuable information from long term monitoring. It is noted that success is only achieved as a result of careful preparation, installation and protection of the instrumentation used. Hence the experience of this project has shown, with two of the three sets of instruments on the nails still providing data.

ACKNOWLDGEMENTS

The authors would like to express thanks to Keller Ground Engineering for financing the research project and EPSRC for supporting the research studentship. Also East Devon District Council, Rust Consulting and Geotechnics Ltd for the data provided and Measurements Group UK for instrumentation of the nails.

References

Geotechnics Limited, *"Madeira Walk, Exmouth Ground Investigation,"* Site investigation report, Geotechnics Limited, Exeter, UK. March 1995.

Jones, A.M., *"The performance of soil nailed slopes,"* PhD thesis, Division of Civil Engineering, University of Wales Cardiff, UK, 1998.

HA68/94, *"Design methods for the reinforcement of highway slopes by reinforced soil and soil nailing techniques,"* Design manual for Roads and Bridges, Volume 4.1, HMSO, London, 1994.

Rust Environmental, *"Design of cliff stabilisation works, Madeira Walk, Exmouth,"* Design Report, Rust Environmental, Bristol, UK, February 1996.

Mark C. Webb,[1] Ernest T. Selig,[2] and Timothy J. McGrath[3]

Instrumentation for Monitoring Large-Span Culverts

Reference: Webb, M.C., Selig, E.T., and McGrath, T.J., **"Instrumentation for Monitoring Large-Span Culverts,"** *Field Instrumentation for Soil and Rock, ASTM STP 1358*, G.N. Durham and W.A. Marr, Eds., American Society for Testing and Materials, 1999.

ABSTRACT: An extensive instrumentation plan was devised to monitor large-span culvert behavior, soil behavior and culvert-soil interaction during backfilling and during live load testing. The goals of the instrumentation plan were to help control construction deformations, including insuring safe working conditions, and to gather sufficient data to advance the state of the art in the design and construction of large-span culverts. The instrumentation plan and sample results are discussed.

A 9.1 m (30 ft) span reinforced concrete arch and a 9.5 m (31 ft) span structural plate corrugated steel arch large-span culvert were tested. Both had shallow cover and live loads. Two tests were conducted, one with dense backfill and one with loose backfill.

Measurements were made of culvert shape, culvert strains, culvert-soil interface pressures, soil density, soil stresses, soil strains, foundation movements, temperature, cracking of the concrete culvert and shear movements between precast concrete segments. The most unique measurement was made by a custom built laser device for measuring movements of specific points on the culvert wall. The accuracy of this device was verified by level survey and tape measurements.

The instrumentation performed well and measured results were within the range expected. Culvert deformations were efficiently measured without impeding construction operations. Culvert-soil interaction effects during backfilling and during live load testing were reasonably defined with the instruments selected.

KEYWORDS: instrumentation, large-span culverts, culvert-soil interaction

---

[1]Senior Engineer, Department of Civil and Environmental Engineering, University of Massachusetts, Amherst, MA 01003.

[2]Professor of Civil Engineering, Department of Civil and Environmental Engineering, University of Massachusetts, Amherst, MA 01003.

[3]Principal, Simpson Gumpertz & Heger, Inc., 297 Broadway, Arlington, MA 02174-5310.

INTRODUCTION

The paper describes the experience gained from instrumenting two large-span culverts for monitoring their behavior during backfilling and during live loading. Two purposes of the instrumentation were to insure that excessive culvert distress did not occur and to provide safety for the workers during construction and during live load testing. In addition, detailed measurements of soil-structure interaction were collected for advancing the state-of-the-art in designing and constructing these large-span culverts. Sample results from the test program are included in this paper.

TEST DESCRIPTION

Two full-scale tests were performed as part of a project called **Recommended Specifications for Large-Span Culverts**, with the objective of developing reliability-based design specifications and construction criteria, for flexible and rigid large-span culverts. Details about the testing program and measured results are given in Webb (1998) and Webb et al. (1998 and 1999).

The tested large-span reinforced concrete arch and large-span structural plate metal arch culvert installed end-to-end in a pre-excavated wide trench are illustrated in Figures 1 and 2. The depth of the trench was 3.0 m (10 ft) so that the culverts were buried with tops about 0.6 m (2 ft) above the existing ground surface (Figure 2). Each culvert was founded on separate reinforced concrete strip footings. The trench was backfilled and a shallow earth embankment was constructed over both culverts in layers. Existing site material was taken from the excavated trench and used for backfilling the culverts and for building the embankment. The in-situ material is a well-graded sand with gravel, per ASTM D 2487, with 1 % fines. The AASHTO M 145 classification is A-1-b, while the Unified Soil Classification is SW. Live load testing with a heavily loaded truck was carried out at various stages of shallow cover.

The installation was done twice. In the first construction sequence, Test 1, structural backfill was compacted and a depth of embankment over the culvert was limited to 0.9 m (3 ft) of soil cover. After the final embankment height was reached a live load test was conducted, which was repeated twice more while the embankment was removed in 0.3 m (1 ft) lifts. Then the culverts were uncovered by removing the remaining embankment and structural backfill material. In the second construction sequence, Test 2, the structural backfill material was placed without compaction up to the top of the arches and then compacted above the top. Live load testing was conducted as in Test 1.

After the completion of Test 2, embankment material was added over the top of the culverts to bring the height of cover to about 1.4 m (4.5 ft). This cover height remained in place for one year to permit monitoring culvert performance over a period of time with constant dead load.

The concrete arch culvert is a 9.1 m (30 ft) span by 3.5 m (11 ft 4 in.) rise (inside dimensions) structure with a constant wall thickness of 254 mm (10 in.). The 12.8 m (42 ft) long structure is assembled from precast segments with a width of 1.82 m (5 ft 11-1/2 in.).

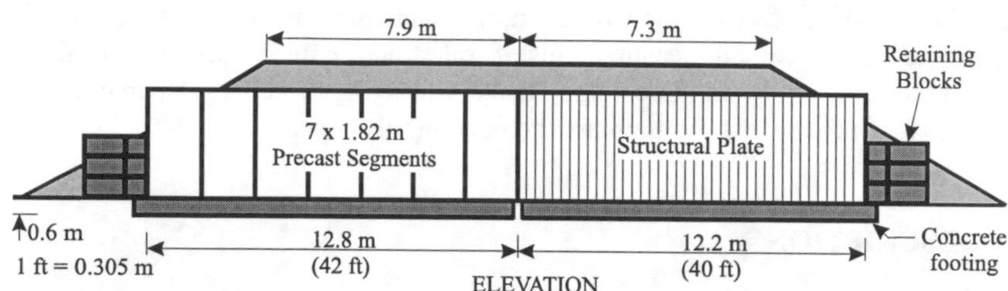

FIG. 1--*End-to-End Installation of Concrete and Metal Culverts*

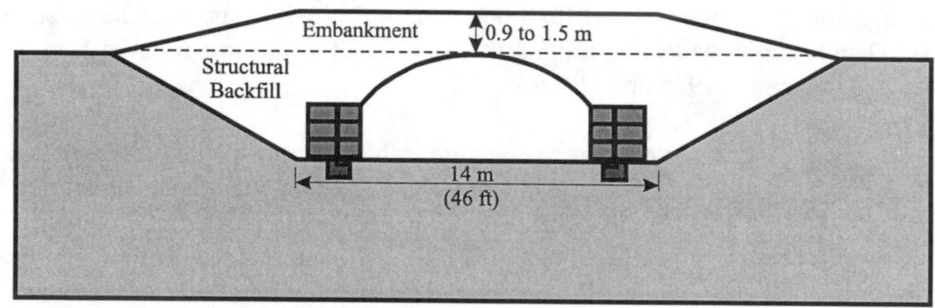

FIG. 2--*Cross-Sectional View of Culvert Installation*

The metal arch culvert is a non-galvanized corrugated steel low-profile arch structure with a 9.50 m (31 ft 2 in.) bottom span and a 3.7 m (12 ft 1 in.) total rise. The culvert is manufactured from structural plates with 152 mm by 51 mm (6 in. x 2 in.) corrugations. The plate thickness is 5.5 mm (0.218 in.) designated as 5 Gauge. The assembled length was 12.2 m (40 ft) long.

The measured parameters and their purpose are summarized in Table 1. A description of each instrument is presented in the following sections.

CULVERT DEFORMATION

Deformation Categories

Culvert deformation measurements were grouped into the following categories:

1) remote readout measurements taken during construction operations,
2) detailed measurements taken when no construction operations were active, and
3) measurements during live load testing.

TABLE 1--*Purpose of Measurements*

Measurement	Purpose
	Metal Culvert
Deformation	Culvert shape and distortion
Strain	Culvert deformations Stresses, moments, and thrusts in culvert wall
Interface Pressure	Culvert-soil interaction Distribution and effects of vehicular loads
Wall Temperature	Correlate temperature change with culvert behavior
	Concrete Culvert
Deformation	Culvert shape and distortion
Interface Pressure	Culvert-soil interaction Distribution and affects of vehicular loads
Crack Length and Width	Culvert performance and distress
Relative Segment Movement	Distribution and effects of vehicular loads
Wall Temperature	Correlate temperature change with culvert behavior
	Foundation
Settlement	Culvert distortion and foundation stability
Transverse Spread	Culvert distortion and foundation stability
Rotation	Settlement and transverse spread
	Soil
Soil Stress	Culvert-soil interaction and live load modeling
Soil Strain	Culvert-soil interaction
Soil Moisture Soil Unit Weight	Correlation with standard field practice Weight of soil over culvert
Soil Stiffness	Settlement predictions and culvert-soil interaction
Surface Elevations	Depth of cover and weight of soil over culvert
	Other
Photographs	Visual documentation
Live Load Magnitude	Document loading
Temperature and Rainfall	Document environmental conditions

Category 1 - Construction Monitoring--Culvert movements during construction operations were monitored to evaluate structural stability and warn of excessive deformations. For this purpose it was necessary to select instruments with remote reading capabilities for safety of workers. Conventional level survey was selected for monitoring vertical deformations while structural extensometers (with electrical displacement transducers) were selected for monitoring horizontal deformations.

Category 2 - Detailed Measurements--Detailed deformation measurements were used for structural analysis and also to expand on, as well as check, readings taken during construction operations. Electrical readings with an infrared laser device at three stations along the length of each culvert were used for this category supplemented by tape extensometer readings. Digital level survey and tilt sensors were used to monitor vertical footing movements at the same stations.

Category 3 - Live Load Monitoring--Evaluating live load effects on the culverts required monitoring of the culvert shape at multiple points quickly. The infrared laser device was selected for this purpose and readings were taken at the three longitudinal stations in each culvert. Level survey was used to obtain the deflection basin of the metal culvert crown during live load testing.

Level Survey

For the digital level readings, photocopies were made of the bar code staff and attached to the inside walls of the culverts to monitor vertical movements at various locations. The paper photocopies were bonded to strips of plastic, about 450 mm (18 in.) long and covered with a transparent polyurethane coating for a moisture seal. Bar codes covered with less transparent coatings were found to be more difficult to read with the digital level even in very bright surroundings. Since the bar codes were installed on the insides of the culverts (in the shade) less transparent coatings would definitely not have worked. To further expedite the reading process all bar codes were extended up or down from the permanent point to the same level to prevent needing to reset the digital level every time that a higher or lower bar code had to be read.

The reported reading accuracy of the digital level was 1.0 mm (0.04 in.) when a foldable bar code staff is used. This was sufficient for monitoring the construction deformations. The digital level was a Model DiNi 10 manufactured by Zeiss.

The digital level together with photocopied bar codes installed on the culverts worked well to monitor culvert behavior without impeding construction operations or risking injury to personnel. One problem that was encountered with this technique, however, was the inability to read the bar codes during vibratory compaction when both the culvert and the foundation soil experienced large vibrations. This problem was most pronounced when a large ride-on vibratory roller compactor was operated close to or immediately above the culverts. However, for smaller walk-behind vibratory plate compactors vibrations caused less problems. This problem was further alleviated by the fact that in most cases it was acceptable to obtain culvert deformation only during placement and after compaction of backfill material layers.

Three reading locations were selected at each of two cross-sections along the length of each culvert as shown in Figures 3 and 4 to monitor construction deformations. For the metal culvert bar codes were attached at the crown and points of curvature change where the small radius side plates meet with the large radius top plates. For the concrete culvert bar codes were attached at the crown and shoulders. The two selected cross-sections for metal culvert were Stations M-P1 and M-S3 (Figure 3) and Stations C-P1 and C-S2 for the concrete culvert (Figure 4). The digital level was set up at an angle to the centerline of the culverts to permit seeing all the bar codes along the length of both culverts. Results of level survey measurements with this arrangement will be discussed in a later section.

The deflection basin under the live load vehicle in the metal culvert was measured with the digital level by holding the inverted bar code staff at 12 longitudinal crown locations.

Structural Extensometers

The structural extensometers consisted of a displacement sensor attached to the structure with tensioned cables using inductance coils as the displacement transducer (Figure 5). These gages have been extensively used in the past and are described by Selig (1975a and b). Locations of the structural extensometers are shown in Figures 3 and 4 for the metal and concrete culverts, respectively.

Inductance Coils--Each transducer consists of a pair of wire coils wrapped around non-conductive disks. An alternating voltage applied to one coil induces a voltage in the second coil which is related to the distance between the coils. The readout device, automatically balances the output and converts it to a direct current signal that could be collected with a data acquisition system.

Gage Description--The coils were mounted on two transfer rods which were each connected to steel cables which were attached to eye bolts inserted in the culvert walls (Figure 5). Therefore, the structural extensometers could be installed at any position on the culvert and spanning any distance. Springs were used on the outside sleeve tubing of the sensor to maintain an adequate tension in the cables. To increase the working range of the inductance coils, steel rings were added or removed to adjust the length of the device when the sensor approached the ends of its range. This aided the monitoring process since, during the stage where backfill is placed at the sides of a flexible culvert, causing the sides of the culvert to move inward, the rings could be removed without cutting or adjusting the steel cables. Conversely, during the stage where backfill material is placed over the top of these flexible culverts the inverse is true as the sides of the structure move outward thus causing the sensor to extend.

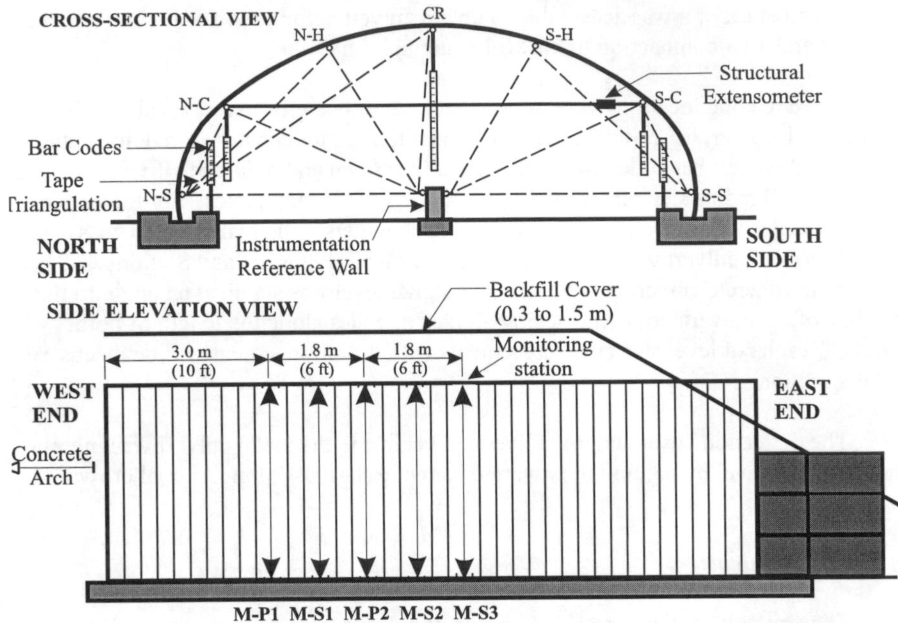

FIG 3--*Monitoring Deformation in Corrugated Metal Culvert*

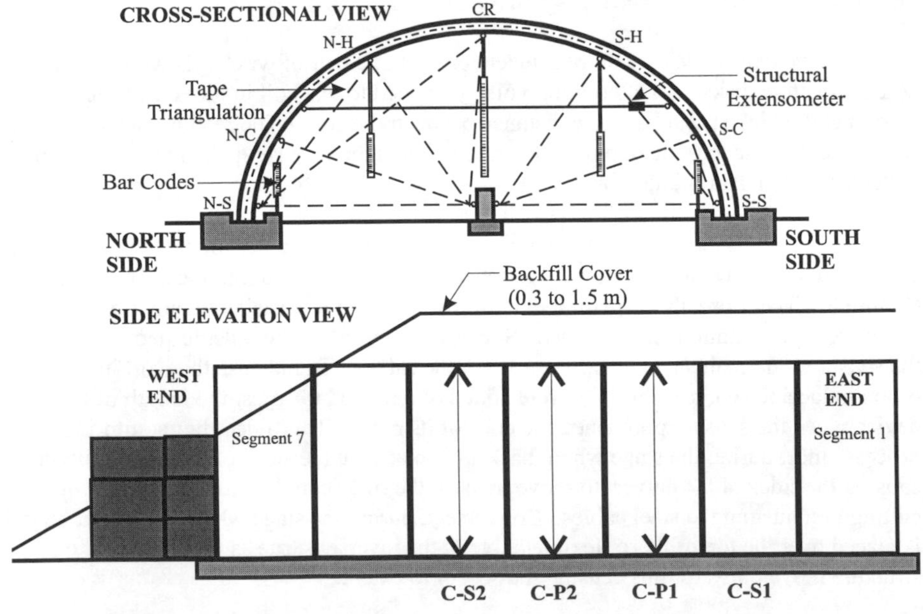

FIG 4--*Monitoring Deformation in Reinforced Concrete Culvert*

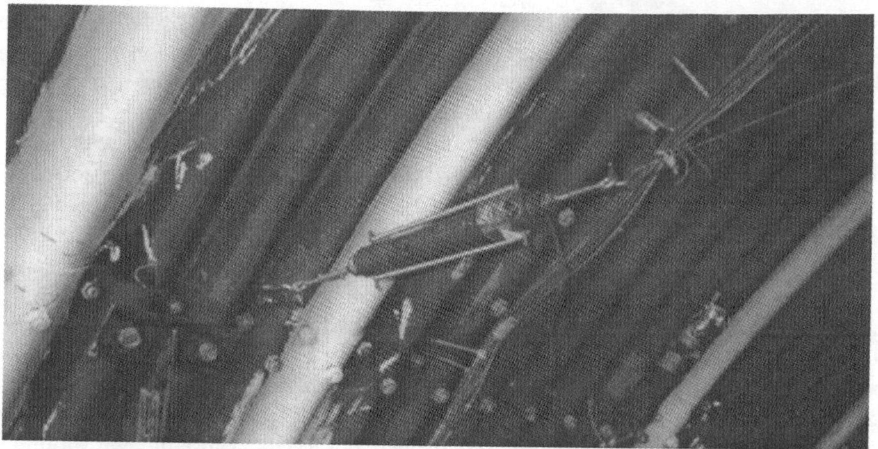

FIG. 5--*View of Structural Extensometer*

The sensitivity and range of the coils is a function of the size and number of turns of wire. For this project 25 mm (1 in.) diameter coils were used which had a range of approximately 75 mm (3 in.), bias of 1.3 mm (0.05 in.) and a resolution of 0.13 mm (0.005 in.) which was controlled by the characteristics of the data acquisition system.

Readout Instrumentation--The readout unit, called the ε-mu and developed at Nottingham University, had only one channel. Thus, manual switch boxes were used to direct the signal from the appropriate sensor into the unit. The processed signal coming out of the readout was collected by a pc-based data acquisition system. The data acquisition program (LabTech Notebook by Laboratory Technologies of Wilmington, MA) for these gages had a manual trigger to allow the operator time to change the switch boxes.

Infrared Laser Device

Infrared Laser--The laser device, an AccuRange Model 4000-LIR manufactured by Acuity Research Inc., Menlo Park, CA emits near infrared light (780 nm wavelength) and can be used for measuring distances up to 15 m (50 ft). Measurement accuracy is affected by three types of noise in different ways and as a function of sample rate. These are detector thermal noise, laser diode noise and a resolution limitation imposed by the maximum sample range. In addition, the reflectivity of the target and therefore the amplitude of the return signal also affects measurement accuracy. Since most of these issues could be controlled by selecting the optimum sample rate, maximum range specification and even improve the reflectivity of targets by painting them broken white or grey colors, the measurement accuracy could be controlled. For the un-galvanized metal culvert (erected from black structural steel plates), white stripes were painted on the inside surface at each measurement station to enhance the reflectivity of the culvert surface. The reflectivity of the concrete culvert surface was sufficient and did not require any painting. Therefore, the laser configuration as used was capable of reading distances up to 6 m (20 ft) with a maximum bias (inaccuracy) of about 2 mm (0.08 in.). Since the laser beam is near infrared, the beam could not always be visually detected to check target alignment. Thus special filter paper had to be used to find the beam in bright light conditions. However, in darker conditions the beam could easily be spotted with the naked eye. Laboratory evaluation showed that the laser device has a warm-up period of about 25 to 50 minutes. Therefore, during field testing the laser device was turned on about 1 hour before any measurements were made and left on during testing.

Pan-Tilt Unit--The laser was mounted on a pan-tilt unit as shown in Figure 6. The pan-tilt unit, Model PTU 46-17.5 manufactured by Directed Perception, Inc., has a tilt range of approximately 78° while it could simultaneously pan through 340°. Specifications of the pan-tilt unit include precise control of position, speed and acceleration with a resolution of 3.086 arc minute (0.051428°) and speeds up to 300° per second. However, too fast acceleration speeds were observed to cause the pan-tilt unit to skip steps causing misalignment of the laser beam. This was a result of a combination of the weight of the laser and restrictions of movement by electrical cables (it was important to ensure that no cables were caught underneath the support fixture). The pan-tilt unit connected to a RS-232 interface on the computer.

FIG. 6--*Laser Device Mounted on Pan-Tilt Unit*

Mounting and Operation--The laser and pan-tilt unit were mounted on a sliding support fixture as shown in Figure 6. In the center of each culvert a concrete wall was constructed extending down the length of the culvert on which steel channels were installed to guide the support fixture. The support fixture could be moved down the length of the culverts to any longitudinal station. The base of the support fixture was designed so that it could be accurately repositioned at the same station location and locked in place during a laser scan.

Using the pan-tilt unit the laser could be pointed at any location on the inside circumference of the culverts fairly quickly and very accurately. This enabled tracking the movements (x, y and z directions) of unique targets around the culvert circumference as the construction and testing progressed. To achieve this, software was developed to control the pan-tilt unit while simultaneously allowing measurements to be made with the laser. Each unit had to be controlled separately through two serial ports on a computer. The basic procedure was to first aim the laser beam to a specific point on the structure by advancing the pan-tilt unit through software and then recording the exact coordinates. This procedure was repeated for each location around the culvert circumference of a specific longitudinal station, therefore creating an input data file for that station. Next, a second software program would read the input data file just created and instruct the pan-tilt unit to advance to each location and allow the laser device to make measurements. Typically, this scanning of the inside circumference was repeated twice to improve the measurement precision which took about two minutes for 26 reading locations around the circumference. On subsequent scans a third software program would read the input data file created during Step 1 (or any other increment) and stop at each location allowing the user to fine adjust the laser beam if any movements occurred. The coordinates were then automatically recorded for use in Step 2. Finally, the measured results were written to hard disk and processed for almost immediate graphical display.

Measurement Locations--A total of 26 locations around the circumference of the metal culvert and 23 locations for the concrete culvert were selected for the detailed laser measurements at each of 3 monitoring stations. All measurements for the metal culvert were made on the inside crest of the corrugations. Example results of laser measurements are discussed in a following section.

Tape Extensometer

Triangulation with distance measurement using a tape extensometer provided a check on the laser measurements. This procedure is highly reliable and accurate, but is time consuming and thus was used only a few times during each test (including the beginning and end of each test) at a relatively few locations. A total of 5 points (Figures 3 and 4) on the culvert wall at each cross-section were measured relative to a fixed reference. Since footings may move and rotate under the applied loads, instrumentation reference walls were constructed as the fixed reference (Figures 3 and 4). Eye bolts installed on the culvert walls and on the reference walls were used to connect the tape.

Measurements were recorded as a combination of a tape measure reading in feet and inches to the nearest 2 inches (25 mm) and a dial gage reading from 0 to 2 inches in 0.001 inch (0.025 mm) increments. All measurements were corrected to account for temperature changes in the steel tape length. For the distances measured the precision was about 0.13 mm (0.005 in.). Distances of up to about 60 m (200 ft) can be measured.

METAL CULVERT STRAIN

Metal culvert wall strains were measured with weldable foil electrical resistance strain gages (Measurements Group, Inc. Type CEA-06-W250A-350). Weldable gages were selected because of their suitability for field installation and long-term stability. Gages with pre-attached leads were not used because of their size and difficulty in waterproofing the Teflon coated leads. However, thin jumper wires were pre-soldered to the gages in the laboratory to speed up actual field installation. Weldable gages are made up of polyimide-encapsulated foil grids pre-bonded in the laboratory to thin (0.13 mm) metal carriers with a high-performance adhesive. However, the effective gage factor of the foil gage on its metal carrier is less than that of the basic strain gage because of the stiffness of the metal shim. Therefore, a steel test beam was instrumented with both conventional and weldable strain gages to obtain the correct properties of the weldable gages.

These encapsulated strain gages had large rugged soldering tabs for easy soldering. Furthermore, the gages could be contoured to culvert corrugation radii as small as 13 mm (0.5 in.). The gages are fully encapsulated for optimum environmental protection however, additional waterproofing was added. The strain gages are self-temperature-compensated for steel, with a resistance of 350 ohms. The gages were wired into a full Wheatstone bridge circuit using the three leadwire system and bridge completion modules supplied by Campbell Scientific, Inc.

Installation

The gages were mounted to the metal surface by spotwelding. Since a reasonably clean surface is required for spotwelding, the un-galvanized and slightly rusted metal surface was de-greased and then cleaned with a small grinder and a cloth. Then grit and grease were removed with a lint-free cloth soaked in a solvent. Capacitor-discharge resistance welding is controlled by the welding time and the probe contact force. Therefore, trials were made with sample gages first before attempting to install actual gages. Furthermore, skill is required to master the spotwelding pattern recommended by the gage manufacturer.

Gage Locations

At each of the 25 locations on the inside wall of the metal culvert (Figure 7), four strain gages were installed (circumferential and longitudinal directions on inside crest and valley locations) in order to compute wall thrust and bending moment. Unfortunately, the technique used to compute hoop strain (and thus wall thrust) from the measured outer fiber strains (assuming a linear strain profile across the wall section) produced erroneous results (Webb et al. 1998). A better technique would have been to measure strain at the centroidal axis of the wall section directly without the need to infer it from the outer fiber strain measurements. Additional strain gages were subsequently added to the centroid of the corrugation profile which produced much better results.

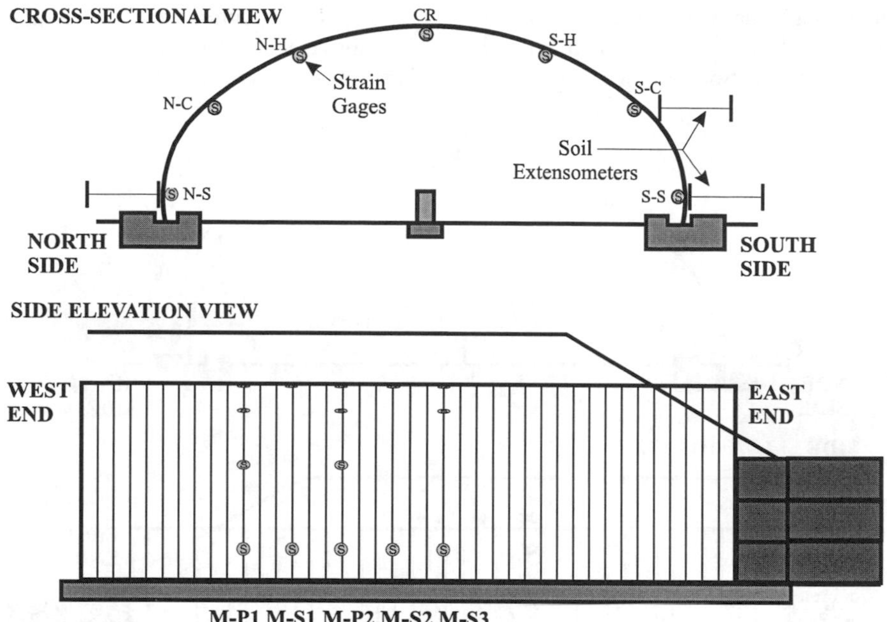

FIG. 7--*Strain Gages for Corrugated Metal Culvert*

INTERFACE PRESSURES

Backfill soil pressures acting on the culvert as a result of culvert-soil interaction were measured with vibrating wire earth pressure cells at locations shown in Figures 8 and 9. A total of 16 earth pressure cells were installed around each culvert. Because of the difficulty in measuring soil stress accurately, cells were installed at duplicate locations.

Gage Description

The pressure cells were 230 mm (9 in.) diameter, fluid filled cells with vibrating wire pressure transducers, manufactured by Geokon, Inc. Each cell had two pressure sensitive faces and also had a built in temperature sensor. The cells were installed radially around each culvert wall 150 to 300 mm (6 to 12 in.) away from the wall. Pressure cells with ranges of 345 kPa (50 psi) and 690 kPa (100 psi), over-range capacity of 150 % full scale, resolution of 0.1 % full scale and reported transducer accuracy to within 0.25 % of full scale were selected. However, accuracy of the earth pressure measurement with the pressure cell is considerably less because of the effects of installation and conformance of the installed cell. The 690 kPa (100 psi) pressure cells were installed above the top of the culverts for measurement of live load pressures while the 345 kPa (50 psi) pressure cells were used at the remaining locations around the culvert wall where lower pressures were expected. In addition, similar cells with a range of 70 kPa (10 psi) were used in regions around the culverts where the lowest pressures were expected. Some of the latter cells only had one pressure sensitive face.

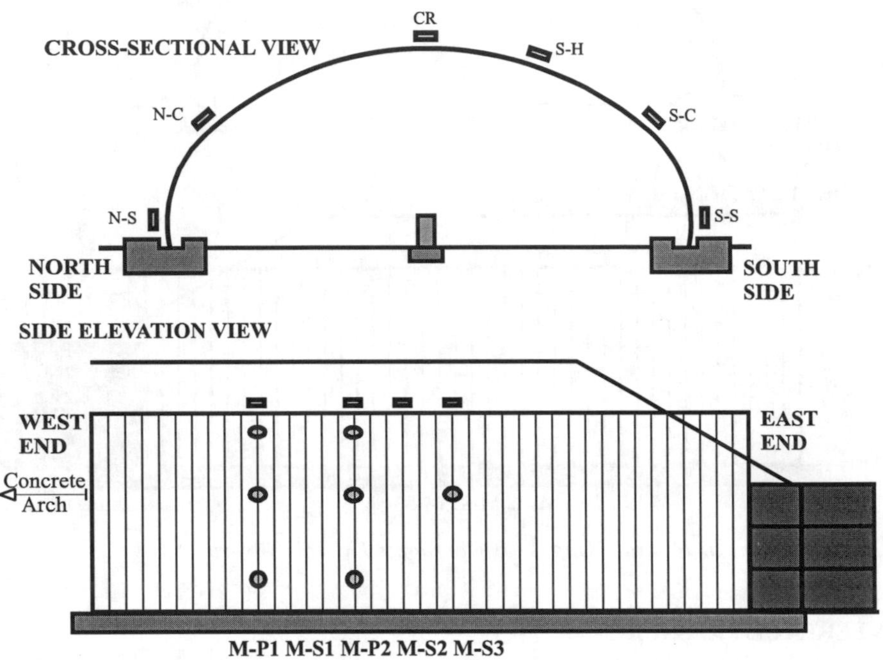

FIG. 8--*Interface Pressure Cells for Corrugated Metal Culvert*

Gage Calibration

The earth pressure cells are all supplied with factory calibration sheets giving pressure and temperature coefficients; however, these coefficients are determined for the vibrating wire transducer itself, and not the complete soil stress cell. Experience with the 230 mm (9 in.) diameter gages shows that the calibration of the complete cell is similar to that of the transducer alone (McGrath and Selig, 1996). This was confirmed by an extensive calibration program.

WALL TEMPERATURE

Temperature of the metal and concrete culvert inside walls were measured using OMEGA Type T copper-constantan thermocouples. Eleven thermocouples were installed on the two culverts at the crown and springline locations. Some of the thermocouples were read automatically by data logger while the rest were read manually.

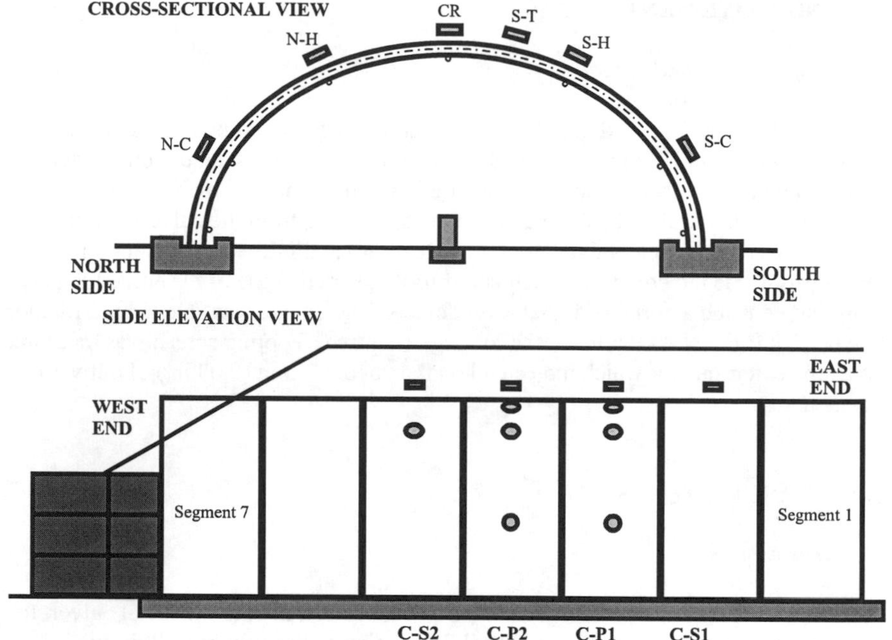

FIG. 9--*Interface Pressure Cells for Reinforced Concrete Culvert*

CONCRETE CULVERT INSTRUMENTATION

Crack Length and Width

Cracks in the concrete culvert were monitored visually and crack length and width dimensions were determined manually using a magnifying crack gage to evaluate culvert performance and distress. Cracks were marked by chalk for identification. Most of the cracks noted in the precast concrete segments occurred during formwork removal, transport and/or installation at the test site.

Relative Segment Movement

Vibrating wire crack gages manufactured by Geokon, Inc. were attached to the concrete culvert wall to determine relative joint movement. Nine gages were installed on the concrete culvert between segments at the shoulder and crown locations to measure relative shear movement. Three gages were installed one each at the crown and both shoulders at the joint between Segments 2-3 (see Figure 4 for segment numbering). Two gages were installed, one each at the South shoulder and crown between Segments 3-4, 4-5, and 5-6. These gages worked well even though very small shear movements (typically less than 2 mm) occurred during testing.

FOOTING MOVEMENT

Settlement, Rotation and Spread

Level survey using a digital level with photocopied bar codes (as discussed in a previous section) were used to measure settlement of the footings under both culverts. Bar code targets were installed near the outer edge of the footing to permit extending them to the height required by the bar code targets hanging from the culvert. Footing settlements were very small (about 4.5 mm maximum) and fairly uniform. Footing rotation was needed to compute settlement of footings at the base of the culvert leg and was measured using a portable digital level device. The device was able to read rotations to the nearest 0.1°. No rotations of the footings occurred. Footing spread was measured with a tape extensometer which indicated less than about 2 mm (0.08 in.) of outward movement.

SOIL MEASUREMENTS

Soil Extensometers

Six soil extensometers were installed in the backfill next to the metal culvert to measure displacement of the structural backfill relative to the culvert wall during backfilling and live load testing. Four of these gages were installed at the springline elevation on both sides of the metal culvert while the other two were installed at the change of curvature on the one side only (Figure 7).

Gage Description–Inductance coils, similar to those used for the structural extensometers, were used for soil extensometers. The coils were mounted on two transfer rods which were each connected to end plate anchors that moved with the soil within which they were embedded causing the coil spacing to change. The portion of the transfer rod near the coil was non-conductive to minimize interference with the inductance field. A schematic drawing of the soil extensometer is shown in Figure 10. The overall length of the installed gage was 1.8 m (6 ft). The guide tubes served two functions. One function was to keep soil away from the transfer rods to minimize any friction that could affect the measurement. The second purpose was to provide a tube for the guide blocks that keep the inductance coils in alignment. The guide tubes were non-conductive because of their proximity to the inductance coils.

Gage Readout Instrumentation--The soil extensometers were read the same way as the structural extensometers.

Other Soil Measurements

In-place soil density and soil moisture content measurements were taken with a direct transmission nuclear density gage (Troxler Model 3411B). A soil compaction modulus gage Model H-4140 manufactured by Humboldt Manufacturing Company was used to measure backfill stiffness for analysis of culvert-soil interaction. Conventional level survey was used to measure backfill layer surface elevations and wheel rutting of the live load vehicle.

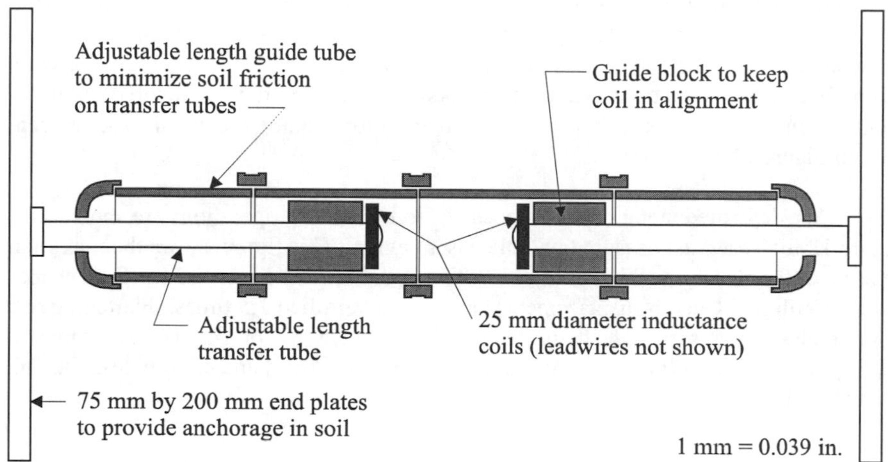

FIG.10--*Schematic of Soil Extensometer*

DATA ACQUISITION SYSTEMS

Due to the variety and number of sensors three different data acquisition systems were used to acquire and process data. The structural and soil extensometers were read with a pc-based data acquisition system. The vibrating wire earth pressure cells and thermocouples were read with a Campbell Scientific, Inc. CR10X data logger, while the strain gages were read with a Campbell Scientific, Inc. CR7 data logger. In order to support the large number of sensors each of the Campbell Scientific data loggers was expanded with three AM416 Campbell Scientific, Inc. multiplexers.

SAMPLE DATA

Sample data are included for illustration purposes. Detailed results can be found in Webb et al. (1998).

Culvert Deformation

Vertical deformations of the metal culvert during backfilling operations, measured with the laser device and with the digital level at the crown and curvature locations, are shown in Figure 11 for Test 2 (loosely placed backfill). The crown continued to rise until earth was placed over the top of the structure at which point the movement was reversed. The relative rate at which the crown rose was about the same as that during downward movement of the crown because of the loose state in which the soil was placed. Maximum measured peaking during Test 2 was about 72 mm (2.8 in.). The South curvature points showed more downward movement than the North points. This trend was also observed by more flattening of the side plates on the South side than on the North side. The metal culvert was top loaded in both tests to prevent excessive peaking during backfilling operations. Concrete blocks were added to the metal culvert with about 2.4 m (8 ft) of backfill in place. The effect of top loading the metal structure can be seen in Figure 11.

The deformed metal culvert shapes due to backfilling operations are shown in Figure 12 for Test 2 as measured with the laser device. This figure shows the maximum peaking of the structure which occurred at 0.3 m (1 ft) of soil cover and the final shape after backfilling. Culvert displacements have been magnified 7.5 times. Flattening of the South plates can be seen. The deformed metal culvert shapes for Test 1 were similar with slightly more peaking (maximum peaking of 80 mm or 3.1 in.) and slightly less flattening of the South plates.

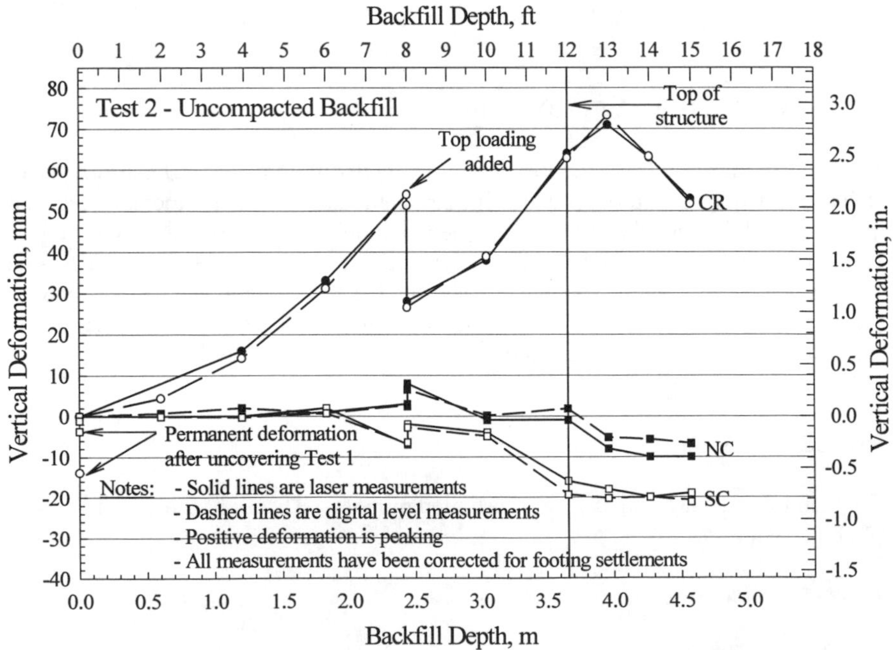

FIG. 11--*Structural Deformations of Metal Culvert During Backfilling*

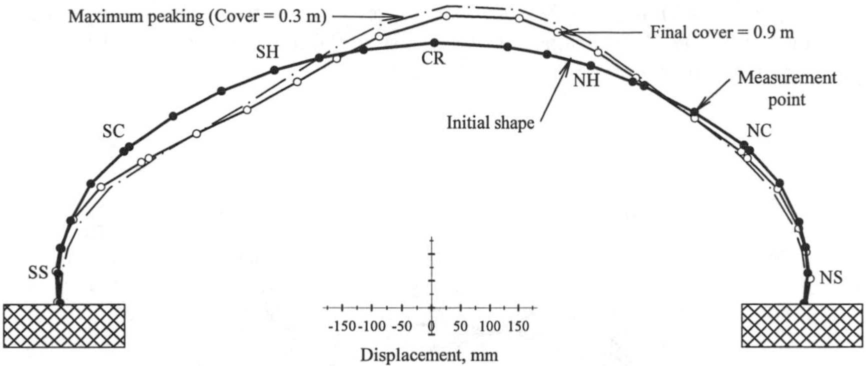

FIG. 12--*Average Metal Culvert Displacements During Backfilling - Laser Device*

Bending Moments During Live Load Testing

The distribution of the change in bending moment due to the tandem axles of the live load vehicle positioned over the culvert crown is plotted in Figure 13 for Test 1. Moment is plotted on the tension side of the structure. Three depths of soil cover and average measurements of the two stations underneath the wheel paths are included. This figure indicates that the live load bending moments at the shoulders at the stations under the wheel paths are inversely related to the depth of soil cover, i.e. decreasing cover results in increasing moment. No significant moments developed at the springline and curvature changes from the live load vehicle. The maximum live load moments did not develop directly underneath the tandem axles but, instead, developed at locations outside the tandem axles (shoulder locations). Bending moments for both tests were similar (Test 1 results are not shown here).

Radial Pressures During Long Term Monitoring

Radial pressures were measured over a 9 month period under constant earth load and the results are plotted in Figure 14 for both culverts. The first set of long term measurements was taken about 3 months after the additional soil cover had been placed and indicates a drop in radial pressure for both culverts and all gages compared to the final cover pressures. However, after the initial drop the pressures generally increased to almost the same values measured at the final cover condition. Temperatures were measured during the long term observation period using thermocouples installed on the inside of the culverts and thermistors located in the earth pressure cells and are plotted in Figure 14a. A comparison between these temperatures and corresponding radial pressures generally suggests a lowering of radial pressures when the culverts contract during the colder months and subsequent increase in radial pressures when the culverts expand again during warmer periods.

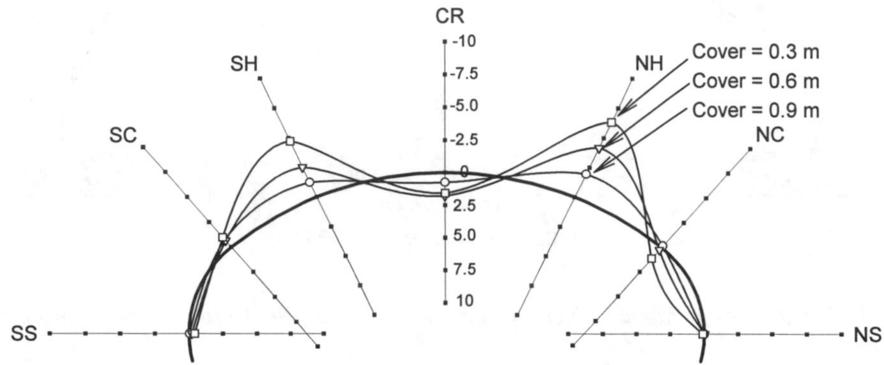

FIG. 13--*Test 2 Bending Moments - Live Load Vehicle Over Culvert Crown*

SUMMARY AND CONCLUSIONS

The instrumentation plan for two large-span culvert tests on a rigid and a flexible culvert has been presented. Experience gained from instrumenting these culverts to monitor construction operations and to obtain detailed measurements for soil-structure interaction analysis has been discussed. The successful procedures followed for monitoring construction deformations using level survey with photocopied bar codes for vertical deformations and inductance coil structural extensometers for horizontal deformations have been documented. The successful use of a laser device mounted on an automated pan-tilt unit for detailed culvert deformation measurements has been described. Problems with the strain gage measurement technique in computing wall thrust from outer fiber strain measurements have been described and a better measurement technique has been given.

Measured vertical deflection results showed uplift of the metal culvert crown until about 0.3 m (1 ft) of soil cover was in place after which movement was reversed during placement of additional soil cover. With the tandem axles of the live load vehicle positioned over the culvert crown measured bending moments at the shoulders are inversely related to the depth of soil cover while moments at the springline and curvature change are small. Long term monitoring of radial pressures around both culverts indicate a lowering of pressures when the culverts contract during the colder months and subsequent increase in pressures when the culverts expand again during warmer periods.

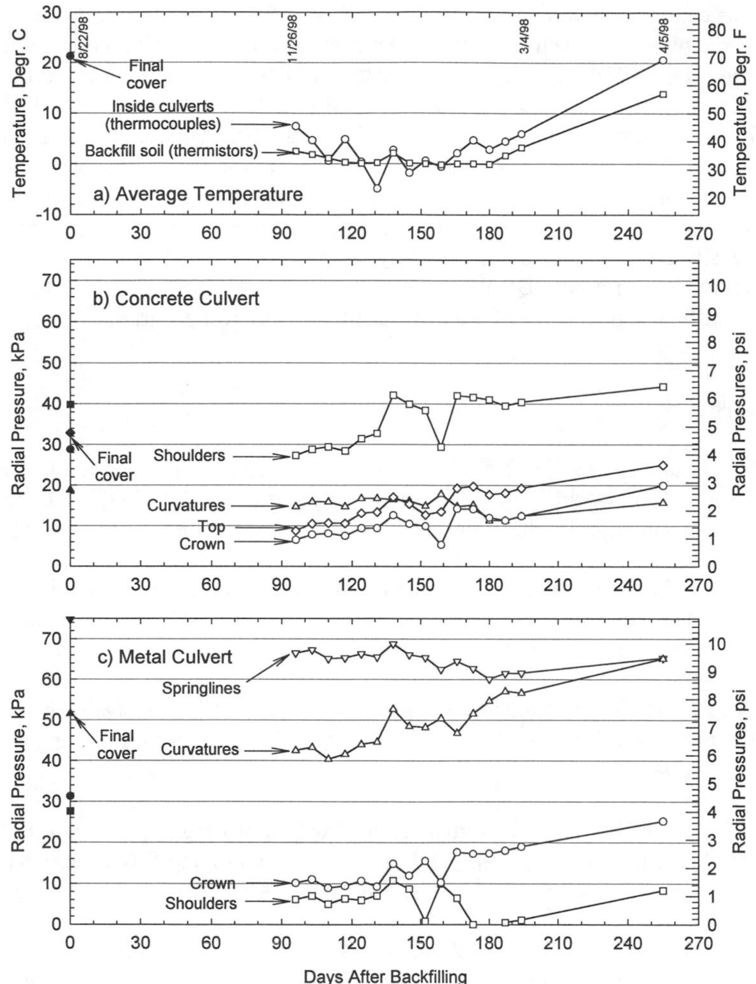

FIG. 14--*Average Long Term Radial Pressures on Culverts*

ACKNOWLEDGMENTS

The full-scale tests were conducted at the University of Massachusetts in Amherst, MA. The project was funded by the American Association of State Highway and Transportation Officials (AASHTO), in cooperation with the Federal Highway Administration (FHWA), and conducted through the National Cooperative Highway Research Program (NCHRP) which is administered by the Transportation Research

Board. The concrete culvert was provided by BEBO of America, Inc. and was manufactured by a local licensee, Rotondo Precast of Avon, CT. The metal culvert was provided by CONTECH Construction Products, Middletown, OH. The culvert tests were performed in a gravel pit owned by Delta Materials Corp., Sunderland, MA. The readout device for the inductance coils was developed at Nottingham University. The soil compaction modulus gage was provided by CNA Consulting Engineers on a trial basis. Val Moser of Campbell Scientific, Inc. provided invaluable assistance with their data acquisition systems. The Massachusetts Highway Department provided the Troxler nuclear density gage. Jeanne Sussmann, Raymond Frenkel, Dan Lovett, Troy Thiele, David Barnett, Zack Barnett, Donna Carver and David Glazier provided substantial assistance with various aspects of the instrumentation and test programs.

REFERENCES

McGrath, T.J., and E.T. Selig., 1996, "Instrumentation for Investigating the Behavior of Pipe and Soil During Backfilling," Geotechnical Report No. NSF96-442P, Department of Civil and Environmental Engineering, University of Massachusetts, Amherst, MA.

Selig, E.T., 1975a, "Soil Strain Measurement Using Inductance Coil Method," *Performance Monitoring for Geotechnical Construction, ASTM STP 584*, American Society for Testing and Materials, pp. 141-158.

Selig, E.T., 1975b, "Instrumentation of Large Buried Culverts," *Performance Monitoring for Geotechnical Construction, ASTM STP 584*, American Society for Testing and Materials, pp.159-181.

Webb, M.C., 1998, "Improved Design and Construction of Large Span Culverts," Geotechnical Report No. NCH98-458D, Department of Civil and Environmental Engineering, University of Massachusetts, Amherst, MA.

Webb, M.C., Sussmann, J.A., and Selig. E.T., 1998, "Large Span Culvert Field Test Results," Geotechnical Report No. NCH98-456F, Department of Civil and Environmental Engineering, University of Massachusetts, Amherst, MA.

Webb, M. C., Selig. E. T., Sussmann, J. A. and McGrath, T. J., 1999, "Field Tests of a Large-Span Metal Culvert," Submitted to Committee A2CO6 for Consideration for Transportation Research Board's 1999 Annual Meeting, National Research Council, Washington, D.C.

Michael Zhiqiang Yang,[1] Eric C. Drumm,[2] Richard M. Bennett[2], and Matthew Mauldon[3]

Measurement of Earth Pressures on Concrete Box Culverts under Highway Embankments

---

REFERENCE: Yang, M. Z., Drumm, E. C., Bennett, R. M., and Mauldon, M., "**Measurement of Earth Pressures on Concrete Box Culverts under Highway Embankments,**" *Field Instrumentation for Soil and Rock, ASTM STP 1358*, G.N. Durham and W.A. Marr, Eds., American Society for Testing and Materials, West Conshohocken, PA, 1999.

ABSTRACT: To obtain a better understanding of the stresses acting on cast-in-place concrete box culverts, and to investigate the conditions which resulted in a culvert failure under about 12 meters of backfill, two sections of a new culvert were instrumented. The measured earth pressure distribution was found to depend upon the height of the embankment over the culvert. For low embankment heights (less than one-half the culvert width), the average measured vertical earth pressures, weighted by tributary length, were about 30% greater than the recommended AASHTO pressures. The measured lateral pressures were slightly greater than the AASHTO pressures. As the embankment height increased, the measured weighted average vertical stress exceeded the AASHTO pressures by about 20%. Lateral pressures which exceeded the vertical pressures were recorded at the bottom of the culvert walls, and small lateral pressures were recorded on the upper locations of the wall. The high lateral pressures at the base of the wall are consistent with the results from finite element analyses with high density (modulus) backfill material placed around the culvert.

KEYWORDS: box culvert, field instrumentation, pressure cell, vibrating wire, highway embankment, earth pressure

---

Introduction

Cast-in-place concrete box culverts are commonly incorporated into highway embankments.To simplify the design of these structures, the earth pressures are usually taken as a function of the equivalent fluid stress due to the overburden. Although the

---

[1]Grad. Res. Asst., Civil & Environ. Engrg. Dept., Univ. of Tenn., Knoxville, TN 37996.
[2]Prof., Civil & Environ. Engrg. Dept., Univ. of Tenn., Knoxville, TN 37996.
[3]Assoc. Prof., Civil & Environ. Engrg. Dept., Univ. of Tenn., Knoxville, TN 37996.

structural response may depend upon the stress level, there is no distinction made between low overburden heights and high overburden heights. The actual loadings experienced by the structures can be complex, and may change during construction and the subsequent service life. Soil-structure interaction effects result in a state of stress around the structure that is dependent upon the stiffness of both the backfill materials and the structure.

Small size circular culverts have been instrumented and studied over the past 60 years (Davis and Bacher, 1968; Corotis and Krizek, 1977; Davis and Semans, 1982). Because circular culverts have equal rigidity and strength in the horizontal and vertical directions, culvert installation methods were developed to reduce the vertical pressures acting on the culvert. These methods divert the vertical stresses from the culvert to the adjacent soil and result in an increase of the lateral pressures on the culvert sides (Spangler and Handy, 1982; Vaslestad, et al., 1994). For culverts built on level ground, the "imperfect trench" (Spangler and Handy; 1982) condition installation method may be used. This method involves spreading a specified thickness of compressible material such as baled straw or plastic foam immediately above the culvert followed by compaction with normal backfill reducing the vertical earth pressures acting on the culverts. A comparison between this method and normal compaction on two instrumented (2.0 m height and 2.55 m width) concrete box culverts under a 10 m silty clay embankment height showed that the imperfect trench method resulted in a significant vertical load reduction (Vaslestad et al. 1994). The earth pressure immediately above the box culvert installed with the imperfect trench condition was 62% of the pressure due to the weight of the soil column above the culvert. The earth pressure on the culvert roof under normally compacted backfill was 125% of pressure due to the soil weight. However, field measurements (Yang et al., 1997) of pressures on an instrumented double cell concrete box culvert (3.66 m in height and 9.91 m in width) under about 12 m backfill height, indicated that the vertical earth pressure was not reduced by placing 2 m loose fill soil around the culvert roof. The average measured vertical pressure was 124% of the soil prism pressure (actual backfill unit weight of 22 kN/m^3) above the culvert and 145% of the current AASHTO recommended pressure(recommended unit weight of 18 kN/m^3). Furthermore, the induced differential settlement due to the loose fill may cause damage to the pavement at the top of the embankment.

Field test data on box culverts with normally compacted backfill are limited. The reported instrumented culverts have been either small size culverts with a width less than 3 m (Russ, 1975; James et al., 1986; Vaslestad et al., 1994) or large size culverts with relatively low backfill heights (Tadros et al., 1989). The suitability of the current AASHTO recommended design pressures for these culverts was examined by monitoring the pressures on an instrumented culvert throughout the backfilling process.

Instrumentation Description

The instrumented box culvert was 99 m long and 3.9 m high by 7.0 m wide. It was a double cell culvert, constructed of cast-in-place reinforced concrete. Typically, the top and bottom slabs were 0.78 m thick and the side walls were 0.41 m thick, although

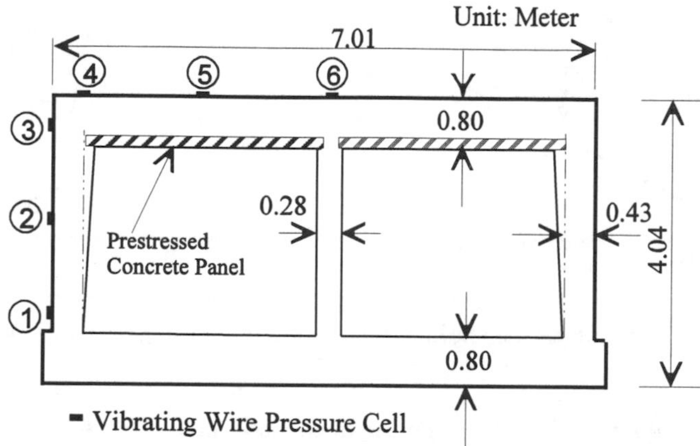

Figure 1 Illustration of Culvert Dimensions and Pressure Cell Location

the dimensions varied along the length of the culvert with the change of the overburden embankment height. The culvert's typical dimensions under the full embankment height are illustrated in Figure 1. The culvert was constructed on relatively level ground at the bottom of a broad valley. The site soil was a soft silty residual clay with shallow outcrops of limestone rock. About 0.6 m of well graded crushed gravel was spread immediately below the culvert in order to level the foundation and adjust for the variable thickness of the bottom slab along the length. Well graded crushed stone was backfilled to a height of 0.6 m above the culvert roof to provide drained conditions for the culvert, then the culvert was backfilled with silty clay, and high plasticity clay with occasional limestone boulders of 0.35m or smaller diameter.

To monitor the earth pressures during the construction, 12 vibrating wire hydraulic type soil contact pressure cells were installed in two separate sections. Upon completion of the embankment, section A would be under about 19 meters of fill, and section B would be under 11.7 meters. The relative locations of instrumented sections are shown in Figure 2. The slope of the embankment was 1:2. Each instrumented section consisted of 6 pressure cells with 3 cells mounted on the surface of the culvert wall and 3 on the roof. The location and pressure gage numbering scheme are shown in Figure 1. The pressure cell (Geokon model 4810) consists of two 230 mm diameter circular plates welded together around their periphery. One of the plates is thicker and designed to bear against the external surface of the structure in order to prevent flexure of the cell. The total thickness of the cell is 6 mm, and the aspect ratio (cell diameter over plate thickness) is 38. This "intermediate" cell size is appropriate for the measurement of soil pressure as suggested by Weller and Kulhawy (1982). Two different cell capacities were chosen: 345 kPa on the roof and 172 kPa on the culvert wall. The cells are capable of operating at up to twice the rated capacity, but the accuracy decreases.

The cell was first fixed to the culvert wall and roof with concrete anchors through 4 mounting lugs around the edge of the plate, and a quick setting high strength grout pad

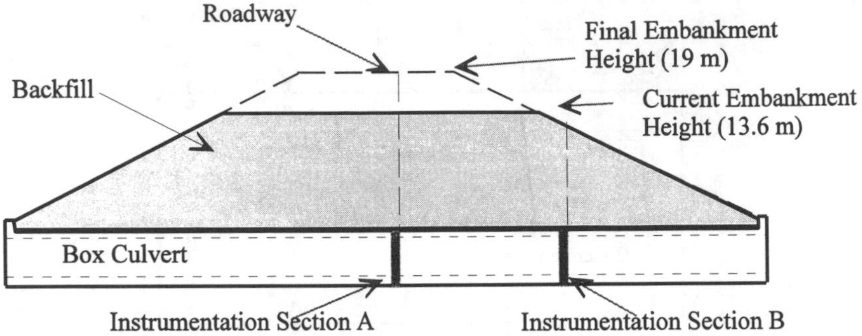

Figure 2 Schematic of Embankment and Culvert Cross-Section with Typical Sections A and B

was used to assure uniform contact between the plate and concrete. Medium sand was used to cover the cell and transducer housing to protect the cell from possible point loads or other stress distortions induced by the large size particles in the crushed gravel. A geosynthetic cover was attached to the concrete with adhesive to separate the gravel and the sand. This installation is illustrated in Figure 3. The backfill was placed with conventional compaction control criteria, with the dry density greater than 95% maximum dry density determined by standard Proctor compaction. The unit weight of the backfill gravel was measured in-situ by the sand replacement method (ASTM D 4914-89) and the average was found to be 22.0 kN/m^3. The unit weights of the silty clay and high plasticity clay were determined by the drive tube method and the average value for both materials was determined to be about 18.0 kN/m^3.

Measured Vertical Stresses on the Roof

The pressure changes during placement of the first 13.6 meters of backfill on the culvert roof were recorded with respect to the backfill height. The recorded vertical pressures and the surveyed backfill height above the cell are shown in Figure 4. The data were collected over about 600 days (20 months), during which there were two periods of about 3 months each when no construction took place. Figure 4 indicates that the

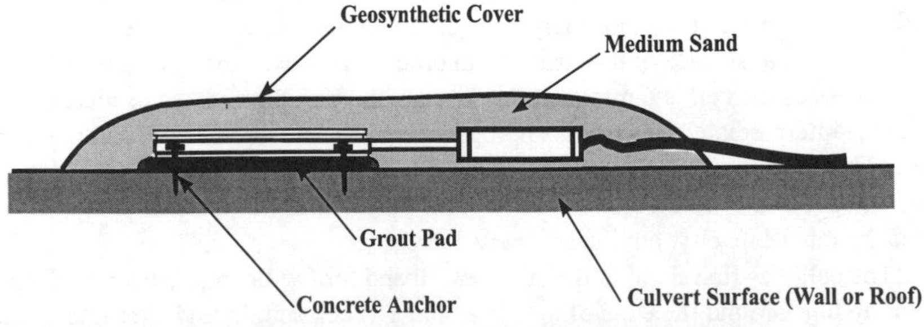

Figure 3 Installation of Contact Pressure Cell on the Culvert

variation in backfill height in both instrumentation sections A and B are similar until about day 580, when section B (under the slope) reached the final embankment height. Up to about 2 m backfill depth, the recorded pressures are similar in sections A and B, and the measured pressures were nearly uniform across the roof. When the backfill height reached about 6 meters, the pressures were found to vary significantly across the roof. The highest recorded pressures were on the culvert corner (gage 4), which corresponds to the location of greatest structural stiffness. This is consistent with results from previous instrumented culverts (Tadros et al., 1989; Yang et al., 1997). The large deviation of vertical pressures may reflect the influence of soil-structure interaction effects, which become more significant as the structural deflections increase with increasing embankment height. At backfill heights greater than about 6 m, the vertical pressures measured at section B are less than those at section A. This is likely a result of the position of section B close to the embankment slope.

Measured Horizontal Stresses on the Walls

The recorded lateral pressures on the culvert wall (Figure 5 and Figure 6) show an increase with backfill height similar to the vertical pressures on the roof. However, the

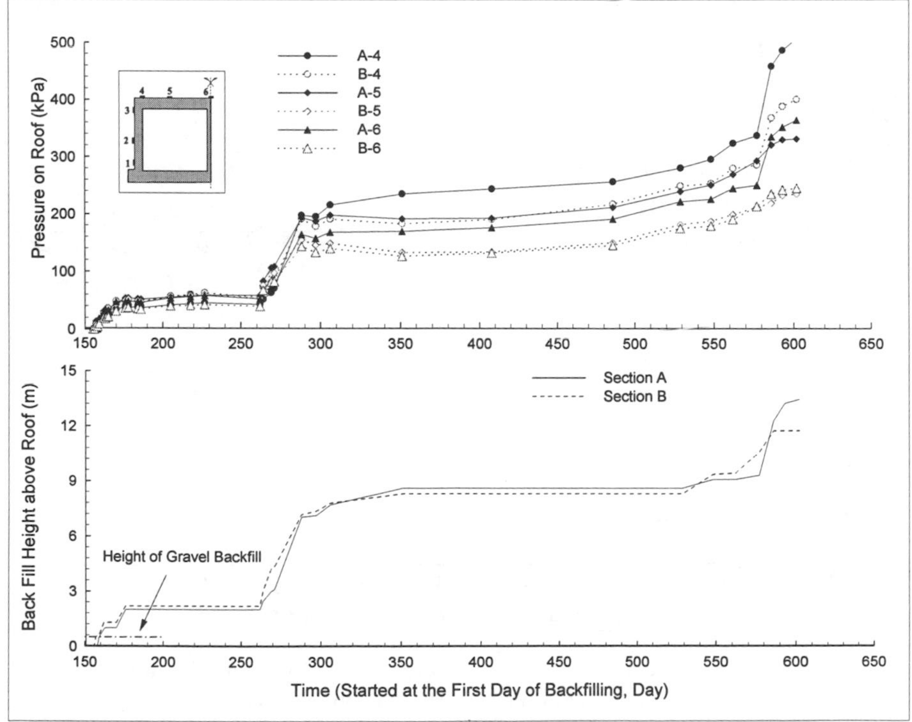

Figure 4 Recorded Vertical Pressures on the Culvert Roof with Backfill Height above Roof

cells at the bottom of the wall (Figure 6) registered pressures exceeding the vertical pressures, whereas the upper cells recorded relatively low lateral pressure. The influence of compaction equipment on the lateral pressure can be identified as the recorded stress peaks in Figure 5 during the 155-170th day, corresponding to the time when the gravel was placed around the culvert. This is followed by a period of stress relaxation with residual stresses differing at sections A and B. This relaxation does not appear to occur in the lower portion of the wall (Figure 6), where the overburden stress is greater. The added lateral pressure contributed by the additional thickness of overburden is negligible with respect to the short term increase caused by compaction. As the backfill height was increased above the culvert roof (about day 155), the upper cells in both sections recorded a decrease in lateral pressure, whereas the pressure cell reading at the bottom of the wall kept increasing and exceeded the vertical pressure. The high horizontal pressures were observed at both section A and section B, and were larger than the manufacturer's rated capacity for the pressure cells.

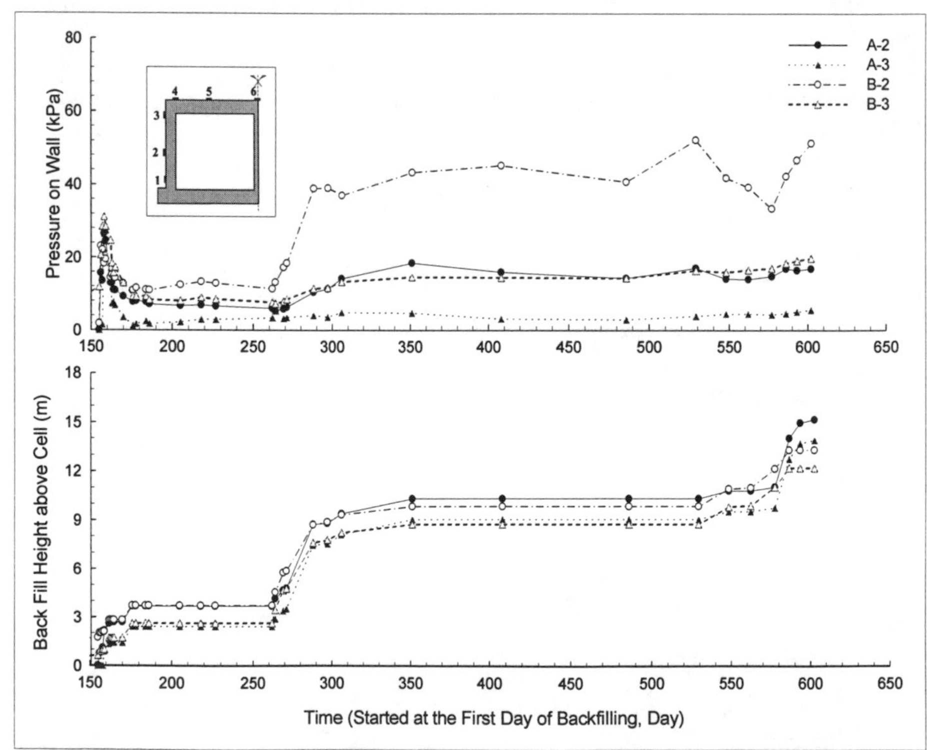

Figure 5 Recorded Lateral Pressures on the Culvert Wall with Backfill Height above Cells

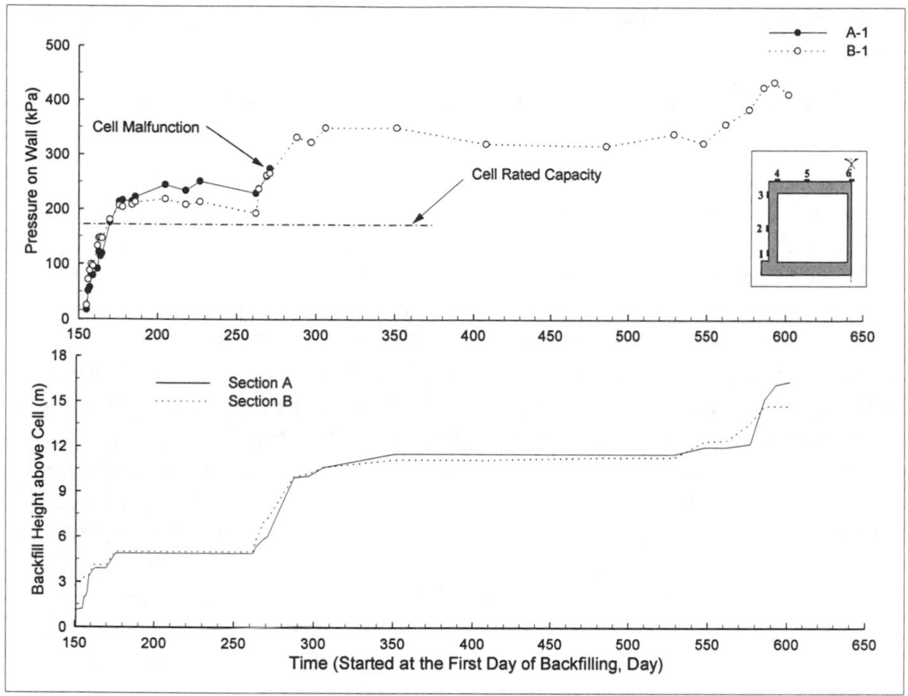

Figure 6 Recorded Lateral Pressures at Bottom of Wall with Backfill above the Cells

Discussion

1. Comparison of Measurements with AASHTO Recommended Vertical Pressure

The instrumentation results were compared with the AASHTO (1996) recommended design pressures. The AASHTO recommended pressure is an equivalent uniform fluid pressure that will give approximately the same internal forces (moments and shears) that are generated by the actual pressure distribution. The AASHTO vertical pressure for the embankment condition is the embankment height times an assumed unit weight of 18.8 kN/m³. To account for soil-structure interaction effects, AASHTO suggests that this pressure be increased by a dimensionless correction factor or soil-structure interaction coefficient, F_{el}

$$F_{el} = 1 + 0.20\frac{H}{B} \leq F_{el,max} \qquad \text{where} \quad F_{el,max} = \begin{cases} 1.15 & \text{for well-compacted side soil} \\ 1.40 & \text{for uncompacted side soil} \end{cases} \quad (1)$$

and H is the embankment height above the box culvert and B is the width of the box culvert. For well-compacted fills at the sides of the culvert, F_{el} should not be taken greater than 1.15, and for uncompacted fills at the sides of the culvert, it should not be exceed 1.40.

The AASHTO recommended vertical pressure σ_v used in design is then:

$$\sigma_v = F_{e1}\gamma H \tag{2}$$

The AASHTO soil-structure interaction coefficient (Equation 1) increases linearly with backfill height, H. For the current instrumented culvert (B = 7 m) with well-compacted fill, the limiting value of 1.15 is reached at a backfill height of 5.3 m.

The results presented in Figure 4 suggest that the vertical pressure distribution on the roof is not uniform and thus cannot directly be compared to AASHTO design pressures. An equivalent uniform pressure can be calculated by assigning a tributary length to each cell and obtaining a weighted average. Based on the predicted distribution of vertical pressure from a finite element analysis, a tributary length was assigned to each pressure cell and the equivalent uniform pressure determined. The tributary width for cells 4 and 6 at the culvert corner and at the centerline was taken as 0.2b and tributary width at cell 5 in the middle of the span was 0.6b, where b is the span of one cell. Figure 7 compares the weighted average vertical pressures with both the uncorrected (F_{e1}=1) and the corrected AASHTO pressures. The weighted average vertical earth pressures were generally greater than the AASHTO design pressures, even at the lower backfill heights (H/B<0.5). Figure 7 shows that the soil pressure recommended by AASHTO is less than the tributary-weighted average pressure observed at the site.

The measured soil-structure interaction coefficient (the measured ratio between the recorded vertical pressure and the calculated overburden pressure, γH), can be compared to the coefficient F_{el} recommended by AASHTO (Eq. 1). Although the AASHTO coefficient is intended to produce an equivalent uniform pressure, it is instructive to compare the coefficient F_{el} determined from individual pressure cells with the AASHTO values. Figure 8 compares the coefficients from the instrumented culvert with the AASHTO coefficients. Of 96 recorded pressure data from the instrumented culvert with backfill height less than 3.5 m (H/B < 0.5), the range of the recorded soil-structure interaction coefficient F_{el} is from 0.73 to 1.85. The average value is 1.30 with a standard deviation of 0.34. The highest recorded F_{e1} values were readings taken immediately after compaction of the backfill.

Also shown in Figure 8 are the soil-structure interaction coefficients calculated from an instrumented culvert in Nebraska reported by Tadros (1989). This was a double cell concrete box structure 8.1 m wide and about 4.3 m high. The permanent backfill height was 2.6 m, but it was temporarily backfilled to a height of 3.7 m for several days. Compacted silty clay was used as fill adjacent to the culvert. This culvert was under a low embankment height (H/B<0.5), with a maximum H/B value of 0.47, and a corresponding F_{el} = 1.06. As noted in Figure 8, the majority of the measured coefficients from the Nebraska culvert also exceed the AASHTO design coefficient. Of 30 recorded vertical pressures from 6 locations on the roof at backfill heights of 1.1, 2.4, 2.6, and 3.7 m, the recorded F_{el} ranged from 0.94 to 2.07. The mean value was 1.36 with a standard deviation of 0.26. Some of the cells recorded slightly less than the soil column pressure γH, when the fill height reached the maximum height of 3.7 m.

Although the materials adjacent to the two culverts were different, the vertical pressures on the roofs under low backfill height were similar. When the embankment height was less than 0.5B, the average recorded vertical pressures were greater than the AASHTO recommended vertical pressure by approximately 30%.

In the current study, when the embankment height exceeded 4 m (H/B > 0.5), some of the recorded soil-interaction coefficients are lower than the AASHTO F_{e1} values. Of 90 recorded pressure data points, the range of the recorded soil-structure interaction

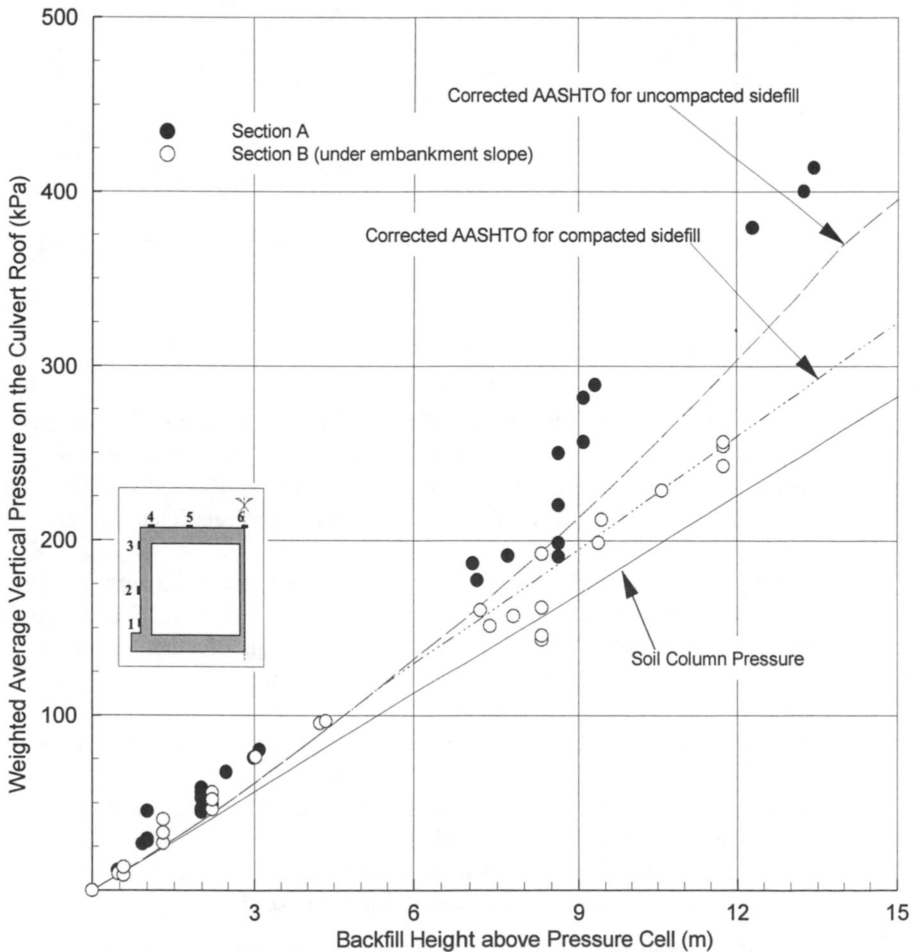

Figure 7 Weighted Average of Recorded Pressures Compared with AASHTO Pressures

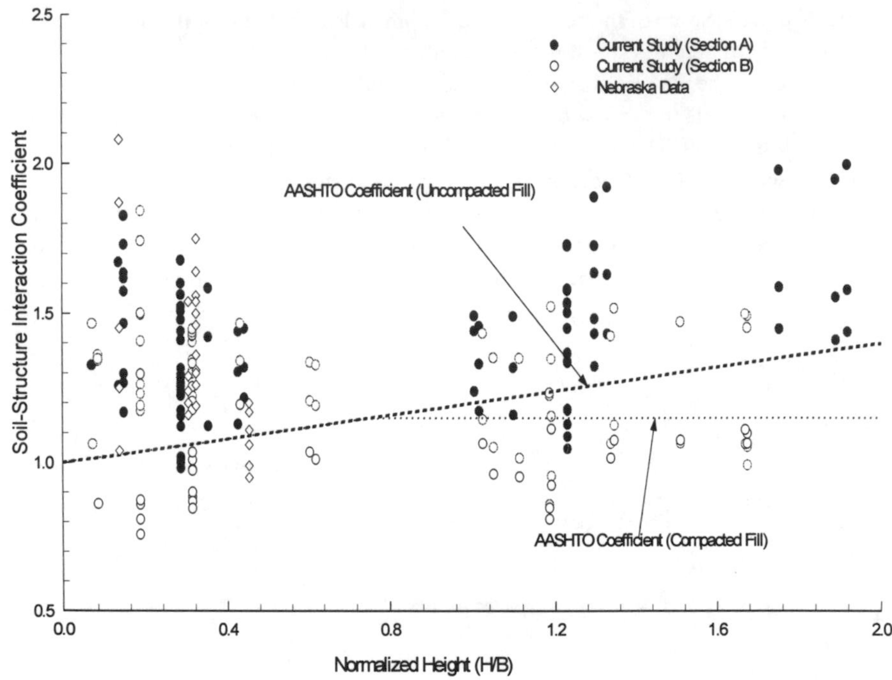

Figure 8 Comparison of the Normalized Vertical Pressures with Normalized AASHTO Pressures

coefficients was from 0.8 to 2.0 , with a mean value of 1.31 and a standard deviation of 0.27 . This suggests that 90% of the measured pressure data were greater than the soil column pressure. The average recorded soil-structure interaction coefficient was still greater than the recommended AASHTO soil-structural interaction coefficient by about 33%.

For any given backfill height, the mean of the weighted average pressures (using the assumed tributary widths of 0.2b for the pressure recorded at cells 4 and 6, and 0.6b for the pressure recorded at cell 5) is greater than AASHTO design pressure. For low embankment heights (H/B<0.5), the mean value is greater than the current AASHTO design pressure by 31%. For high embankment heights (H/B>0.5), the mean value is greater by 19%. The AASHTO upper limit of F_{el}=1.15 does not seem justified based on these measurements.

In Figure 8, all of the recorded soil-structure interaction coefficient values less than 1.0 from the current study are at locations in section B, which are close to the edge of the slope (Figure 2). Lower vertical pressures recorded in this section may be attributed to the lower lateral constraint compared with that in section A.

2. Comparison of Measurements with AASHTO Recommended Lateral Pressure

The current AASHTO (1996) recommended lateral pressure is obtained by the

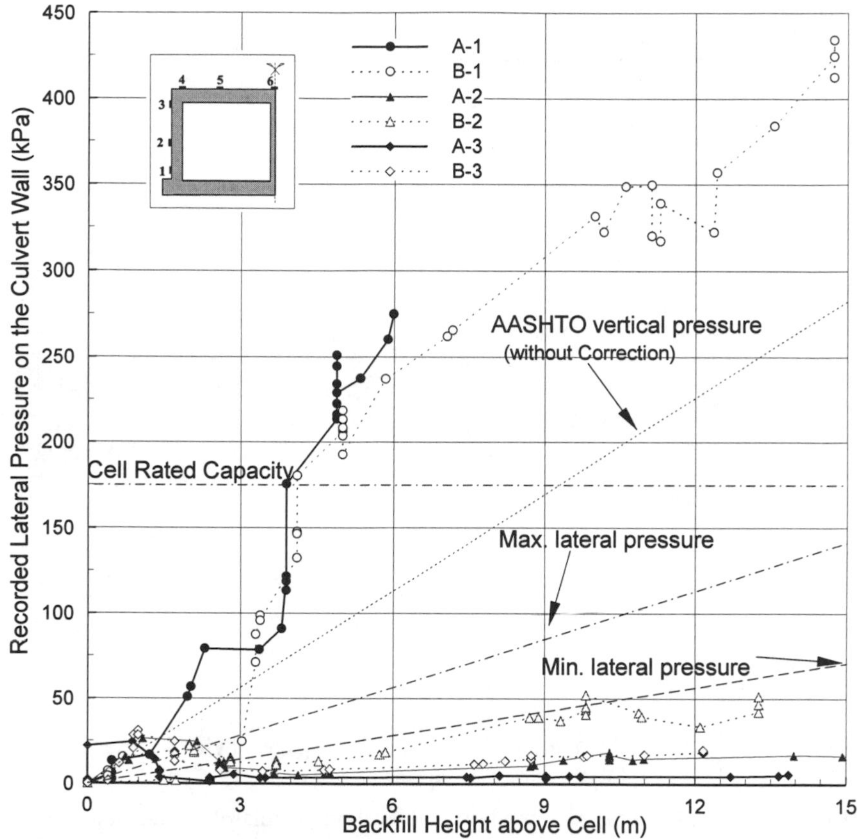

Figure 9 Recorded Lateral Pressures Compared with AASHTO Pressures

embankment height multiplied by an equivalent liquid unit weight. For the maximum lateral pressure, this equivalent unit weight is 9.4 kN/m^3, and for the minimum pressure, the equivalent unit weight is 4.7 kN/m^3. Based on the unit weight of 18.8 kN/m^3 used for the vertical pressure, this corresponds to lateral earth pressure coefficients of 0.5 and 0.25, respectively. Figure 9 compares the recorded lateral pressures at different fill heights with the AASHTO recommended pressures. The bottom cells (Number 1) recorded larger than vertical pressure and the intermediate and upper cells (Numbers 2 and 3) recorded pressures below the AASHTO minimum pressure. This was observed at both instrumented sections A and B.

Very high lateral earth pressures were observed for some combinations of backfill modulus during a parametric study using the finite element method (Yang et al., 1997). It was assumed that the backfill modulus was proportional to compaction energy, and lateral pressures against a 9.9 m high, 3.7 m wide double cell culvert under 11.7 m fill

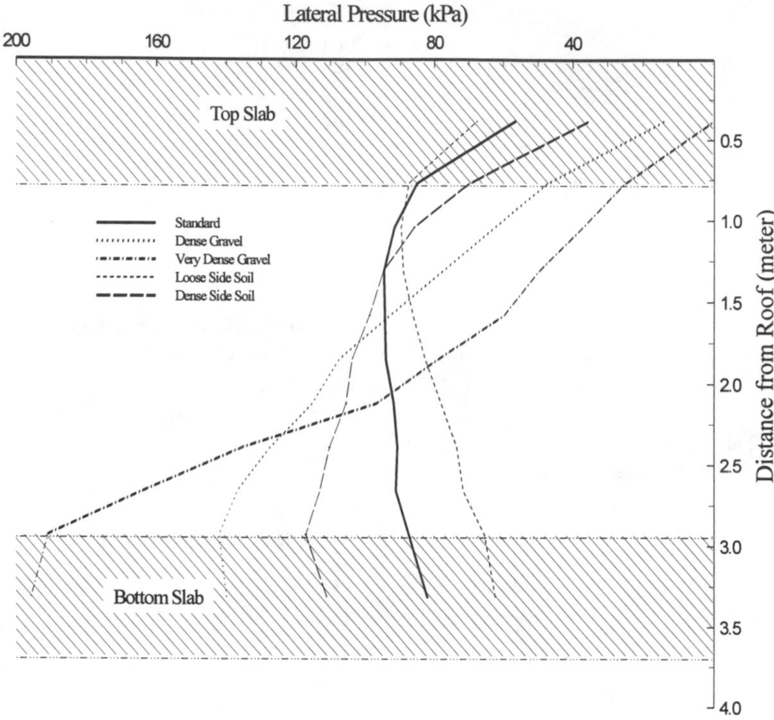

Figure 10 Parametric Study of Earth Pressures under Different Compaction Conditions (Yang et al., 1997)

were investigated. For a range of modulus values, both the distribution and the magnitude of the lateral pressure were found to depend strongly on the modulus of backfill. As shown in Figure 10 (Yang et al., 1997), very large lateral pressures at the bottom of culvert walls were obtained with the very dense gravel (high modulus) surrounding the culvert. The elastic modulus of the very dense gravel was 74 MPa, or 5 times the modulus of gravel in the "standard analysis" which used a value of 14.8 MPa (Penman et al., 1975). These analytical results suggest that high lateral pressures may be induced at the bottom of the culvert wall when the backfill is well compacted. These high lateral pressures may result in significant shear forces in the bottom of the culvert wall.

Summary and Conclusions

Pressure measurements on an instrumented culvert during backfill construction provided a record of the change in vertical and lateral pressure with increasing embankment heights. The pressure acting on two different sections along the culvert changed in a similar manner as the height of the embankment increased. At lower backfill height ($H/B < 0.5$), the vertical pressures acting on the roof are consistent with the results of other instrumented culverts of similar size. Based on the results from a numerical analysis on a similar box culvert, the tributary width of pressure cells at different locations can be determined and then a weighted average vertical pressure can be obtained. The recorded weighted average vertical pressures are about 30% greater than the recommended AASHTO pressure at low backfill heights. At high backfill heights (*H/B>0.5*) the weighted average vertical pressures exceed the AASHTO recommendation by about 20%. Although there are no other reported test results for box culverts at high values of H/B, consistent measurements were obtained at two different instrumented sections along the box culvert.

The lateral pressures acting on the culvert wall vary in a complex manner with the increase of backfill height. At low values of *H/B*, the AASHTO recommended pressures were found to slightly underestimate the actual lateral pressure on the culvert. At higher *H/B* ratios, or *H/B>0.5*, the recorded lateral pressure became very large at the base of the wall, exceeding the vertical pressure. The pressure at the top and mid-height of the wall experienced some relaxation after compaction, and thereafter remained below the AASHTO recommendation. The high lateral pressure at the base of the wall was consistent with results from finite element analyses reported previously, in which high modulus, a well compacted gravel backfill was found to result in large lateral pressures at the base of the culvert wall.

The reported field measurements of earth pressures on large concrete culverts suggest that the current AASHTO recommended pressure does not reflect field observations . Furthermore, the current soil-structure interaction coefficient may not account for important features such as the effect of backfill compaction effort.

Acknowledgment

This investigation was supported by the Tennessee Department of Transportation, contract #CUT123RES1085. This support, and the input from William D. Trolinger, Division of Materials and Tests, and Billy R. Burke, Structures Division, are appreciated.

References

Corotis, R. B. and Krizek, R. J., "Analysis and Measurement of Soil Behavior Around Buried Concrete Pipe," *Concrete Pipe and the Soil-Structure System*, ASTM STP 630, 1977, pp. 91-104.

Davis, R. E. and Bacher, A. E., "California's Culvert Research Program-Description, Current Status, and Observed Peripheral Pressures," *Highway Research Record 249*, 1968, pp. 14-23.

Davis, R. E. and Semans, F. M., "Rigid Pipe Proof Testing under Excess Overfills with Varying Backfill Parameters," *Transportation Research Record 878*, 1982, pp. 60-82.

James, R. W., Brown, D. E., Bartoskewitz, R. E. and Cole, H. M. "Earth Pressures on Reinforced Concrete Box Culvert" *Research Report 294-2F*, Texas Trans. Ins., The Texas A&M Univ. Sys. College Station, 1986.

Penman, A. D. M., Charles, J. A., Nash, J. K. and Humphreys, J. D. "Performance of Culvert under Winscar Dam," *Geotechnique*, Vol. 25, No. 4, 1975, pp. 713-730.

Russ, R. L., "Loads on Box Culverts under High Embankments: Positive Projection, without Imperfect Trench," *Research Report No. 431*, Division of Research, Department of Transportation, Lexington, KY., 1975.

Spangler, M. G. and Handy, R. L., "*Soil Engineering*" 4th edition, Harper & Row, New York, 1982.

Tadros, M. K., Benak, J. V., Abdel-Karim, A. M. and Bexten, K. A., "Field Testing of a Concrete Box Culvert," *Transportation Research Record 1231*, 1989, pp. 49-55.

Vaslestad, J., Johansen, T. H. and Holm, W., "Load Reduction on Rigid Culverts Beneath High Fills: Long-Term Behavior," *Transportation Research Record 1415*, 1994, pp. 58-68.

Weller, W. A., Jr., and Kulhawy, F. H., "Factors Affecting Stress Cell Measurements," *Journal of Geotechnical Engineering Division*, ASCE, Vol. 108, No. 12, 1982, pp. 442-449.

Yang, M. Z., Drumm, E. C., Bennett, R. M. and Mauldon, M., "Influence of Compactive Effort on Earth Pressure on a Box Culvert," *Proceedings, 9th International Conference of the Association for Computer Methods and Advances in Geomechanics*, Wuhan, China, 1997, pp. 2021-2026.

Standard Specifications for Highway Bridges. 16th Ed., The American Association of State Highway and Transp. Officials (AASHTO), Washington, D. C. 1996.

Timothy J. McGrath,[1] Ernest T. Selig,[2] and Mark C. Webb[3]

Instrumentation For Monitoring Buried Pipe Behavior During Backfilling

---

Reference: McGrath, T.J., Selig, E.T., and Webb, M.C., **"Instrumentation for Monitoring Buried Pipe Behavior During Backfilling,"** *Field Instrumentation for Soil and Rock, ASTM STP 1358*, G.N. Durham and W.A. Marr, Eds., American Society for Testing and Materials, 1999.

ABSTRACT: An extensive instrumentation plan was devised to monitor buried pipe behavior, soil behavior and pipe-soil interaction during backfilling. The emphasis of the instrumentation plan was to monitor these parameters under different installation techniques without impeding construction operations.

Different types and sizes of pipe were selected for installation in trenches excavated in undisturbed in situ soil conditions. Installation variables included in situ soil conditions, trench widths, backfill material (including controlled low strength material), haunching effort, and compaction methods. A total of fourteen tests, each including reinforced concrete, corrugated steel, and corrugated HDPE, were conducted. Eleven of the installations were conducted with 900 mm inside diameter pipe and three with 1500 mm inside diameter pipe. The pipes were buried to a cover depth of 1.2 m.

Measurements of pipe shape, pipe strains, pipe-soil interface pressures, soil density, soil stresses, and soil strains were collected. Pipe shape changes were measured by a custom built profilometer. Custom designed bending beam pressure transducers were used in the steel pipe to measure interface pressures.

Most of the instrumentation performed well and measured results were within the range expected. Pipe-soil interaction effects were effectively measured with the instruments selected. Pipe shape changes were a very valuable parameter for investigating pipe-soil interaction.

KEYWORDS: instrumentation, buried pipe, pipe-soil interaction, measurements

---

[1]Principal, Simpson Gumpertz & Heger, Inc., 297 Broadway, Arlington, MA 02174-5310.

[2]Professor of Civil Engineering, Department of Civil and Environmental Engineering, University of Massachusetts, Amherst, MA 01003.

[3]Senior Engineer, Department of Civil and Environmental Engineering, University of Massachusetts, Amherst, MA 01003.

INTRODUCTION

The interactions between a buried pipe and the soil surrounding it are complex and challenging to monitor. Difficulties arise from the nature of soil, which is widely variable in terms of particle size, and stiffness, and due to the presence of the pipe which causes significant redistribution of soil stresses. The interactions between pipe and soil during backfilling when soil is placed in a very loose state and then compacted, sometimes under high impact loads, can be even more complex and have not been well documented. This paper documents the instrumentation and data acquisition hardware and software used to collect data during extensive field tests conducted to study the behavior of buried pipe during the placement and compaction of backfill. It provides details of instrument development, describes the performance of the instruments used in these tests, and provides suggestions and guidance for similar measurements.

Scope

Instrumentation used in the field studies included both new devices developed specifically for this project and commercially available instruments. This paper provides construction details of the new devices as well as information on accuracy and precision. Four groups of equipment are discussed:

- computer hardware and software to perform the tasks of data collection,
- instruments for monitoring pipe behavior,
- instruments for monitoring soil behavior, and
- instruments to monitor pipe-soil interface pressures.

Background

The project formed part of **Fundamentals of Buried Pipe Installation** (McGrath, 1998) evaluating current and proposed construction practices for achieving quality installations. Part of this task was accomplished by conducting laboratory tests (Zoladz et al. 1996) and field tests (Webb et al. 1996, 1998) on different types of pipe using a variety of soil types, equipment and techniques. However, only the field test instrumentation is discussed in this paper. Detailed information on the instrumentation for both laboratory and field tests is given in McGrath and Selig (1996).

The pipes used in the field tests were corrugated steel (metal), reinforced concrete (concrete), corrugated high density polyethylene pipe with 900 mm (36 in.) inside diameter, and profile wall high density polyethylene pipe with 1500 mm (60 in.) inside diameter. The two polyethylene pipes are referred to as the plastic pipe in this paper. Each test included all three types of pipe (Figure 1). Field tests were conducted by installing an access manhole and then conducting four tests, two in each direction (narrow trench condition, wide trench condition) from the manhole. Thus every test was conducted in undisturbed in situ soil conditions. After the four tests were conducted the manhole was retrieved and moved to the location of the next series of tests. A total of fourteen field tests was completed.

Instruments used to make measurements, and the purpose of the measurements, are summarized in Table 1. Figure 2 is a typical instrument layout for the field tests. The figure shows the instruments for the metal pipe; however, the instruments for the concrete and plastic pipe were similar, except as noted in the subsequent sections.

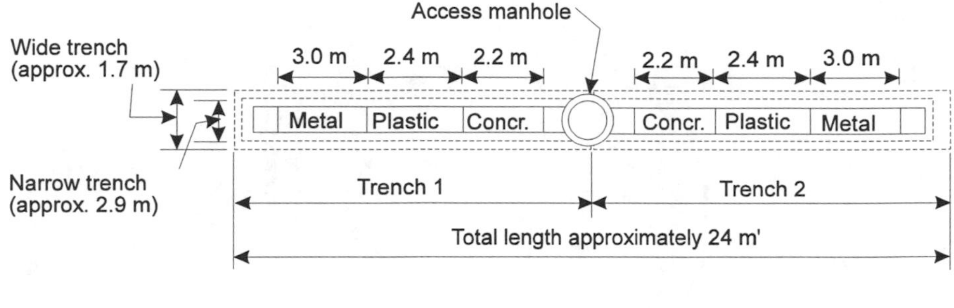

FIG. 1--*Typical Field Test Trench Layout*

TABLE 1--*Summary of Measurements*

Measurement	Instrument	Purpose
Pipe profiles	Profilometer	Pipe deflections and distortions
Pipe wall strains	Metal and plastic - resistance strain gages RCP - none	Structural deformations, moments and axial forces in pipe wall
In-place density & moisture	Nuclear gage	Correlation with standard field practice, relationship to soil stiffness
Soil strains	Inductance coil soil strain gages	Relative motions of pipe and trench wall
Horizontal soil stresses at backfill-in situ soil interface	230 mm (9 in.) earth pressure cells	Trench width affects, in situ-backfill soil interaction
Vertical soil stresses	230 mm (9 in.) earth pressure cells	Soil loads on pipe (arching)
Soil strength & stiffness	Penetrometers	Quality of haunching technique, uniformity of haunch support
Pipe-soil interface pressures	Concrete - 100 mm (4 in.) earth pressure cells embedded in pipe wall Metal - bending beam cells Plastic - none	Pipe-soil interaction, quality of haunching

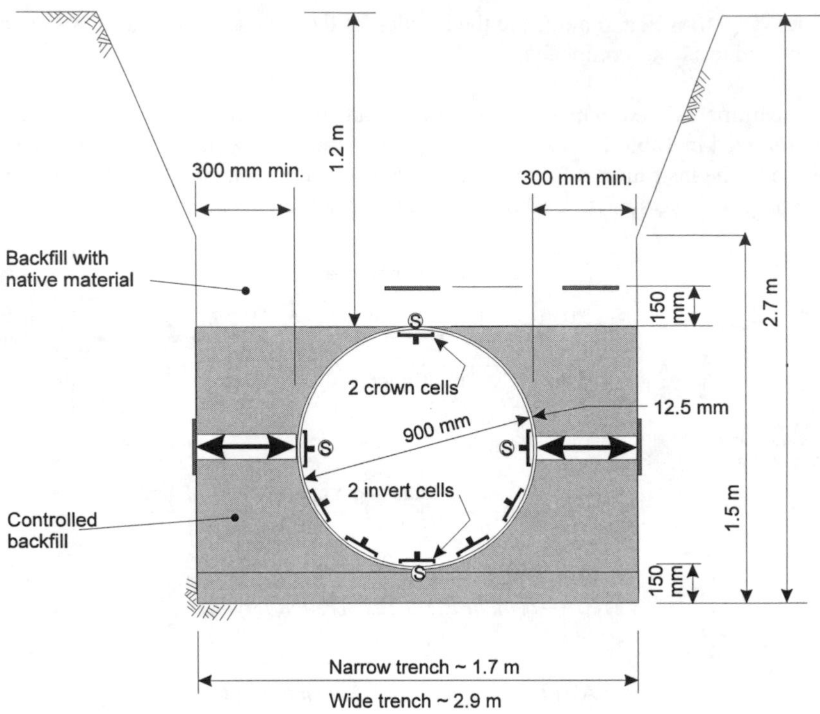

Notes:
1. Instruments occurring at the same locations in this figure were offset longitudinally in the ground.
2. The 900 mm metal pipe is shown in this figure. Details of instrumentation for other pipe and diameters are similar.

1 in. = 25.4 mm
1 ft = .305 m

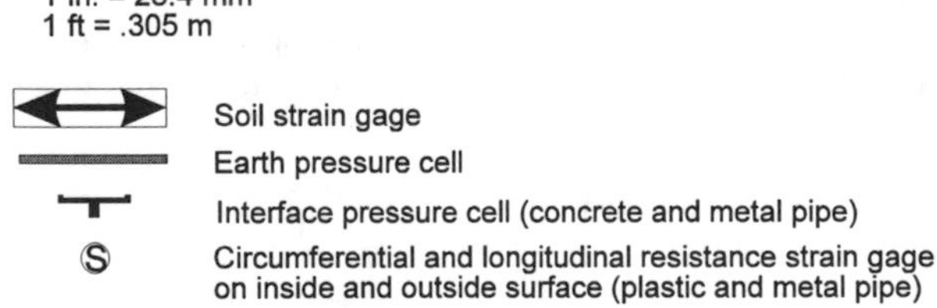

FIG. 2--*Typical Field Test Trench Cross-Section with Instrumentation*

DATA ACQUISITION SYSTEM

A computerized data acquisition system was assembled to acquire and process data for the project. The measurement instruments were connected by cabling to a signal conditioning board which amplified the signal as required. From the signal conditioner the signal was sent to an analog to digital (A/D) conversion board installed in the computer. After conversion the signal was processed into engineering units, displayed, and stored for future analysis. Several software programs were used to read the different instruments. An external power supply was used to power the instruments and part of the data acquisition system. The system was installed in a van.

A/D Board

A Model CIO-DAS08 A/D board manufactured by Computer Boards, Inc. was used. Conversion resolution was 12 bits, thus the selected ±5 volt input range could be divided into 4,096 increments of 2.44 millivolts. Readings were all static in nature, therefore actual sampling rates were not an issue.

Signal Conditioner

Two 32-channel signal conditioning boards were used. On-board multiplexer units coordinated the data from sixteen signal conditioner channels through a single channel of the A/D board. Thus the 64 channels available on the signal conditioner board used four channels on the A/D board. The remaining four channels on the A/D board were not used. The signal conditioners, Model CIO-EXP32, were manufactured by Computer Boards, Inc. On board amplification (gain) was available in increments of 1, 10, 100, 200, 500, or any sum of these increments. Strain gage based instruments were read with a gain of 300 and the other instruments were read with a gain of 1.

Software

Software used for data acquisition was LabTech Notebook (Notebook) by Laboratory Technologies of Wilmington, MA. Notebook had capabilities including complex programs, channel triggers, and direct conversion of voltage signals to engineering units. The last feature allowed real-time assessment of the data.

Power Supply

Three external voltage supplies were required to operate the system. The signal conditioner boards were operated with supply voltages of +12 v and -12 v, while the profilometer, pipe strain gages, and soil strain gages were powered with +4.76 v. This voltage level was constantly monitored using one of the data acquisition channels.

PIPE INSTRUMENTATION

Profilometer

The changes in the pipe shape during installation and backfilling are indicative of the pressure distribution on the pipe and of the stresses in the pipe wall. A device called a profilometer was developed to measure the distance from a point near the center of the pipe to the pipe surface at one degree intervals around the entire pipe circumference. This data was processed to compute the pipe shape and change in vertical and horizontal diameter and to estimate the radius of curvature of the pipe wall at all stages of the installation process.

Design and Construction--Several criteria were developed for the performance of the profilometer. These are:

- Mobility - The profilometer was designed to be easily installed and then removed to avoid causing pipe restraint and to allow measurements at several locations in a single test.

- Centering in pipe not required - The mobility requirement introduced a second requirement that the center of rotation of the profilometer would not need to be located in exactly the same position for every measurement of a single pipe, nor would it need to be located at the center of the pipe.

- Measurement of Corrugated Pipe - The profilometer was designed with a roller to bridge the corrugations of the metal pipe and provide a smooth curve.

- Range, Precision and Accuracy - It was estimated that the pipe deflections during a test could be plus or minus 2.5 percent of the inside diameter and that the center of measurement might be up to 25 mm (1 in.) away from the center of the pipe. Thus, for the 1500 mm diameter pipe, the range of the measurement arm had to be greater than 100 mm. Radial measurements accurate to the nearest 0.05 percent of the inside diameter were desired. Precision similar to the accuracy was also required.

The major elements of the profilometer design, based on the above criteria are shown in Figure 3 and include:

- a three legged support arrangement to hold the device in a pipe during measurements with adjustability of the legs achieved through threaded feet on the ends,

- a bearing to allow free rotation of the measurement arm,

- a rotary optical encoder to monitor the orientation of the measurement arm, and

- the measurement arm, which consisted principally of a LVDT, to measure the distance from the center of rotation of the profilometer to the inner surface of the pipe wall.

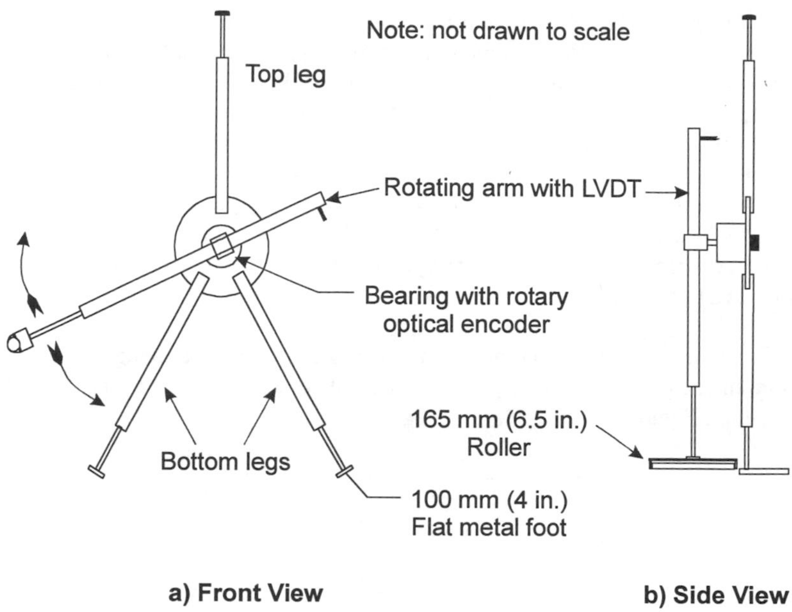

FIG. 3--*Schematic of Pipe Profilometer*

Electronic Components--The principal measuring elements of the profilometer were the optical encoder for rotational position and the LVDT for radial measurements.

The optical encoder was Model No. K15DM-360-5-SE-4A purchased from Lucas Ledex Inc. The profilometer used two channels on the encoder. Channel M provided a 5 volt signal once per revolution. This channel was used to start the series of profilometer readings. Channel A provided a separate 5 volt signal once for every degree of rotation. This channel was used to signal when to read the LVDT.

The LVDT was actually a DC-LVDT because it incorporated the necessary circuitry to allow DC input and output. The LVDT was a model LDC/3000/A purchased from RDP-Electrosense Inc. and had a range of ± 75 mm (±3.0 in.). The manufacturer's specified calibration coefficient was 33.9 mm/v (1.33 in./v); however, when assembled and operated by the data acquisition system the coefficient was 39.9 mm/v (1.57 in./v). Resolution of the LVDT is infinite, thus the A/D card resolution of 2.44 mv limited the profilometer resolution to 0.10 mm (0.004 in.).

Operation--A data acquisition program, written in Notebook, monitored the optical encoder, read the LVDT at the appropriate times, and stored the data on a computer disk. Both instruments were read with a gain of 1.0.

Calibration--The calibration of the assembled profilometer was completed using a 750 mm inside diameter concrete pipe. The profilometer was installed in the pipe and two complete 360 degree profile measurements were taken. The profilometer was then removed from the pipe, reinstalled and two more complete profiles were taken. This procedure was repeated until a total of four installations, and thus eight profiles, were taken. The results of these tests showed that for a second set of readings without removing the profilometer from the pipe the average change in reading is less than 0.1 mm (0.004 in.), the resolution of the A/D board. When the profilometer is removed and reinstalled between tests the average change increases to 0.5 mm (0.02 in.) with a standard deviation of 0.9 mm (0.04 in.).

Sample Data--The profilometer was used to determine the change in vertical and horizontal diameter, and the complete pipe shape at several stages of a test. Sample data is presented in Figure 4.

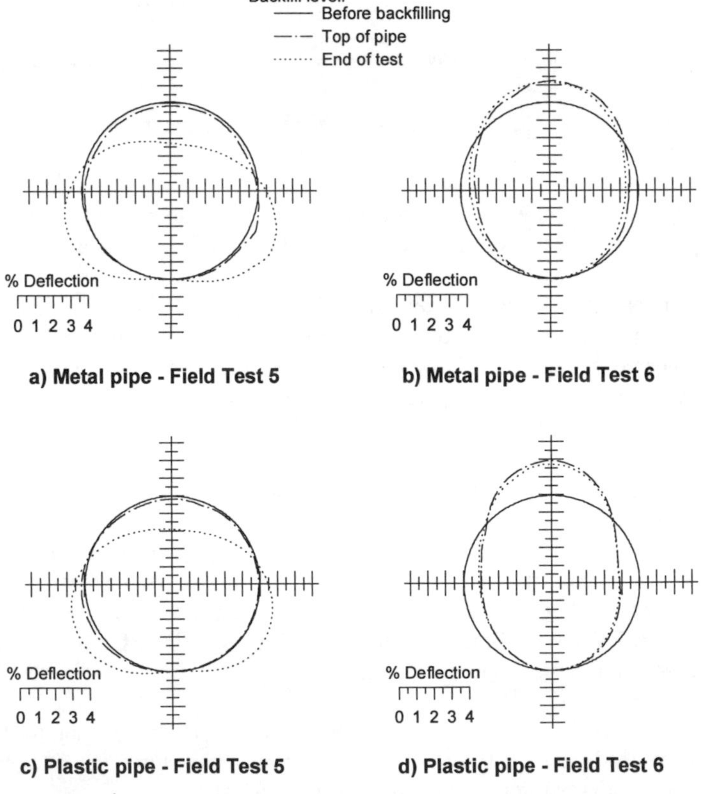

FIG 4--*Sample Profilometer Data, Magnified 10 Times*

Strain Gages

Strain gages were used to monitor the plastic and metal test pipe. The concrete pipe was not expected to deform sufficiently under 1.2 m of cover to warrant gaging. Strain gages were mounted on the inside and outside surface at the crown, invert, and springlines of each pipe. Two gages, one each oriented in the circumferential and longitudinal directions, were mounted at each location, thus each pipe had a total of 16 gages (Figure 5).

Strain gages were also used to build the interface pressure cells for the metal pipe. Details presented here on gage selection, installation and waterproofing are also applicable to that instrument.

Gage Selection and Installation--All strain gages, adhesives, and waterproofing materials were purchased from Micro-Measurements Division of Measurements Group, Inc. (MM).

The gages selected were standard constantan foil faced, polyimide backed, resistance gages. All gages had a gage length of 6.4 mm (0.25 in.). Selection of specific features of the gages was based on the characteristics of the test pipe materials. Steel is a good heat sink and gage heating is not a problem during operation; therefore standard 350 ohm gages were used (Type EA-06-250BF-350). Steel is also a common material and self-temperature-compensating gages can be purchased directly from gage manufacturers. Polyethylene is not a good heat sink, thus 1,000 ohm gages were used (Type EA-06-250BK-10C) to reduce the power consumption and associated heat output. Thermal properties in polyethylene can vary widely; thus no attempt was made to purchase self-temperature-compensating gages.

All strain gages were bonded using M-M M-Bond 200, a cyanoacrylate adhesive. M-Bond 200 is an easy to use, fast setting, room-temperature curing material. This adhesive is somewhat sensitive to moisture; however, the short, two month, life of the project, and the extensive waterproofing system planned allowed its use rather than requiring slower curing epoxy adhesives. All gages were waterproofed with M-M's M-Coat F.

Procedures for bonding the strain gages generally followed M-M recommendations; however, the steel and plastic pipe each created some problems. Polyethylene is difficult to bond to with any adhesive because of the waxy nature of the surface of the material, and the galvanized surface of the metal pipe is difficult to bond to because of the acidic nature of the surface after the galvanizing process. For both types of pipe high humidity interfered with successful bonding. All the gaging was completed in air conditioned laboratories.

For bonding gages to the plastic pipe the surface was treated with several cycles of scrubbing with a household cleanser (Ajax with bleach), rinsing with distilled water, and abrading with sandpaper, and then degreasing with isopropyl alcohol.

a) Plastic pipe - outside gages (914 mm diameter pipe)

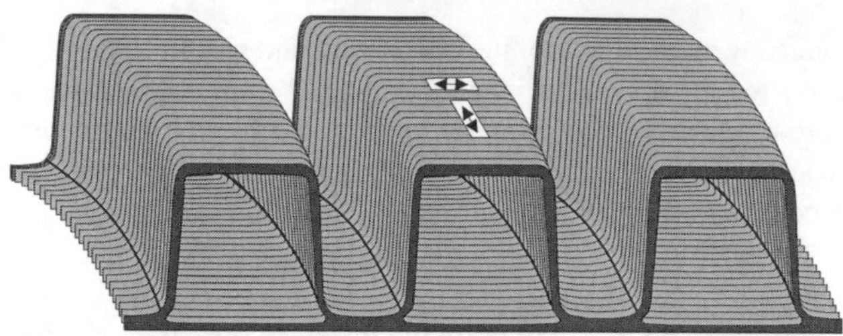

b) Plastic pipe - inside gages (914 mm diameter pipe)

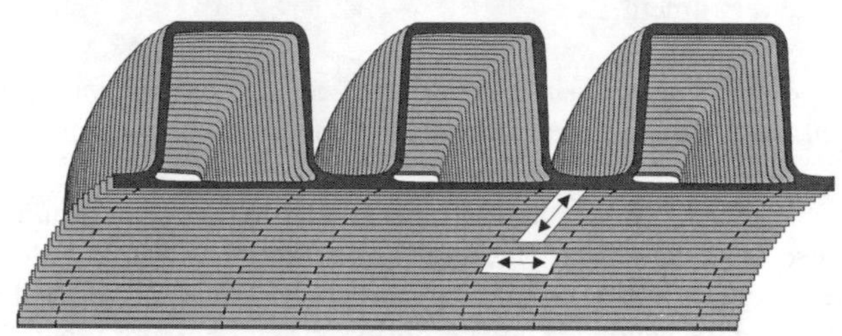

c) Metal pipe - inside gages

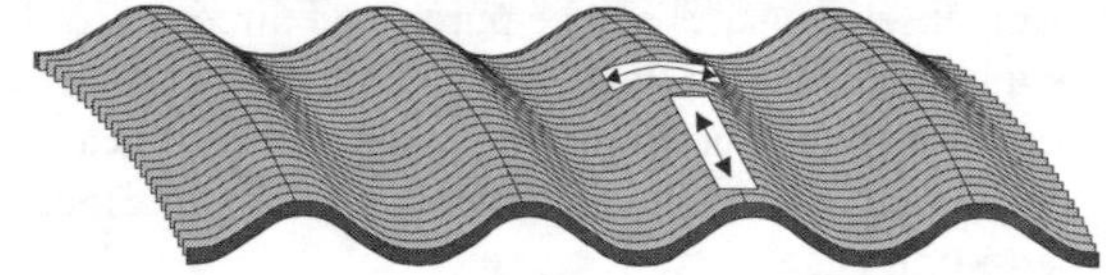

↔ Strain gage location and direction

Notes: Inside gages for metal pipe mounted on inside corrugation crest.
Gage locations for 1524 mm (60 in.) diameter pipe are similar.

FIG. 5--*Strain Gage Locations for Plastic and Metal Pipe*

Procedures for bonding the strain gages to the metal pipe involved the following steps:

- The surface was degreased with isopropyl alcohol.
- Cleaning with M-Prep Conditioner A was omitted.
- The surface was warmed slightly with a heat gun prior to bonding. This step was added to protect against humidity forming on the surface of the pipe.

Wheatstone Bridge Circuits--All strain gages were read using standard Wheatstone bridge circuits. To minimize lead wire effects, all bridges were completed in the test pipe, as close to the active strain gage as possible. After the test pipe were set in place, the lead wires from the data acquisition system were brought into the pipe and the lines were connected.

The gages for the plastic pipe were not self temperature compensated for polyethylene, therefore dummy gages had to be provided. This was accomplished by cutting out sections of pipe (from spare pieces) and bonding gages to the same portion of the pipe profile as the actual sensing gage. The remaining half of the bridge was completed using precision resistors. Wiring for the metal pipe was similar to that of the plastic pipe except that temperature compensating strain gages were not required.

There was some random noise in the system when the data acquisition system was set to a gain of 300. This was resolved by reading each gage ten times and averaging the ten readings.

SOIL INSTRUMENTATION

Nuclear Soil Density Gage

In-place soil density and soil moisture content measurements were taken with a Troxler, Model 3411B nuclear density gage on loan from the Massachusetts Highway Department.

Soil Strain Gages

The soil strain gages (Selig, 1975) were used in the field tests to monitor the change in the sidefill width, which is the dimension between the trench wall and the pipe (see Figure 2). Six gages were used, one on either side of each of the three test pipe.

Gage Description--The soil strain gages operate on the principle of electromagnetic inductance and are sometimes referred to as inductance coil strain gages. A schematic drawing of the soil strain gage as used is shown in Figure 6.

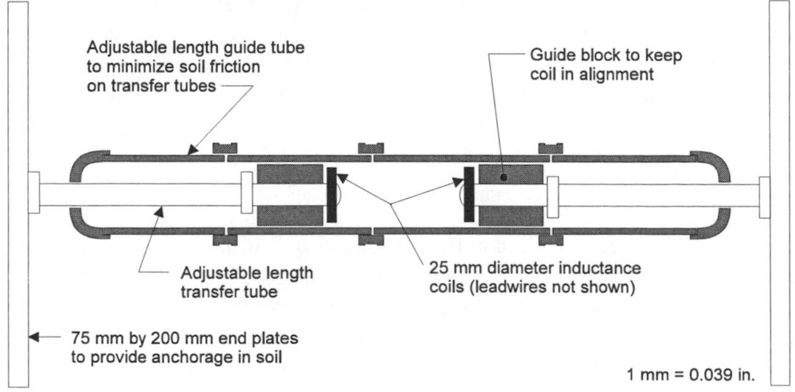

FIG. 6--*Schematic of Soil Strain Gage*

The measuring element of the soil strain gage is the pair of inductance coils, each consisting of a non-conductive disk wrapped with wire conductor. Driving one with an alternating current induces a voltage in the second coil that is a function of the distance between the two coils. The output of the gages is non-linear and is an alternating current signal that must be balanced for phase and amplitude to accurately determine the distance between the two coils. A signal processing unit, called the ε-mu, developed at Nottingham University, was used to automatically balance the phase and amplitude, and to convert the output to a direct current signal. This is a redesigned version of the original Bison Instruments, Inc. device. This allowed measurements to be taken through the data acquisition system.

The sensitivity and range of the coils is a function of the size and number of turns of wire. For this project 25 mm (1 in.) diameter coils were used which had a range of approximately 75 mm (3 in.). As with many of the instruments, the sensitivity of the gages was limited by the data acquisition system.

End plates were sized to insure movement with the soil within which they are embedded. The transfer rods translate any motion of the end plates to the inductance coils. Thus the measured change in coil spacing is the change in the end plate spacing. The transfer rods were adjustable to allow for the different trench widths that were being tested. The portion of the transfer rod near the coil was non-conductive to minimize interference with the inductance field.

The guide tubes served two functions. One function was to keep soil away from the transfer rods to minimize any friction that could affect the measurement. The second purpose was to provide a tube for the guide blocks that keep the inductance coils in alignment. The guide blocks control the lateral movement of the transfer rods to keep the inductance coils in alignment. The guide tubes and guide blocks were non-conductive because of their proximity to the inductance coils.

The soil strain gages were installed at the springline level of the pipe. A shallow cross-trench was dug into the loose backfill and the gage was placed with one end plate against the springline of the pipe and the other against the trench wall. The trench wall was smoothed as much as possible and any spaces that were left were filled with a medium sand.

Earth Pressure Cells

Soil stresses were measured at twelve locations, four for each pipe. As shown in Figure 2, two earth pressure cells were set at an elevation 150 mm (6 in.) above the top of the pipe to measure the vertical soil stresses directly above the pipe and over the trench sidefill, while the other two gages measured horizontal soil stresses at the interface between the trench backfill and the in situ material at the elevation of the pipe springline.

The earth pressure cells were 230 mm (9 in.) diameter, fluid filled cells with vibrating wire pressure transducers, manufactured by Geokon, Inc. Each cell also had a built in temperature sensor. The cells for measuring vertical soil stress were built to read pressure on both faces. The cells to measure horizontal stresses were designed with only one pressure sensitive face. The other side of these gages was a 6.25 mm (0.25 in.) thick heavy plate. The heavy plate was set against the in situ material in the trench wall to protect against erroneous readings if the trench wall was not smoothed adequately. The Geokon cells have a range of 70 kPa (10 psi), accuracy of 0.25% of full scale, precision of 0.2% of full scale, and a resolution of 0.1% of full scale. Resolution of the temperature sensors is 0.1°C.

The gages were monitored with a GK-403 vibrating wire readout by Geokon, Inc. The GK-403 reads both the vibrating wire transducer and the temperature sensor.

PIPE-SOIL INTERFACE INSTRUMENTATION

Interface pressures between the pipe wall and the soil were measured for the concrete and metal pipe. Different approaches were used for the two pipe because of the difference in flexural stiffness and wall profile of the pipe.

Concrete Pipe Interface Pressure Cells

Pipe-soil interface pressures for the concrete pipe were measured with 100 mm (4 in.) diameter fluid filled cells with vibrating wire transducers. The cells were custom made for the project by Geokon, Inc. The pressure cell consists of two diaphragms welded at the edges, with the fluid tube to the vibrating wire pressure transducer coming off of the back diaphragm, as shown in Figure 7a. The transducers had a range of 170 kPa (25 psi), accuracy of 0.25% of full scale, precision of 0.2% of full scale and a resolution of 0.1% of full scale.

Design and Construction--The needs of the project imposed several criteria on the design of the cells:

- Cell diameter - The size of the measuring face was restricted because of the curvature of the outside diameters of the test pipe. A cell diameter of 100 mm (4 in.) was considered an acceptable compromise as being large enough to minimize the effects of local nonuniform stresses on the face of the cell, yet small enough to minimize deviation from the curved surface of the pipe.

- Transportability - The test program was designed to use two sets of test pipe. This approach allowed backfilling of one test trench while the next trench was being prepared for a subsequent test (Webb, 1996). This sequence required that the interface pressure cells be removable so that they could be transported from one set of test pipe to the next.

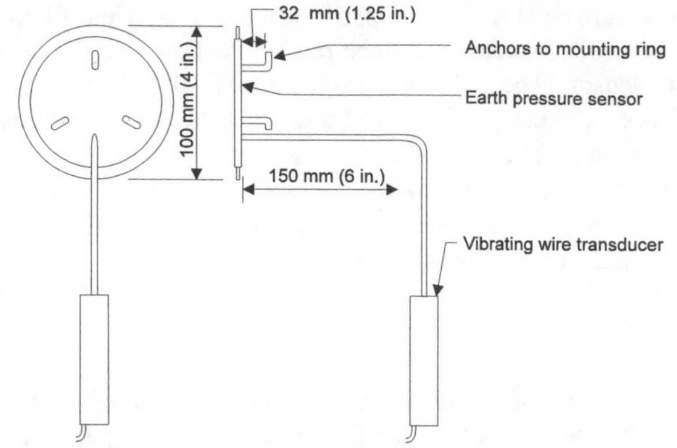

a. Cell as Purchased

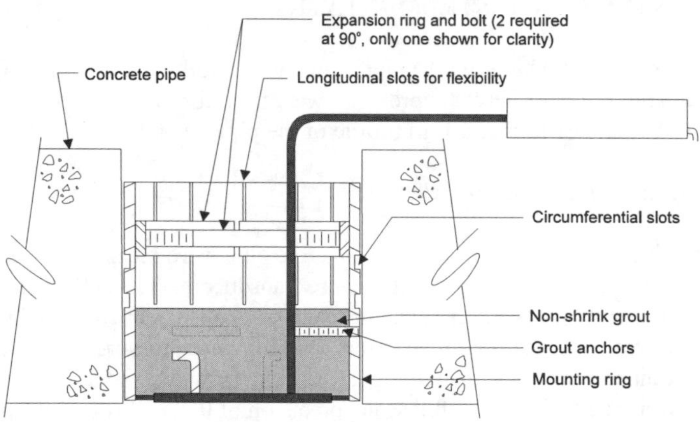

b. Cell in Mounting Ring and Set in Concrete Pipe Wall

FIG. 7--*Interface Pressure Cell for Concrete Pipe*

A mounting ring and clamp system were developed to achieve transportability. The complete interface pressure cell assembly, shown in Figure 7b, includes the pressure cell, a mounting ring, and two expansion rings with bolts. The mounting ring and pressure cell were cast together with non-shrink gout. The mounting ring was manufactured with circumferential and longitudinal slots to create flexible tabs that would push out against the concrete pipe wall when expanded with the expansion clamp.

Calibration--After being cast into the ring clamp, each gage was calibrated under fluid pressure and for changes due to temperature. The fluid pressure calibrations showed that the special construction did not change the response of the cell to externally applied pressure. The calibration factors were within 3% of the transducer calibrations supplied by Geokon. The small size of the cells, however, did result in a significant increase in sensitivity to temperature. The average transducer temperature correction factor supplied by Geokon was 0.04 kPa/°C (0.006 psi/°C) while the average temperature correction factor for the complete interface pressure cell was 1.24 kPa/°C (0.18 psi/°C).

Installation in Pipe--Ten interface pressure cells were used, two at both the crown and invert, and the remaining cells were located at 30°, 60° and 90° from the invert on each side of the pipe (Figure 2). This provided redundancy for all measurements.

Corrugated Metal Pipe Interface Pressure Cells

Measuring the soil pressure on the surface of the corrugated metal pipe is complicated by the profile surface. Standard pressure cells have flat surfaces which, if installed in the corrugated profile, create edges and sudden geometry changes that result in conformance problems, causing arching of load onto or off of the cell. To minimize this problem, custom interface pressure cells were designed to preserve the exterior profile of the pipe. This was accomplished by using a close fitting cutout of the pipe wall as a sensing element, supported by a beam instrumented with strain gages to serve as a transducer.

Cell Configuration--The sensing element of the metal pipe interface pressure transducer was a cutout of the pipe wall as shown in Figure 8.

The cutout was supported on a 140 mm long by 25 mm wide by 6.35 mm thick (5.5 in. by 1 in. by 0.25 in.) aluminum transducer beam and secured by three screws, one at the center of the cutout into the transducer beam and two at the edges of the cutout into a transverse cross brace (see Figures 8b and 9). The ends of the transducer beam were secured to the pipe by screws spaced at 127 mm (5 in.).

Calibration--The cells were calibrated by dead weights and through soil. The dead weight calibrations were in general agreement with design calculations (correlation coefficients r^2 of about 0.99 and the reproducibility was typically 0.6 kPa).

Subsequent testing showed that the cells were less reliable in measuring actual interface pressures than indicated by the dead weight tests because: 1) the screws securing the cells to the pipe wall restrained the ends from free rotation and introduced an apparent pressure as a result of pipe deflection, and 2) as the pipe deflected under soil load the change in curvature of the pipe would cause the cutout to project out from the face of the pipe or withdraw back from the face of the pipe, causing arching of load onto or off of the cutout. These two effects were demonstrated in calibration tests.

Overall the two issues discussed above raise questions about the accuracy of the metal pipe interface pressure cells. While the cells were used in the tests and the data is reported, the results should be considered qualitative rather than quantitative in demonstrating interface pressures.

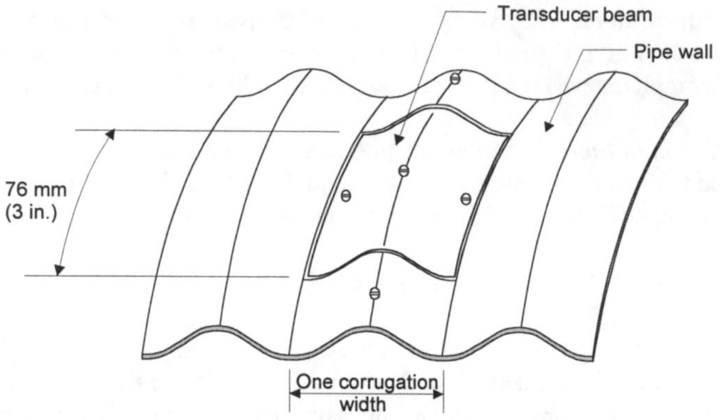

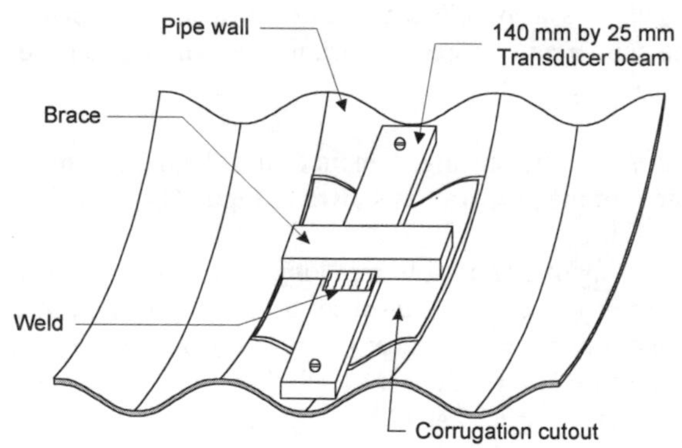

FIG. 8--*Corrugated Metal Pipe with Interface Pressure Cell*

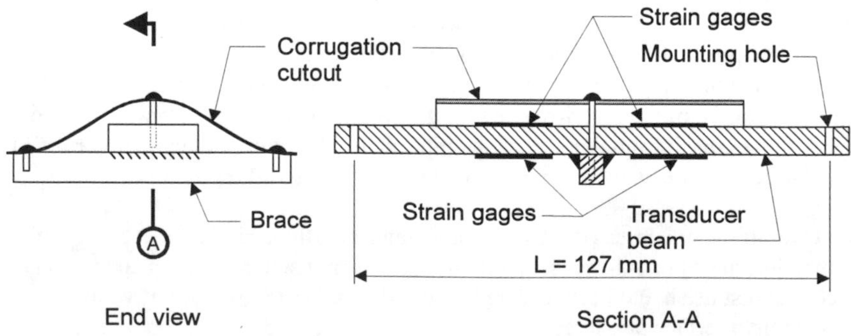

FIG. 9--*Metal Pipe Interface Pressure Cell*

CONCLUSIONS

This paper documents the selection and application of instruments used to monitor soil-culvert interaction during backfill placement. A wide array of instruments were used to monitor the pipe strains and deformations, the soil stresses and strains and the pressures at the soil-culvert interface. Most instruments were read using a computer data acquisition system that is also documented.

Most of the instrumentation functioned as desired; however, some of the instruments and sensors required special care and pertinent issues were discussed. In addition, the interface pressure cells that were developed for the corrugated metal pipe provided good qualitative information but the results may not represent actual values.

ACKNOWLEDGMENTS

The work reported in this paper was supported by funding from the National Science Foundation, the Federal Highway Administration, the states of California, Iowa, Kansas, Louisiana, Massachusetts, Minnesota, New York, Ohio, Oklahoma, Pennsylvania, and Wisconsin, and the Eastern Federal Lands Highway Division of the Federal Highway Administration. The pipes used in the tests were donated by Contech Construction Products, CSR/New England, Hancor Inc., and Plexco/Spirolite Inc. UMass graduate students Glen Zoladz and George Costa provided substantial assistance with the tests. The Massachusetts Highway Department provided a nuclear density gauge for use during the field tests.

REFERENCES

McGrath, T. J., 1998, "Pipe-Soil Interactions During Backfill Placement," Geotechnical Report NSF98-444D, Department of Civil and Environmental Engineering, University of Massachusetts, Amherst, MA.

McGrath, T. J. and Selig, E. T., 1996, "Instrumentation for Investigating Behavior of Pipe and Soil During Backfilling," Geotechnical Report NSF96-443P, Department of Civil and Environmental Engineering, University of Massachusetts, Amherst, MA.

Selig, E. T., 1975, "**Soil Strain Measurement Using Inductance Coil Method**", *Performance Monitoring for Geotechnical Construction, ASTM STP 584*, American Society for Testing and Materials, pp. 141-158.

Webb, M. C., McGrath, T. J., and Selig, E. T., 1998, "**Field Test of Buried Pipe with CLSM Backfill**," "*The Design and Application of Controlled Low-Strength Materials (Flowable Fill)*," *ASTM STP 1331*, A.K. Howard and J.L Hitch, Eds., American Society for Testing and Materials.

Webb, M. C., McGrath, T. J., and Selig. E. T., 1996, "Field Tests of Buried Pipe Installation Procedures," Transportation Research Record 1541, TRB, National Research Council, Washington, D.C., pp. 97-106.

Zoladz, G. V., McGrath, T. J., and Selig. E. T., 1996, "Laboratory Tests of Buried Pipe Installation Procedures," Transportation Research Record 1541, TRB, National Research Council, Washington, D.C., pp. 86-96.

Instrumentation Support Construction Activities

Dar-Hao Chen,[1] John Bilyeu,[1] and Fred Hugo[2]

Monitoring Pavement Response and Performance Using In-Situ Instrumentation

---

REFERENCE: Chen, D., Bilyeu, J., and Hugo, F., **"Monitoring Pavement Response and Performance Using In-Situ Instrumentation,"** *Field Instrumentation for Soil and Rock, ASTM STP 1358*, G. N. Durham and W. A. Marr, Eds., American Society for Testing and Materials, West Conshohocken, PA, 1999.

ABSTRACT: The purpose of this paper is to present the effectiveness of in-situ instrumentation on diagnosing the pavement layer conditions under full-scale accelerated traffic loading. The test section is an in-service pavement (US281) in Jacksboro, Texas. Multi-Depth Deflectometers (MDDs) are used to measure both permanent deformations and transient deflections, caused by accelerated traffic loading and Falling Weight Deflectometer (FWD) tests. Four different FWD loads of 25, 40, 52, and 67 kN were applied in close proximity to the MDDs at various traffic loading intervals to determine pavement conditions. It was found that the majority of rutting occurred in the newly recycled asphalt mix. The aged (> 40 years) underlying base and subgrade layers contributed less than 30% to overall rutting. Only the top recycled asphalt layer underwent notable deterioration due to traffic loading. Up to 1.5 million axle repetitions, the test pad responded to FWD load almost linearly, not only over the whole pavement system but also within individual layers. However, under higher FWD loads, the percentage of total deflection contributed by the subgrade increased.

KEYWORDS: accelerated pavement testing, deflectometer, remix, instrumentation, load simulator

---

[1]Data Analysis Engineer and Assistant, respectively, Texas Department of Transportation, Design Pavement Section, 4203 Bull Creek Road, #37, Austin, Texas 78731, Tel (512) 467-3963, Fax (512) 465-3681.

[2]Research Fellow, Center for Transportation Research, University of Texas at Austin, 3208 Red River, Austin, Texas and the Institute for Transport Technology, University of Stellenbsoch, South Africa, 7600.

Accelerated Pavement Testing (APT) provides an avenue for authorities to find fast and reliable solutions to pavement problems. The Texas Mobile Load Simulator (MLS), an accelerated pavement testing device, was employed in this investigation. The MLS is a machine that rolls tandem axles with standard truck tires over a 3m by 12m test pad. The load on each tandem axle is set to carry a legal load of 151 kN at a speed of approximately 19 kph. More details of the MLS can be found in references [Metcalf, 1996; Chen et al., 1997a].

Since the testing was conducted under closely watched conditions, the specific pavement response and performance information could be obtained. As with all APT, a link must be established between the data obtained from the pavement responses and the pavement performance models being investigated. This link is generally established via the pavement instrumentation and elastic-layer modeling. The effects of accelerated traffic on the pavement was measured using in-situ instrumentation installed at different depths and locations within the pavement structure.

In this investigation, the FWD was used to determine the response of the pavement structure to four different loads, which are approximately 25 kN, 40 kN, 52 kN, and 67 kN. The FWD testing was done at regular traffic loading intervals. Other analyses included the comparison of these responses to those from the loading applied by the MLS. Performance was measured in terms of rutting relative to the load applications. Also the rutting was also analyzed in terms of permanent deformation in the respective pavement layers.

Test Section

The test section is an in-service pavement located on US281 near Jacksboro in the Fort Worth District. US281 is a two-lane highway in each direction. The Fort Worth District Pavement Engineer indicated that there was an average of 3,100 vehicles per day (1550 per direction) in 1994. The percent of trucks is approximately 17.4%. Approximately 10% of traffic falls on the inside lane. Since the pavement was rehabilitated in 1995, an estimated 9,850 trucks had traveled on the inside lane (or approximately 10,000 to 19,700 ESALs of traffic, depending on the conversion factor used) before the MLS was moved onto the test site.

The first asphalt layer of the test section was constructed in 1957. There were four major upgrades/rehabilitations that were completed in 1971, 1976, 1986, and 1995. Figure 1 shows the complete pavement history. TxDOT forecasts that the pavement section in the outside lane will be subjected to 2 million ESALs over a twenty-year period. The last major rehabilitation was done in 1995 with 50mm of recycled ACP using the Remix process. Prior to that there was a major rehabilitation, using 76 mm of lightweight aggregate ACP, in 1986. The inside southbound lane of US281 was closed to traffic in April of 1997 for testing. The MLS was then moved onto the test section in May. The outside lane remained open for public use. At the time of this analysis, 1.5 million axle repetitions had been applied to the test pavement by the MLS. Each axle had been set to carry a load of 75.6 kN.

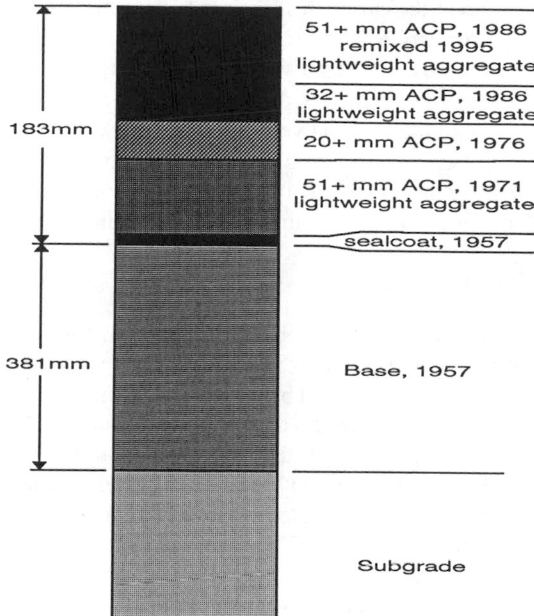

FIG. 1 -- *Pavement Section of Southbound US281*

The wander pattern was set such that the maximum percent coverage would be 83% over a cumulative width of 100 mm, centrally located, while 405 mm of the section would be covered with 67% of the MLS "traffic", as shown in Figure 2. This calculation was based on a 230 mm wide tire and 100 mm spacing between dual tires, which were measured from the MLS. The total width of the wander pattern was approximately 860 mm.

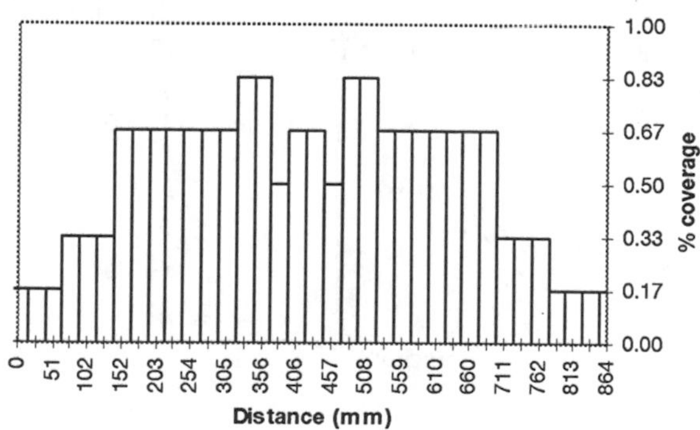

FIG. 2 --*Wander Pattern Applied by the MLS*

Objective

The main objective of the current testing was to determine the effectiveness of the rehabilitation process performed in 1995. It was important to make sure that the pavement failure would not be caused by a subsurface layer(s) within the 40-year-old pavement. Several measures were taken to prevent incorrect adjudication of the rehabilitation process, since failure of an overlay is sometimes caused by a subsurface layer and not the overlay itself. Thus, it was imperative to have in-situ instrumentation with the capacity to identify the condition of the various layers under traffic loading. The Multi-Depth Deflectometer (MDD) was found to be a very suitable tool for this purpose.

Multi-Depth Deflectometer (MDD)

The Texas Transportation Institute (TTI) was awarded the contract to install MDDs. Two MDDs were installed, one in the center of each wheel path. MDDs reveal not only the transient responses under load but also the accumulated permanent depth deformation. Each MDD hole contains three LVDTs to measure deflections at three different depths, as shown in Figure 3. Ideally, an LVDT should be placed at each layer interface, so that the deflection contribution by each layer could be directly measured. Mechanical limitations made this impossible, so the MDD sensors were installed as close as possible to the ideal depths. MDD-T, MDD-M, and MDD-B has been used to refer to the LVDTs at the depths of 90 mm, 320 mm and 570 mm, respectively. Figure 3 shows the MDD depths and their relation to the pavement layers.

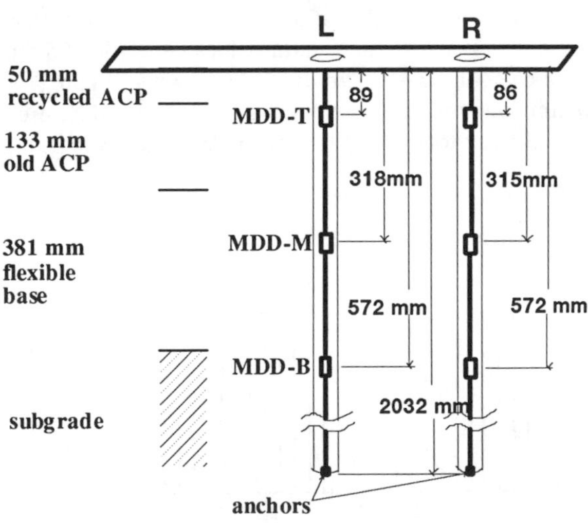

FIG. 3 --*MDD and Pavement Layer Depths*

FWD-MDD Interrelationship

Deflection profiles from FWD tests provide valuable information to assess pavement layer moduli through a back-calculation process. The back-calculated moduli are often obtained from a best fit to the measured deflection profile. The verification of back-calculated moduli has been done in two ways: (1) engineering judgment; and (2) comparison with other test results such as laboratory and field seismic testing. Kim et al. [1992] and Uzan and Scullion [1990] applied FWD tests on top of an MDD and measured the resulting surface deflections and depth deflections simultaneously using both the FWD geophones and the MDDs, respectively. They found that the depth deflections measured by the MDD can be a powerful tool in evaluating the accuracy and dependability of back-calculated moduli values from FWD data. The authors also used the MDD data for analyzing the two-dimensional pavement structure of the pavements tested in Victoria. In addition, FWD drops on MDDs provide valuable information to determine if different pavement layers respond to FWD loads in a linear or nonlinear fashion, thus validating the appropriateness of using linear elastic theory in the calculation.

Four different loads, 25, 40, 52, and 67 kN, were applied in close proximity to the MDDs with each load repeated three times to examine the repeatability of the results. In the previous study [Chen et al. 1997b] applied four different loads with each load repeated 10 times and found that the repeatability was high for the MDD readings and for most of the FWD readings. The results indicated that errors were observed in the FWD W1 measurement because the first sensor was in direct contact with the MDD cap, which generated slippage that caused erratic readings. Therefore, two changes were made in this study.

1. The center of the FWD loading plate was offset approximately 220 mm from the center of the MDD cap, as shown in Figure 4. This was done so that the FWD loading plate would not be in direct contact with the MDD cap.

2. The number of repeated tests was reduced from 10 to 3 because of the excellent repeatability shown in the previous study. The results show insignificant variation among the 3 tests.

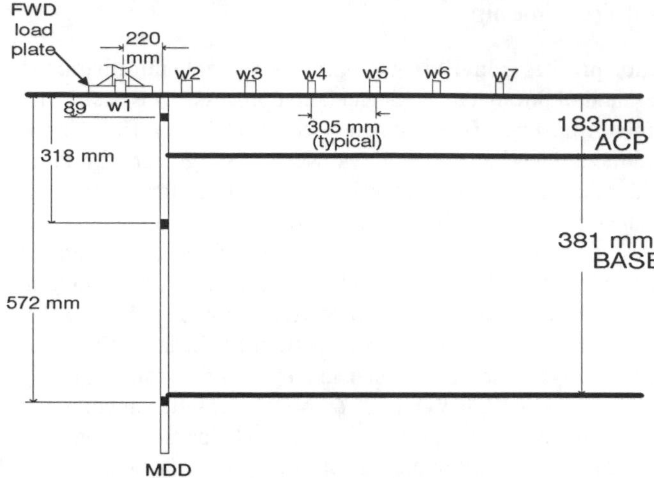

FIG. 4 --*MDD-FWD Schematic*

FWD-MDD data was collected prior to testing and after 2500, 5000, 20k, 40k, 80k, 160k, 300k, 450k, 600k, 750k, 900k, 1050k, 1200k, 1350k and 1.5 million axle repetitions. Analyses were performed on the data to determine pavement degradation due to axle loading. Figure 5 shows the history of depth deflections under 40 kN load. It is encouraging to see that the MDD registered clear signals under FWD load at all three depths. All collected FWD-MDD signals were high in quality. Figure 5 shows the duration of the stress pulse to be approximately 28 milliseconds; and that the subgrade contributed up to approximately 50% of total deflection.

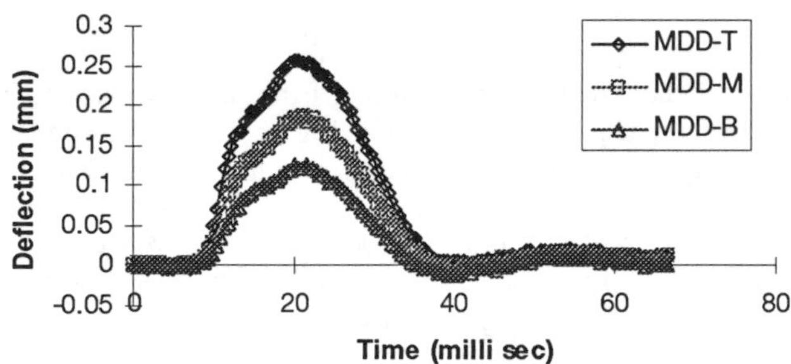

FIG. 5 --*Deflection Response History on Left Wheel Path for 42 kN FWD Load*

Figure 6 illustrates the MDD depth deflections vs. axle repetitions for the left wheel path. Only the averaged results from 40 kN FWD loads are presented in Figure 6. Figure 7 shows the top MDD deflection (MDD-T) measurements under four different load levels. Note that the pavement temperature was not controlled but was recorded. Asphalt pavement temperatures were collected from three different depths: top (13 mm from surface), middle (90 mm from surface), and bottom (165mm from surface). The middle pavement temperature data were also incorporated into figures 6 and 7. The deflections increased not only due to the traffic loading but were also affected by pavement temperature. The MDD deflections increased up to 1 million axles due to increased temperatures but decreased after 1 million because of decreased temperature. This trend for deflections is the same at all three depths.

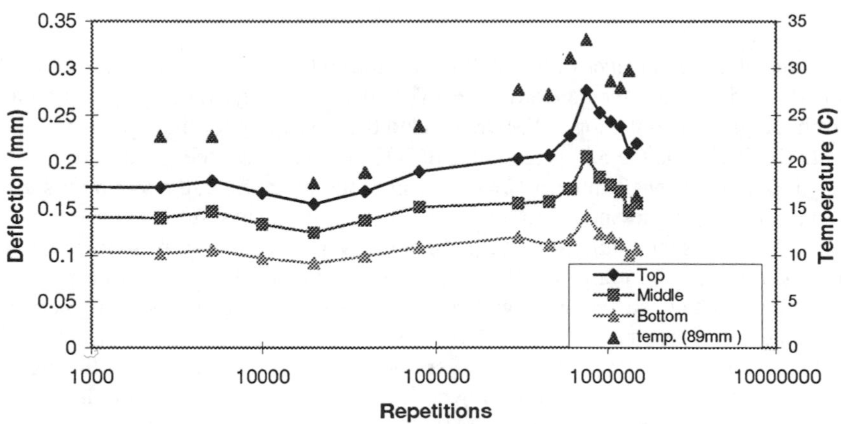

FIG. 6 --*Depth Deflections at Various Loading Stages on Left Wheel Path*

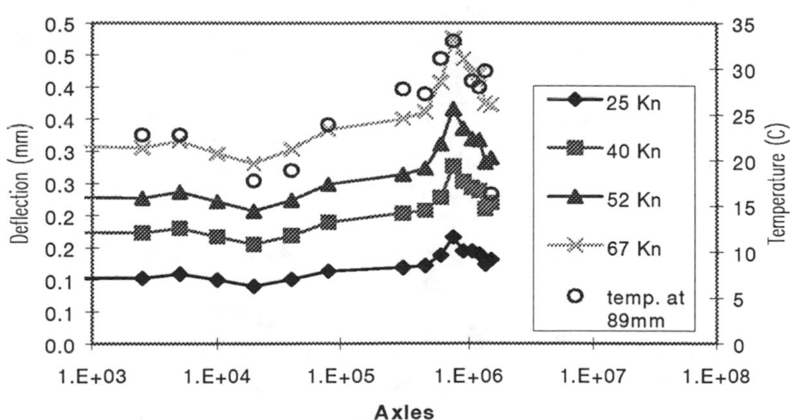

FIG. 7 --*Pavement Responses Under Four Levels of FWD Loads on Left Wheel Path*

The depth deflections at MDD-M and MDD-B at the end of 1.5 million axles were approximately of the same magnitudes as at zero axles. However, there was some deflection increase at MDD-T. No surface cracking that was not already present at the beginning of the test (two thermal cracks), was observed at the end of 1.5 million axle repetitions. It was thus concluded that no significant damage had occurred in the base and the subgrade at the end of the test. A prime reason for this was the fact that very little water (if any) could have filtered into the pavement layers, because the pavement test section was covered by the MLS. Also, the base and subgrade had been in service and consolidated for 40 years.

One of the advantages of the MDD is its ability to measure the permanent depth deformation within the pavement structure. The surface rutting (S) or total rutting minus the rutting registered at the top (T) MDD yielded the net rutting for the top 90 mm of asphalt. Similarly, the top minus the mid MDD (M) reading was the rutting for the bottom 100 mm of asphalt and top 127 mm of granular base. Also, the mid minus the bottom (B) reading represent the bottom 254 mm of base material rutting. The B reading in Figure 8 denotes the rutting from the subgrade layer. It was no surprise to see from Figure 8 that the major rutting (> 60%) was contributed by the top 90 mm of asphalt. The top 50 mm of asphalt is the newly overlaid (1995) layer. The aged underlying pavement layers had reduced their rate of consolidation. In view of Figure 8, the rate of rutting in the newly overlaid ACP layer increased after 300,000 reps partly because of the high temperature (mid June) during that testing period. The underlying pavement layers also began to contribute some rutting after 300,000 axles, because the soft surface layer resulted in higher stress penetrating into the subsurface. After 1 million reps, the rate of rutting for underlying pavement layers decreased, because the temperature was lower during this testing period. The test passed 1 million repetitions in mid September.

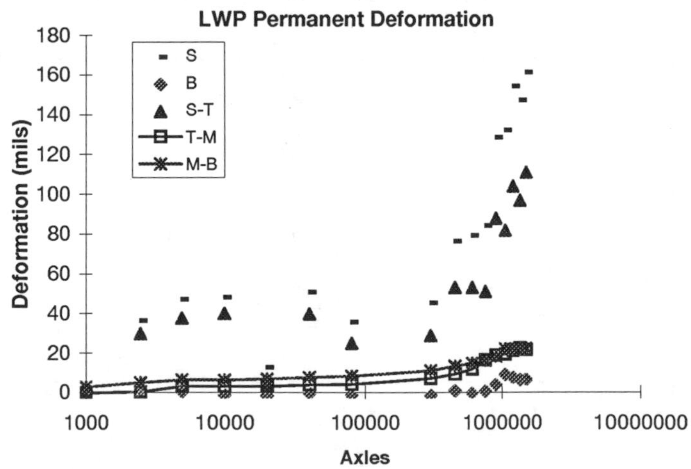

FIG. 8 --*Layer Permanent Deformation For Left Wheel Path*

At the end of 1.5 million repetitions, the surface rut depth, on average, had reached 3.4 mm. No surface cracking was observed. Both the MDD transient responses and permanent deformations indicated that the majority of material property changes occurred in the top asphalt layer. Since the subsurface layers have been in service since 1957, their material characteristics were assumed to have remained fairly constant during the test.

The test pavement had carried the equivalent of 8 year's traffic in accelerated form without severe distress. Of course, the pavement had not been subjected to the full environmental impact. Nevertheless, the district pavement engineer was satisfied with the performance of the rehabilitation process.

The test pad is currently being subjected to a 25% overload to study the effect of overload and to investigate whether it would bring about more visible distress. Subsequently, it is planned to continue the test with the surfacing subjected to water infiltration during testing.

BISAR Analysis and Anchor Movement Monitoring

The measured peak surface deflections were used to back-calculate the moduli of the pavement layers using the MODULUS software. For this step, loads of approximately 40 kN (from the 2nd FWD drop height) were used. The moduli from MODULUS were then used in BISAR to forward-predict deflections both at the surface and at depth. The moduli obtained from MODULUS were used only as a starting point. Several iterations within BISAR were needed to obtain a set of moduli that would estimate both the depth and surface deflections accurately at any load.

The accuracy of pavement analysis is greatly affected by the estimation of the bedrock depth. Particularly, when shallow bedrock is encountered, the back-calculated subgrade modulus can often be overestimated. Before the test site was selected, ground-coupled ground penetrating radar was employed to assess the subsurface structure. From the drilling and ground penetrating radar, bedrock was found in the range of 0.76m to 3.5m from the surface, for the 182 m long section which includes the MLS site. During the MDD anchor installation, no bedrock was encountered up to a 2 m depth, where the anchors were placed.

It is believed that the bedrock depth under the MLS test section is between 2.5 m and 3.5 m. Since the anchors were not located in the fixed bedrock, any anchor movement will affect the true MDD deflection readings. Therefore, anchor movement monitoring becomes important to assure the integrity of the test data. In addition, anchor movement information could be used to estimate the subgrade stiffness and depth to bedrock. The lesser anchor movements mean that there may be shallow bedrock and/or stiffer subgrade. Anchor movement data under FWD load was collected by attaching the 7th FWD geophone to the MDD core rod, which is connected directly to the anchor.

Tables 1 and 2 present the FWD deflections W1 (at the center), W2 (300 mm) from the center of the load, and MDD deflections MDD-T, MDD-M, MDD-B. The FWD load plate was offset approximately 220 mm from the MDDs. The magnitudes of anchor movement are also given in tables 1 and 2. As shown in tables 1 and 2 and Figure 6, the subgrade deflection constitutes slightly less than 50% of the total deflection. It was found that at higher loads, a slightly higher percent of deflection was contributed by the subgrade.

TABLE 1 --*MDD Deflections Including the Anchor Movement for the Left Wheel Path at the End of 1.2 Million Axle Repetitions*

Load (kN)	W1(mm)	W2 (mm)	MDD-T (mm)	MDD-M (mm)	MDD-B (mm)	Anchor (mm)
25	0.201	0.150	0.189	0.145	0.113	0.044
40	0.343	0.259	0.319	0.251	0.195	0.076
52	0.454	0.345	0.421	0.337	0.264	0.101
67	0.600	0.460	0.558	0.456	0.359	0.136

TABLE 2 --*MDD Deflections Including the Anchor Movement for the Right Wheel Path at the End of 1.2 Million Axle Repetitions*

Load (kN)	W1(mm)	W2 (mm)	MDD-T (mm)	MDD-M (mm)	MDD-B (mm)	Anchor (mm)
25	0.203	0.139	0.185	0.139	0.112	0.015
40	0.338	0.239	0.310	0.240	0.192	0.025
52	0.444	0.318	0.408	0.323	0.258	0.031
67	0.588	0.428	0.543	0.437	0.351	0.040

In view of tables 1 and 2, the anchor movements were three times higher in the left wheel path than in the right wheel path. However, the deflections recorded by the MDD-T, MDD-M, and MDD-B were only 2% higher in the left than the right wheel path. Also, the surface FWD deflections are approximately the same between the left and right wheel paths, as shown in Figure 9. It is the authors' opinion that the left wheel path has deeper bedrock and higher moduli values for asphalt, base and subgrade. Bedrock in the left and right wheel path was estimated to be at the depth of 3.2 m and 2.7 m, respectively. The variation in bedrock depth and subgrade moduli is possible (even within 1.8 m) in this case. Since the test section was built in a fill section, variation of the subsurface properties is not uncommon. The confirmation of bedrock depth through drilling will be performed once the test is complete. Figure 9 also shows that both surface and depth deflections responded linearly to increased load. This observation was supported by the fact that the data fit linear trendlines, with R-squared values exceeding 0.99.

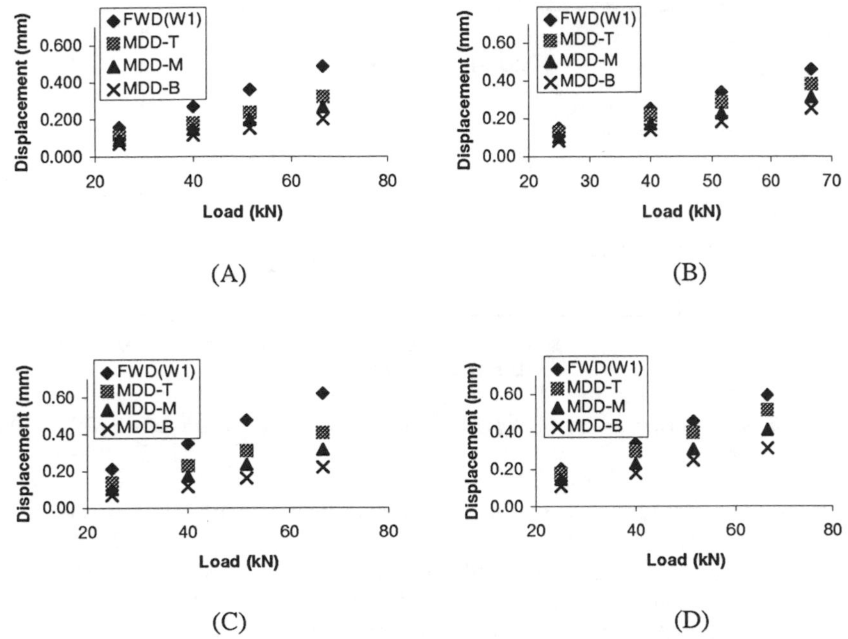

FIG. 9 --*FWD-MDD Response Data (A) at 0 Axle Repetition Left Wheel Path (69 F mid-Depth Pavement Temperature) (B) at 0 Axle Repetition Right Wheel Path (69 F temperature) (C) at 600 000 Axle Repetitions Left Wheel Path (94 F) (D) at 600 000 Axle Repetitions Right Wheel Path (94 F)*

Additional efforts were made to place the MDD between the second and third sensors of the FWD with a spacing of 480 mm (center of FWD load plate to center of MDD hole). Graphical displays of the measured surface and depth deflections at a single load level (but two different offsets) are shown in Figure 10. At an offset between 220 mm and 480 mm from the load, the deflections measured within the asphalt layer (MDD-T) are equal to or greater than those measured on the surface.

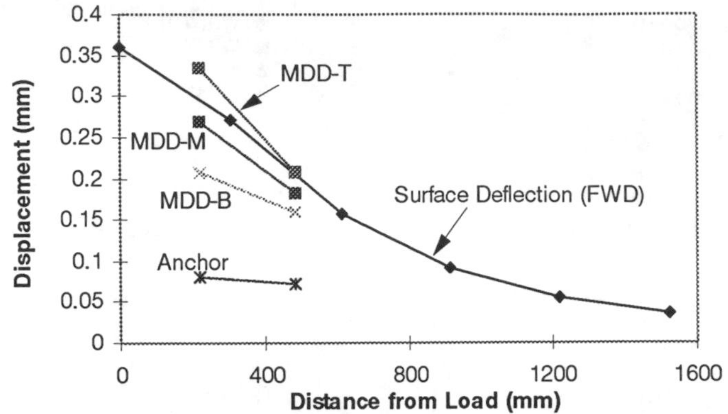

FIG. 10 --*Measured Surface and Depth Deflections on Left Wheel Path for 40 KN FWD Load*

There are four major layers shown in the cores taken from the MLS test site. Thus, the asphalt was divided into 4 different layers in the BISAR analysis. The total thickness of ACP is 183 mm, as determined from the cores. The moduli and layer thicknesses used in the analyses are presented in Table 3. The analyses included two different load offsets (220 and 480mm) and four levels (25, 40, 52 and 67 kN) of loads. After running MODULUS to obtain surface deflection-based layer moduli, these values were adjusted (using BISAR) to better fit the actual measurements taken at various offsets, depths, and loads. The comparison between measurements and predictions for surface deflection from FWD and depth deflections from MDD are given in Table 4. As shown in Table 3, the moduli are higher in left than right wheel path but the bedrock is deeper, as explained in the previous section. It should be noted that the analyses in tables 3 and 4 represent the pavement condition at the end of 1.2 million axle repetitions. The moduli variation in relation to the axle repetitions will be discussed in subsequent publications.

TABLE 3 --*Layer Properties Used in the BISAR Program*

LEFT WP			RIGHT WP		
Modulus (Mpa)	Thickness (mm)	Poisson's Ratio	Modulus (Mpa)	Thickness (mm)	Poisson's Ratio
3101 (AC)	51	0.35	2963 (AC)	51	0.35
2412 (AC)	25	0.35	2274 (AC)	25	0.35
1723 (AC)	25	0.35	1723 (AC)	25	0.35
1860 (AC)	81	0.35	1585 (AC)	81	0.35
289 (BASE)	381	0.38	276 (BASE)	381	0.38
76 (SUBG)	508	0.4	69 (SUBG)	508	0.4
48 (SUBG)	2159	0.4	55 (SUBG)	1651	0.4
1723 (BEDROCK)	-	0.38	1723 (BEDROCK)	-	0.38

TABLE 4 --*Left Wheel Path Measured Vs. Predicted Deflections, mm (Predicted in Parentheses)*

Left WP	Offset =	220mm			
Load	**W1**	**W2**	**MDD-T**	**MDD-M**	**MDD-B**
25 KN	0.224	0.166	0.211	0.163	0.127
	(.253)	(.177)	(.201)	(.181)	(.156)
40KN	0.360	0.272	0.335	0.268	0.207
	(.381)	(.267)	(.302)	(.272)	(.234)
52KN	0.467	0.355	0.437	0.353	0.276
	(.48)	(.335)	(.381)	(.343)	(.295)
67 KN	0.625	0.479	0.586	0.484	0.378
	(.63)	(.439)	(.5)	(.452)	(.386)
	Offset =	480mm			
Load	**W1**	**W2**	**MDD-T**	**MDD-M**	**MDD-B**
25 KN	0.213	0.162	0.126	0.119	0.099
	(.254)	(.177)	(.143)	(.14)	(.129)
40KN	0.342	0.263	0.204	0.182	0.160
	(.384)	(.267)	(.216)	(.211)	(.194)
52KN	0.444	0.343	0.264	0.241	0.208
	(.48)	(.335)	(.272)	(.267)	(.244)
67 KN	0.602	0.465	0.363	0.329	0.288
	(.63)	(.439)	(.356)	(.348)	(.32)

Conclusions

The purpose of this paper is to present the effectiveness of the in situ instrumentation on diagnosing the pavement layer conditions under accelerated traffic loading. It was found from this study that the MDD is a valuable tool in assisting pavement engineers in characterizing the pavement material properties. Pavement response and performance information of underlying layers under traffic loading were evident from the MDD data. Several observations were derived from the analyses:

1. The majority (> 60%) of rutting occurred in the top new recycled asphalt mix. The aged (40 years) underlying base and subgrade layers contributed less than 30% of overall rutting.

2. Only the top recycled asphalt layer suffered notable deterioration due to accelerated traffic loading. Deflections, at any given load, increased in this layer with increased axle repetitions.

3. At intervals up to 1.5 million axle repetitions, the test pad responded to FWD load linearly. Increased FWD loads caused proportionately larger deflections not only within the whole pavement system, but also within individual layers.

4. High loads caused a slightly higher percent of total deflection to be contributed by the subgrade. The subgrade exhibited this non-linear behavior while the asphalt layers were still in their linear range.

5. The test pavement carried 8 years of equivalent traffic in accelerated form without severe distress. The pavement engineer was satisfied with the performance of the rehabilitation process.

Acknowledgments

The authors thank Mr. Ken Fults, Dr. Andrew Wimsatt and Mr. Mike Murphy for their input and suggestions. In addition, the authors also thank Mr. Tom Scullion and Mr. John Ragsdale for installing the MDDs. Thanks and appreciation are extended to Mr. Sherwood Helms and the MLS field operation personnel for conducting the field tests and collecting data.

References

Chen, D., Fults, K., and Murphy, M. (1997a) "The Primary Results for the First TxMLS Test Pad," Transportation Research Board. Record No.1570. p. 30-38.

Chen, D., Meilahn, N., Murphy, M. and Fults, K. (1997b) "Evaluation of the Change of Material Properties Under Accelerated Trafficking of Texas Mobile Load Simulator," *Proceeding CD-ROM, 13th International Road Federation (IRF) World Meeting.* Toronto 1997.

Kim, Y. R., Khosla, N.P., Satish, S., and Scullion, T. (1992) "Validation of Moduli Backcalculation Procedures Using Multi-depth Deflectometers Installed in Various Flexible Pavement Structures," Transportation Research Board. Record 1377. p. 128-142.

Metcalf, J. B., (1996) "Application of Full-Scale Accelerated Pavement Testing," *Synthesis of Highway Practice 235*. National Cooperative Highway Research Program.

Uzan, J., and Scullion, T. "Verification of Backcalculation Procedures, " *Proceeding Third International Conference on Bearing Capacity of Roads and Airfields*. 3-5 July, 1990. Trondheim, Norway, p. 447-458.

Ali Regimand[1] and Alan B. Gilbert[2]

Apparatus and Method for Field Calibration of Nuclear Surface Density Gauges

---

REFERENCE: Regimand, A. and Gilbert, A.B., "**Apparatus and Method For Field Calibration of Nuclear Surface Density Gauges,**" *Field Instrumentation for Soil and Rock, ASTM STP 1358*, G.N. Durham and W.A. Marr, Eds., American Society for Testing and Materials, West Conshohocken, PA. 1999.

ABSTRACT: Nuclear gauge density measurements are routinely used for compliance verification with specifications for road and construction projects. The density of construction materials is an important indicator of structural performance and quality. Due to speed of measurement, flexibility and accuracy, nuclear gauge density measurement methods are becoming the preferred standard around the world.

Requirements dictate that gauges be verified or calibrated once every 12 to 18 months. Presently, there are no field portable devices available for verification of the gauge calibration. Also, the density references used for calibration of gauges, are large and not designed for field portability. Therefore, to meet the present standards, users are required to ship gauges back to a service facility for calibration.

This paper presents results obtained by a newly developed device for field verification and calibration of nuclear density gauges from three different manufacturers. The calibrations obtained by this device are compared to the factory calibration methods and accuracies are reported for each gauge model.

KEYWORDS: verification, calibration, density, moisture, nuclear density gauge, density gauge, backscatter, direct transmission

---

Nuclear density gauges are used for measurement of wet density and moisture in the construction industry. Due to the large errors in the measurements of the density gauges, a conference was held in Virginia to discuss and examine the variations in gauges (Hughes and Anday 1967). Later in 1969, a symposium was held at North Carolina State University to address improvements for nuclear gauges (McDougall et al. 1969).

Since their initial development in the 1950's, nuclear gauges have undergone many improvements. Several variations of gauges are now available in the market to address different needs in the industry (Regimand 1987). In the last twenty years, the major focus has been on the accuracy of the calibration references and the methods used. In the last version of ASTM Test Method on Density of Soil and Soil-Aggregate in Place by Nuclear

[1]Chief Nuclear Engineer, InstroTek, Inc., 3201 Wellington Ct. Ste. 101 Raleigh, NC 27615.
[2]Product Manager, InstroTek, Inc., 3201 Wellington Ct. Ste. 101 Raleigh, NC 27615.

Methods(D 2922) and Test Method for Density of Bituminous Concrete in Place by Nuclear Methods(D 2950), the five block calibration method was specified as the preferred calibration method. However, in the new revision of these standards the calibration requirements were changed to allow flexibility in gauge calibration methods used by the manufacturers. This was done to encourage research and development innovations in the gauge calibration area.

A weakness that has always existed in the nuclear gauge application has been the field verification of gauges to determine the amount of drift in calibration. For years, gauge standard counts were used to determine the gauge calibration accuracy. However, for density measurements, standard counts can not be relied on for calibration accuracy check. Standard counts are primarily utilized in the calculations to correct for radioactive source decay. The geometrical relationship of source to detectors in the standard count position is completely different as compared to the measurement positions used on the gauge. For this reason, changes in counts at a particular position is not necessarily reflected in the standard count position taken on a reference block provided with each gauge.

ASTM standards D 2922 and D 2950 require that gauges be verified or calibrated once every 12 to 18 months. But due to the expense and difficulties related to shipment of nuclear products, gauge owners, in many cases, continue to use these devices for years without a current verification or calibration.

For best results, a field device with a known density is needed to allow the users an opportunity to collect density readings at each source position and determine the status of the gauge calibration.

Nuclear Gauge Theory

Nuclear gauges are used for measurement of density and moisture of soil, asphalt and concrete. The material density is measured by employing a small radioactive source such as Cesium-137 and one or more photon detectors. There are two modes of operation routinely referred to as "backscatter" and "direct transmission". In the backscatter mode, the source and the detector are placed on the test material in the same plane. Enough shielding is provided in the base to stop the majority of direct photons from the source to the detector. The photons from the source penetrate the test material and a fraction will scatter back to the detectors. In the construction density range of interest, the number of photons scattered back to the detector is inversely proportional to the material density. Therefore, the higher the number of detected counts, the lower the density.

In the direct transmission mode, a hole is formed in the test material and the source is placed into the material to a predetermined depth. The photons detected in this case are representative of the density of the material in the path between the source and the detectors. Again, in direct transmission mode, the number of photons detected is inversely proportional to the material density. Both backscatter and direct transmission densities are calculated from the relationship in (Eq 1).

$$CR = A\exp(-BD) - C \qquad (1)$$

where A,B and C are gauge parameters determined at the time of calibration, D is the material density and CR is the ratio of counts on the test material to standard counts taken on a reference block with the radioactive source in the shielded position.

Moisture is generally measured by employing a fast neutron source such as an americium-241:Beryllium and a thermal neutron detector such as a helium-3 tube. Both the source and the detector in most gauges are placed in the base of the gauge. The fast neutrons from the source penetrate the material and interact with the hydrogen nucleus in water (H_2O). Multiple interactions reduce the energy of the incident neutron to a level known as thermal energy. The detector in the gauge base is primarily sensitive to thermal neutrons and there is a direct proportionality between the number of counts detected by the tube and the amount of water in the material. The relationship used for calculating moisture from the gauge counts is given by

$$MCR = a + bM \quad (2)$$

where a is the intercept and b is the slope of the line determined at the time of calibration, M is the moisture content and MCR is the ratio of counts on the test material to the standard count taken on the reference block.

In most manufactured gauges, the source and detector position for moisture measurement is fixed inside the gauge base and for both the standard count and the measured count the source to detector relationship is the same. The advantage of this method is that once the gauge is calibrated at the factory, the moisture count ratio (MCR), if taken at relatively the same time as the measured count will correct for all variations in the gauge. However, in the density measurement geometry, the measurement position is different from the standard count position and the density count ratio (DCR) does not correct for variation in gauge readings.

Nuclear Gauge Calibration

Gauge density calibration is performed by using large blocks with known densities. The gauge is placed on these blocks and a count is collected. A minimum of three blocks are needed to find the constant parameters in (Eq 1).

There are two methods of choice for gauge calibration. Namely the "Three" and "Five" block method.

In both these methods, magnesium (mg), magnesium/aluminum laminate (m/a) and aluminum (al) blocks are used. The gauge is placed on each of these blocks and a count is collected. The known block densities are normalized by using the ratio of the metal mass attenuation coefficient to that of an average soil. The detectors used in the manufactured gauges are Geiger Mueller tubes and not capable of energy discrimination. The normalization factors are calculated by assuming an average detected energy by the gauge.

The counts along with the normalized metal densities are used with an appropriate curve fitting routine to determine the A,B, C constants in (Eq 1). This method is referred to as three block calibration. It is important to note that in some cases, other combinations of materials are used as the three calibration blocks to determine the A, B and C constants.

In the five block method, the A, B and C constants are calculated using the metallic blocks used for the three block method. In addition, counts on limestone and granite blocks are used to adjust the B value. This forms the basis for the five block calibration and the fundamental argument for this approach is that soil composition is half way between granite and limestone. Whereas, three block calibration is based on historical five block calibration data to determine the position of the soil curve and to adjust the metal block densities accordingly. Again, this is done because accurate detected photon energy information is not available for these gauges.

Moisture calibration is performed by using at least two reference blocks, with different hydrogen densities. Counts on the two blocks with the known moisture content are used to calculate the slope and intercept of (Eq 2).

All gauges are calibrated prior to shipment from the factory. However, due to electronic and mechanical variations with time, calibrations have to be performed periodically. The main problem is the determination of calibration frequency. While some gauges can continue operation with accurate calibration for many years, others might need adjustments right after shipment from the factory. To effectively address this problem, a portable device is required to determine the status of the calibration and to allow accurate on site calibration.

The details on moisture calibration and density verification is outside the scope of this paper. The main emphasis of this paper is comparisons of factory to the newly developed field calibration method for the manufactured gauges.

Gauge Errors

There are three errors inherent in gauges manufactured. Namely, nuclear precision, surface roughness and composition error. Gauge precision is the function of source to detector geometry and the source intensity. Composition error is the effect of elemental composition of the material to the density measured by the gauge. Surface roughness error is the effect of photon streaming from the source to the detectors between the base of the gauge and the material surface. Since test material surface varies for each test location, surface roughness is the most difficult error parameter to define.

Composition error in most gauges can be minimized by appropriate filtering of the detection system or adjustment of the source to detector distance. Even though this error can be controlled, the adjustments to reduce this error usually causes an increase in the surface roughness error. Manufacturers optimize the gauge geometry to balance these errors and to hold these effects within a tolerable limit.

There are other external factors that significantly influence the gauge readings. Walls or structures in the proximity of gauges can change the gauge measurements. Manufacturers recommend gauge measurements to be taken at least one meter from any walls. However, due to space restrictions and production schedules, most calibration facilities, including some of the manufacturers, ignore this critical restriction. Also, radiation background from other sources will reduce the gauge measured density. The effects from the wall and background are easy to control and should be avoided during calibration and field measurements. Significant errors can be introduced in calibration and field results between gauges can vary, if these effects are not controlled.

Field Verification and Calibration Method

Recently a new development in the area of field verification and calibration was introduced to the market. The portable engineered block (PEB) is a portable device that can be used to determine the gauge calibration status and to generate new calibration constants for the gauge. This device weighs approximately 18 kg and works with a number of different manufactured gauges.

A combination of materials is used within a same block to create densities within the construction range of interest. Composition error, surface roughness and boundary conditions for each manufactured gauge is considered when determining the PEB density. The block densities are traceable to NIST by standard procedures used by the manufacturer. The gauge can be placed on the block and gauge density readings can be compared to the densities provided with the PEB. The comparison may serve as the basis for acceptance of the gauge calibration accuracy.

If calibration verification fails, the added benefit of this system is its capability of field calibration. Counts taken on the PEB along with a PC program can be used to simulate counts for blocks employed during the gauge factory calibration. Since the density and the composition of the calibration blocks are well established, one can accurately predict the counts based on the present status of the gauge. This routine can be used to generate counts for a new three block or a five block calibration.

Test Method

The purpose of this test was to compare the results obtained on construction materials using calibrations provided by the conventional factory methods and the one performed on the PEB.

Test Procedure

Three different manufactured gauges, Gauge 1 to Gauge 3, were used for this study. All gauges were calibrated at InstroTek according to the standard methods used by the manufacturer of each gauge. Wall and background effects were eliminated in the calibration facility by isolating the calibration area. Twenty minute calibration counts were taken at each depth on known NIST traceable and ASTM required standard blocks . Gauges 1 and 3 were calibrated with the three block method. Blocks of mg, m/a and granite were used for Gauge 1 and mg, m/a and al for Gauge 3. Gauge 2 was calibrated with the five block method using mg, m/a, al, limestone and granite blocks. For a direct comparison, the PEB calibration method performed on each gauge used the same calibration method as used by the factory. The PEB calibration involves taking counts on one block, and simulating the counts for the blocks used in the conventional factory method.

Measurements were taken in the field on six different materials with the density range of approximately 1922 to 2563 kg/m^3(120 to 160 lb/ft^3). The field counts were used to calculate densities by using the constants generated from the following:

1- Factory Method- Calibration normally provided with the gauge, when shipped from the factory.
2- PEB Method- Calibration performed using the PEB device.

Test Materials

The test material consisted of three different types of asphalt mixtures and three soils. The asphalt mixture readings were performed in the backscatter mode on a base, binder and a surface mixture. Two sites with North Carolina red clay, a pure limestone block and a pure granite block were used for backscatter and direct transmission measurements.

The asphalt projects were located on Interstate 40 between Raleigh and Research Triangle Park, North Carolina. The clay site was located on Highway 401 and U.S. 1, north of Raleigh, North Carolina. The limestone and granite block measurements were performed in the laboratory. These two blocks along with the other blocks used during this study are traceable to National Institute of Standards and Technology (NIST).

Since gauge calibrations are optimized for construction materials, the best comparison between the calibration methods for different manufactured gauges should be based on construction materials similar to the ones selected in this study.

a- Asphalt Surface Mixture (ASM)- This material is generally used for riding surface material on asphalt construction projects. The gauge density results presented in the tables are averages of ten 4 minute readings, for each manufactured gauge. The actual density (AD) for this material is taken as the average of ten core density tests performed by the contractor.
b- Asphalt Intermediate Mixture (AIM)- This material is used as an intermediate layer on asphalt construction projects. The gauge density results presented in the tables are averages of ten 4 minute readings, for each manufactured gauge. AD for this material is taken as the average of eight core density tests performed by the contractor.
c- Asphalt Base Mixture (ABM)- This mixture is used as the base layer for new construction sites with larger proportion of large size aggregates. The gauge density results presented in the tables are averages of ten 4 minute readings, for each manufactured gauge. AD for this material is an average of twenty five core density tests performed by the contractor.
d- Clay site1 and Clay site 2- This material was compacted in preparation for a new road. The gauge density results presented are averages of two 4 minute readings, one from each site, for depths backscatter (BS), 2,4,6,8, 10 and 12 of each manufactured gauge.
e- Limestone Block- These counts were collected in the laboratory at depths BS, 2, 4, 6, 8, 10 and 12 for each manufactured gauge. The limestone block density used in this study is NIST traceable.
f- Granite Block- These counts were collected in the laboratory at BS, 2, 4, 6, 8, 10 and 12 for each manufactured gauge. The granite block density used in this study is NIST traceable.

Results

Four minute count time was utilized throughout this study to minimize the statistical variations in each gauge.

The average count for all test sites for each manufactured gauge, 1, 2 and 3, is reported in (Table 1) through (Table 3).

Table 1--Accumulated counts for all projects for Gauge 1

Depth Inches	ASM	AIM	ABM	Clay	Limestone	Granite
DS[3]	35065	35065	34798	35114	34520	34526
BS	15998	15618	14352	20027	15201	12572
2	...	...	...	86336	68950	51777
4	...	...	...	71216	56565	40398
6	...	...	...	48734	38963	25493
8	...	...	...	28854	23002	14164
10	...	...	...	16291	12514	7652
12	...	...	...	9408	7131	4759

To calculate the density of the test material, it is necessary to use the corresponding gauge constant parameters A, B and C in (Table 4) to (Table 6) calculated by utilizing the constants provided by the Factory Method and the ones performed by the PEB method.

Table 2-- Accumulated counts for all projects for Gauge 2

Depth Inches	ASM	AIM	ABM	Clay	Limestone	Granite
DS	3469	3469	3496	3508	3460	3449
BS	993	934	881	1293	951	761
2	...	...	...	3908	3113	2276
4	...	...	...	3220	2544	1671
6	...	...	...	2365	1866	1123
8	...	...	...	1575	1224	703
10	...	...	...	1000	747	414
12	...	...	...	606	436	256

Table 3-- Accumulated counts for all projects for Gauge 3

Depth Inches	ASM	AIM	ABM	Clay	Limestone	Granite
DS	2901	2901	2909	2902	2930	2874

[3] DS is the daily density standard count.

BS	813	783	705	1134	797	596
2	...	...	...	3590	2728	1862
4	...	...	...	3342	2575	1673
6	...	...	...	2456	1889	1140
8	...	...	...	1526	1189	662
10	...	...	...	892	661	356
12	...	...	...	497	342	180

Table 4-- Gauge constant parameters for Gauge 1, for each method

	Depth	Factory Method			PEB Method	
	A	B	C	A	B	C
BS	2.9734	57.40371	0.19588	2.3828	67.80975	0.15349
2	11.4374	74.89223	0.22950	9.6361	97.31010	-0.28891
4	14.0418	63.03570	0.12731	11.6818	73.46961	-0.07307
6	17.3612	48.11310	0.16631	12.8271	57.27014	-0.00009
8	19.8939	38.10837	0.14181	13.4902	44.74362	0.05611
10	20.3064	31.94563	0.10585	12.8428	37.20080	0.06355
12	20.6020	26.98492	0.09367	13.4682	30.25024	0.08053

Table 5-- Gauge constant parameters for Gauge 2, for each method

Depth	Factory Method			PEB Method		
	A	B	C	A	B	C
BS	2.8789	1.24874	-0.10566	2.9389	1.25607	-0.10771
2	6.9728	0.92231	-0.01640	7.1775	0.93334	-0.02671
4	9.0004	1.07577	0.07625	9.3703	1.08689	0.070326
6	10.5446	1.27141	0.06881	11.0273	1.28290	0.065651
8	11.9348	1.53628	0.02167	12.4651	1.54835	0.019517
10	13.1583	1.84295	-0.01086	13.5364	1.85564	-0.01216
12	15.1417	2.20183	-0.02693	14.9782	2.21519	-0.02651

Table 6-- Gauge constant parameters for Gauge 3, for each method

Depth	Factory Method			PEB Method		
	A	B	C	A	B	C
BS	5.6871	1.59553	-0.12106	4.8485	1.49027	-0.10700
2	16.9252	1.42566	-0.25673	14.0234	1.30207	-0.18828
4	21.6530	1.53094	-0.20321	17.9546	1.40740	-0.13323
6	24.8264	1.71560	-0.13758	21.0438	1.61328	-0.09670
8	26.8041	1.95809	-0.08181	23.5309	1.88292	-0.06487
10	28.4525	2.25320	-0.04735	24.6028	2.17730	-0.04061
12	26.1565	2.50666	-0.02743	27.5310	2.54279	-0.02888

Densities calculated for each gauge uses the constants and the counts at each depth. The density equation (Eq 3) is used for calculating the densities for Gauge 1 and (Eq 4) is used for Gauge 2 and Gauge 3.

$$D = B\ln\left(\frac{A}{CR - C}\right) \tag{3}$$

$$D = \frac{1}{B}\ln\left(\frac{A}{CR + C}\right) \tag{4}$$

where CR is the count ratio.

The densities for each gauge for all test sites are given in (Table 7) to (Table 9).

Analysis

The objective of this study is to determine if the factory and the PEB methods of calibration yield similar density results. The data in (Table 7) shows that in all but one case, the percent difference between the factory and the PEB methods were less than $\pm 1.0\,\%$.

For Gauge 2, at depth 12, the percent difference for clay was -1.19 %. The depth 12 reading for limestone using the PEB method is in line with the measurements with the other gauges. Also, assuming the granite block is homogenous, the reading at depth 12 is consistent with the other depths when using the PEB method. Therefore, the factory method readings at depth 12, in this case can be considered to be slightly off. Since counts at depth 12 are lower than the other depths, it is possible that the higher statistical variation in the counting rate resulted in a slight deviation in factory calibration.

To determine if the calibration performed by the PEB method is accurate, it is necessary to compare the densities measured by this method to an acceptable value for each material. If one accepts the fact that the gauges similar to the ones used in this study are used daily for control of density on job sites, then the average of the density measurements from the three gauges on each material at each depth should serve as a good estimate for the density of the tested materials. Therefore, the overall standard error of estimate for each gauge at all depths will provide accuracy values for the two methods of calibration.

The data in (Table 8) is a comparison of the individual measurements taken with the factory calibration method for each gauge, at all depths for each material and (Table 9) represents the same data with the PEB method.

Table 7—Comparison of densities (kg/m^3) for all test sites, Factory Method (FM) and PEB Method (PEBM)

	Gauge 1			Gauge 2			Gauge 3		
Depth	FM	PEBM	%diff	FM	PEBM	%diff	FM	PEBM	%diff
Asphalt surface Mix: Avg. core density 2273 kg/m^3									
BS	2239	2241	0.07	2216	2229	0.56	2240	2235	-0.25
Asphalt Intermediate Mix: Avg. core density 2419 kg/m^3									
BS	2279	2281	0.09	2296	2309	0.57	2282	2276	-0.28
Asphalt Base Mix: Avg. core density 2513 kg/m^3									
BS	2409	2424	0.63	2385	2388	0.12	2410	2397	-0.57
Clay: N/A									
BS	1905	1892	-0.72	1916	1915	-0.06	1910	1903	-0.34
2	1962	1954	-0.40	2004	2009	0.26	1997	1990	-0.35
4	2019	2017	-0.09	2047	2059	0.58	2042	2038	-0.22
6	2045	2039	-0.33	2085	2097	0.55	2072	2066	-0.29
8	2061	2055	-0.28	2104	2111	0.37	2093	2087	-0.27
10	2066	2066	0.00	2099	2097	-0.13	2083	2077	-0.29
12	2063	2070	0.36	2108	2083	-1.19	2075	2069	-0.27
Limestone: NIST traceable density 2225 kg/m^3									
BS	2297	2300	0.10	2271	2284	0.56	2273	2267	-0.27
2	2240	2242	0.11	2248	2265	0.75	2260	2255	-0.18
4	2251	2260	0.42	2236	2257	0.93	2264	2260	-0.18
6	2229	2230	0.01	2244	2263	0.83	2267	2260	-0.30
8	2219	2219	0.00	2243	2257	0.63	2254	2248	-0.28
10	2237	2241	0.18	2264	2268	0.14	2250	2245	-0.24
12	2251	2264	0.57	2283	2262	-0.91	2265	2259	-0.25
Granite: NIST traceable density 2629 kg/m^3									
BS	2641	2635	-0.22	2556	2571	0.60	2624	2601	-0.88
2	2637	2625	-0.44	2585	2603	0.68	2641	2624	-0.67
4	2625	2637	0.42	2584	2604	0.78	2641	2620	-0.82
6	2630	2619	-0.43	2587	2605	0.69	2658	2634	-0.93
8	2628	2609	-0.74	2593	2607	0.54	2652	2632	-0.78
10	2644	2620	-0.90	2597	2601	0.15	2625	2612	-0.53
12	2656	2645	-0.41	2626	2601	-0.95	2636	2635	-0.03

Table 8—Comparison of densities (kg/m^3) of the Factory Method densities and the average density for each test site.

Material	Avg. Dens.	Gauge 1	Gauge 2	Gauge 3	%diff Gauge 1	%diff Gauge 2	%diff Gauge 3
ASM	2232	2239	2216	2240	0.33	-0.70	0.37
AIM	2286	2279	2296	2282	-0.29	0.45	-0.16
ABM	2401	2409	2385	2410	0.32	-0.68	0.36
Clay BS	1910	1905	1916	1910	-0.28	0.30	-0.02
Clay 2	1988	1962	2004	1997	-1.29	0.82	0.47
Clay 4	2036	2019	2047	2042	-0.83	0.54	0.29
Clay 6	2067	2045	2085	2072	-1.08	0.85	0.23
Clay 8	2086	2061	2104	2093	-1.20	0.86	0.34
Clay 10	2083	2066	2099	2083	-0.80	0.78	0.02
Clay 12	2082	2063	2108	2075	-0.91	1.25	-0.34
Lime BS	2280	2297	2271	2273	0.73	-0.41	-0.32
Lime 2	2249	2240	2248	2260	-0.41	-0.06	0.47
Lime 4	2277	2251	2236	2264	-1.13	-1.79	-0.57
Lime 6	2247	2229	2244	2267	-0.79	-0.12	0.91
Lime 8	2239	2219	2243	2254	-0.88	0.19	0.68
lime 10	2250	2237	2264	2250	-0.59	0.61	-0.01
lime 12	2266	2251	2283	2265	-0.68	0.74	-0.06
Granite BS	2607	2641	2556	2624	1.30	-1.96	0.65
Granite 2	2621	2637	2585	2641	0.61	-1.37	0.76
Granite 4	2617	2625	2584	2641	0.32	-1.25	0.93
Granite 6	2625	2630	2587	2658	0.19	-1.45	1.26
Granite 8	2624	2628	2593	2652	0.14	-1.19	1.05
Granite 10	2622	2644	2597	2625	0.84	-0.95	0.11
Granite 12	2639	2656	2626	2636	0.63	-0.51	-0.13
Accuracy					±0.79	±0.98	±0.57

The standard error of estimate, defined here as accuracy, indicates that for all gauges the two calibration methods will approximately produce the same results. For Gauge 1, the accuracy of the factory calibration is 0.04% better than the PEB method. However, for Gauge 2 and Gauge 3, the PEB method is 0.12% and 0.21% better than the factory method.

Table 9—Comparison of densities (kg/m^3) of the PEB method densities and the average density for each test site.

Material	Avg. Dens.	Gauge 1	Gauge 2	Gauge 3	%diff1 Gauge 1	%diff2 Gauge 2	%diff3 Gauge 3
ASM	2232	2241	2229	2235	0.42	-0.12	0.15
AIM	2286	2281	2309	2276	-0.20	1.02	-0.42
ABM	2401	2424	2388	2397	0.94	-0.56	-0.18
Clay BS	1910	1892	1915	1903	-0.96	0.24	-0.38
Clay 2	1988	1954	2009	1990	-1.69	1.07	0.12
Clay 4	2036	2017	2059	2038	-0.93	1.13	0.10
Clay 6	2067	2039	2097	2066	-1.37	1.44	-0.06
Clay 8	2086	2055	2111	2087	-1.49	1.20	0.05
Clay 10	2083	2066	2097	2077	-0.80	0.69	-0.27
Clay 12	2082	2070	2083	2069	-0.58	0.05	-0.62
Lime BS	2280	2300	2284	2267	0.86	0.16	-0.58
Lime 2	2249	2242	2265	2255	-0.33	0.70	0.25
Lime 4	2277	2260	2257	2260	-0.73	-0.86	-0.75
Lime 6	2247	2230	2263	2260	-0.74	0.73	0.59
Lime 8	2239	2219	2257	2248	-0.88	0.82	0.42
Lime 10	2250	2241	2268	2245	-0.41	0.79	-0.24
Lime 12	2266	2264	2262	2259	-0.10	-0.19	-0.32
Granite BS	2607	2635	2571	2601	1.07	-1.38	-0.23
Granite 2	2620	2625	2603	2624	0.15	-0.69	0.11
Granite 4	2617	2637	2604	2620	0.78	-0.48	0.13
Granite 6	2625	2619	2605	2634	-0.23	-0.76	0.34
Granite 8	2624	2609	2607	2632	-0.58	-0.66	0.29
Granite 10	2622	2620	2601	2612	-0.08	-0.80	-0.38
Granite 12	2639	2645	2601	2635	0.21	-1.45	-0.16
Accuracy					±0.83	±0.86	±0.36

Conclusion

Results of this study indicates that the PEB calibration method produces density results comparable in accuracy to the factory calibration method. Tests conducted on six different construction materials with three different nuclear density gauges calibrated with the factory calibration and PEB calibration methods produced accuracy values of ±0.79%, ±0.98% and ±0.57% for the factory method and ±0.83%, ±0.86% and ±0.36% for the PEB method. Direct density comparisons between the two methods showed that 71 out of 72 densities had percent differences of less than ± 1%. Based on this study, the PEB calibration method can be used as a reliable and accurate alternative to the factory calibration methods used for nuclear density gauges.

References

[1] Hughes, C. S., and Anday, M.C. (1967), Correlation and Conference of Portable Nuclear Density and Moisture Systems. Highway Research 177,.

[2] Gardner, R. P., and Ely, R. L. (1967), Radioisotope Measurement Applications in Engineering, Reinhold Publishing Corporation, New York,.

[3] McDougall, F. H., Dunn, W. L., and Gardner, R. P. (1969), Report on Nuclear Soil Gauge Calibration Workshop-Symposium, Nuclear Engineering Department, North Carolina State University, Raleigh,.

[4] Regimand, A. (1987), "A Nuclear Density Gauge for Thin Overlays of Asphalt Concrete," Transportation Research Board, Record 1126, Washington, D. C.,.

Peter Deming,[1] and David Good[2]

Two Weight System for Measuring Depth and Sediment in Slurry-Supported Excavations

REFERENCE: Deming, P. W. and Good, D. R., **"Two Weight System for Measuring Depth and Sediment in Slurry-Supported Excavations,"** *Field Instrumentation for Soil and Rock, ASTM STP 1358,* G. N. Durham and W. A. Marr, Eds., American Society for Testing and Materials, West Conshohocken, PA, 1999.

ABSTRACT: This paper describes a two weight system using bar and flat shaped weights for measuring depth and detecting sediment at the bottom of slurry-supported excavations. Currently there are no standard depth measurement weights or methods for reliably identifying bottom sediment. Two weights and a procedural system for using the weights is described. Details suitable for manufacture are provided.

KEYWORDS: Slurry, Sediment, Field Inspection, Bentonite, Slurry Trench, Slurry Wall, Drilled Pier

Introduction

This paper describes two sounding weights and procedures for measuring depth and thickness of sediment at the bottom of slurry-supported excavations. At present, there is no standard weight or procedure for depth measurement. There is also no standard means for measuring sediment, although "bobbing" a weight to discern bottom drag is often performed to infer the presence of sediment. The current typical practice of bottom sounding, particularly for sediment control, is inconsistent, inaccurate and subjective. Sounding with a heavy weight provides little quality assurance information to the field engineer even though proper depth control and sediment removal are crucial to the performance of foundations and hydraulic barriers constructed using the slurry method.

We use a two weight system for quality control inspection of slurry-supported trench, panel and drilled shaft excavations. One weight is a solid "bar" which readily penetrates loose bottom sediment. The other weight is a "flat" weight which rests on the sediment surface. The thickness of sediment at the bottom of the excavation at any one location is determined by comparing depths carefully measured by each weight, from a uniform reference.

With the ability to reliably measure bottom sediment, the engineer can determine the suitability of the excavation bottom for backfill or concrete placement. Contract specifications should include a provision for using the two weight system to determine sediment thickness and should define criteria for bottom acceptance.

In addition to the shape and mass of the sounding weights, the ability to detect sediment depends on other factors. These include: the time readings are taken relative to trench events, slurry viscosity, character of sediment, excavation depth, reliability of the fixed reference and most notably inspection technique.

[1] Partner, Mueser Rutledge Consulting Engineers, 708 Third Ave., New York, NY 10017
[2] Associate, Mueser Rutledge Consulting Engineers, 708 Third Ave., New York, NY 10017.

Description of Weights

The two weights which we have used are detailed for manufacture in Figures 1 and 2. The weights are attached to the end of fiberglass reinforced graduated survey tapes. Use of a light cable or rope to support the weight has been shown to allow the measuring tape to wrinkle, giving falsely deep and non-repeatable readings. The weight should be raised and lowered only by the survey tape. In our excavation specifications, the foundation contractor is required to provide the weights and graduated tapes. Several weights should be provided, as weights are lost when the graduated tape breaks.

The "flat" weight consists of a 345 mm (14 in.) long piece of 19 mm (3/4 in.) square or round bar stock welded to the center of a 160 mm (6 in.) diameter steel disk cut from 12 mm (1/2 in.) thick steel plate. A 32 mm (1-1/4 in.) wide band is welded around the edge of the disk to create a 20 mm (3/4 in.) high lip. The lip retains sediment when the weight is removed from the slurry column, giving the operator the ability to recover a specimen of sediment or thick slurry. A slot in the bottom of the weight receives the survey tape grip. The survey tape is attached to the shaft of the weight with duct tape. The flat weight weighs 3.1 kg (6.9 lbs) and has a bearing intensity of 1.5 kPa (0.015 tsf), less than 1/30th that of the bar weight.

The "bar" weight is a 610 mm (24 in.) long piece of 25 mm (1 in.) square or 29 mm (1-1/8 in.) diameter round bar stock. It weights approximately 3.1 kg (6.8 lbs) and has a tip bearing pressure of 47 kPa (0.5 tons per square foot). We have used bar weights ranging from approximately 3 to 5 kg (7 to 12 lbs), and find equivalent performance of these for sinking the measurement tape and penetrating through sediment-laden slurry. The grip at the end of the survey tape is extended and placed over the bottom of the bar, and duct tape is wrapped around the tip and shaft of the weight to attach it to the graduated measuring tape.

Causes of Sediment

Slurry is used to resist earth pressure and groundwater, enabling excavations to progress safely to great depths without structural bracing. When initially prepared, slurry is a mixture of water, bentonite clay and additives which enhance the colloidal suspension. Freshly prepared slurry is only slightly heavier than water, because it contains only a small amount of solids. As the excavation progresses, clay, silt and sand particles become suspended in the slurry. Suspended soil particles increase the slurry weight, which enhances the stability of the excavation.

The ability of slurry to suspend particles depends primarily on its viscosity and gel strength. Most construction slurries will support fine sand sized particles in suspension for long periods of time. However, viscous slurry is required to suspend coarse sand and fine gravel particles. Solids which cannot be suspended by the slurry fall to the bottom. Bottom sediment is a deposition of these solids within a semi-fluid slurry. The sediment can include sand and gravel particles, clay balls, debris, and silt or clay-laden slurry.

Theoretical Bearing Capacity

The pointed weight fails the sediment in end bearing as it passes through the soft sediment to reach the bottom of the excavation or the top of tightly packed granular sediment. Because of its larger bearing area, the flat weight rests near the top of the sediment when it attains a factor of safety of 1.0 against bearing failure. The bearing intensity of the tip of the bar weight is approximately 30 times that imposed by the flat weight. If the flat weight is considered to be a simple circular footing statically bearing on the surface of a soft clay, a shear strength greater than 0.24 kPa (5 psf) should support the flat weight. However, the flat weight may be supported by weaker sediments because of the following:

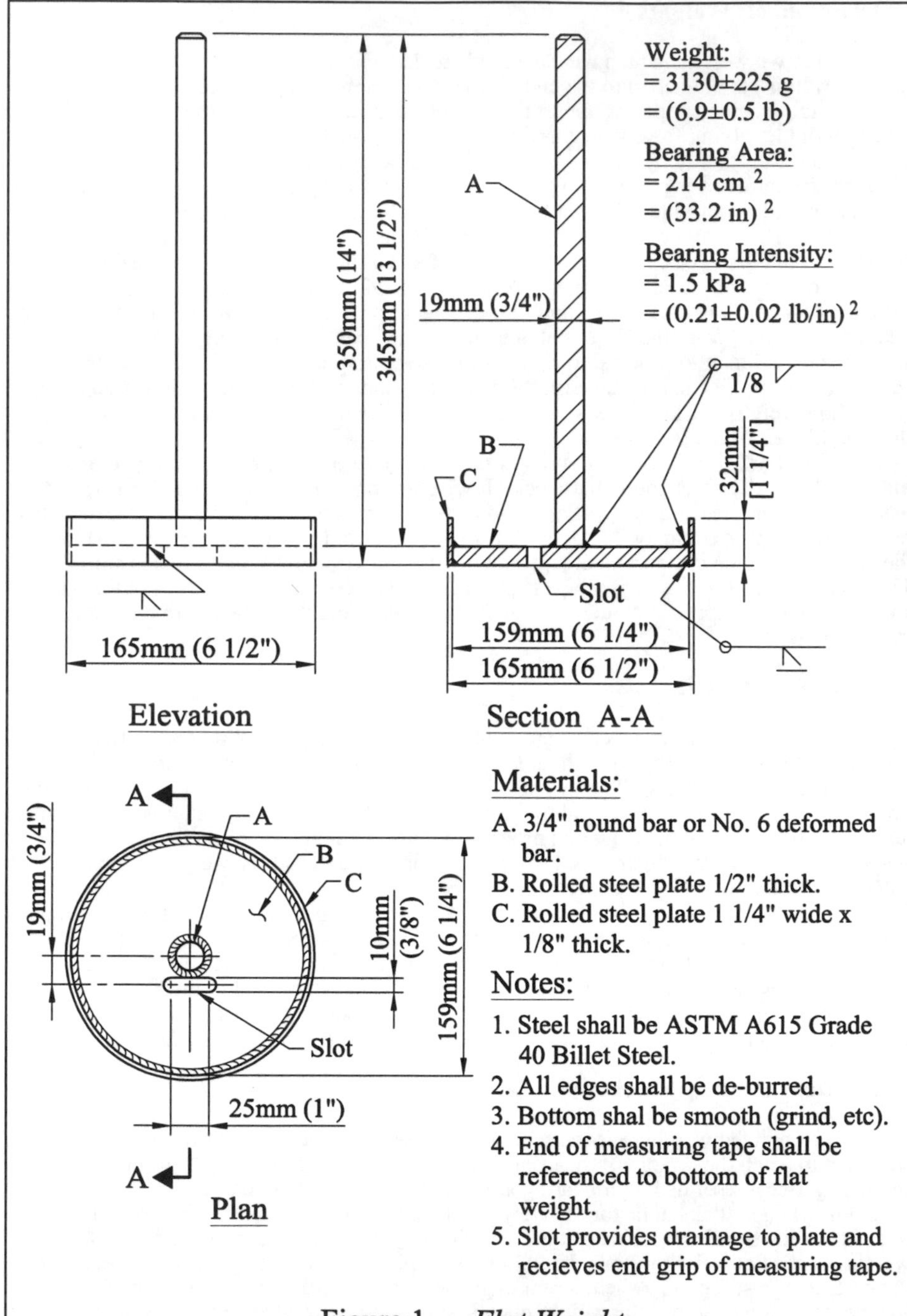

Figure 1 - *Flat Weight*

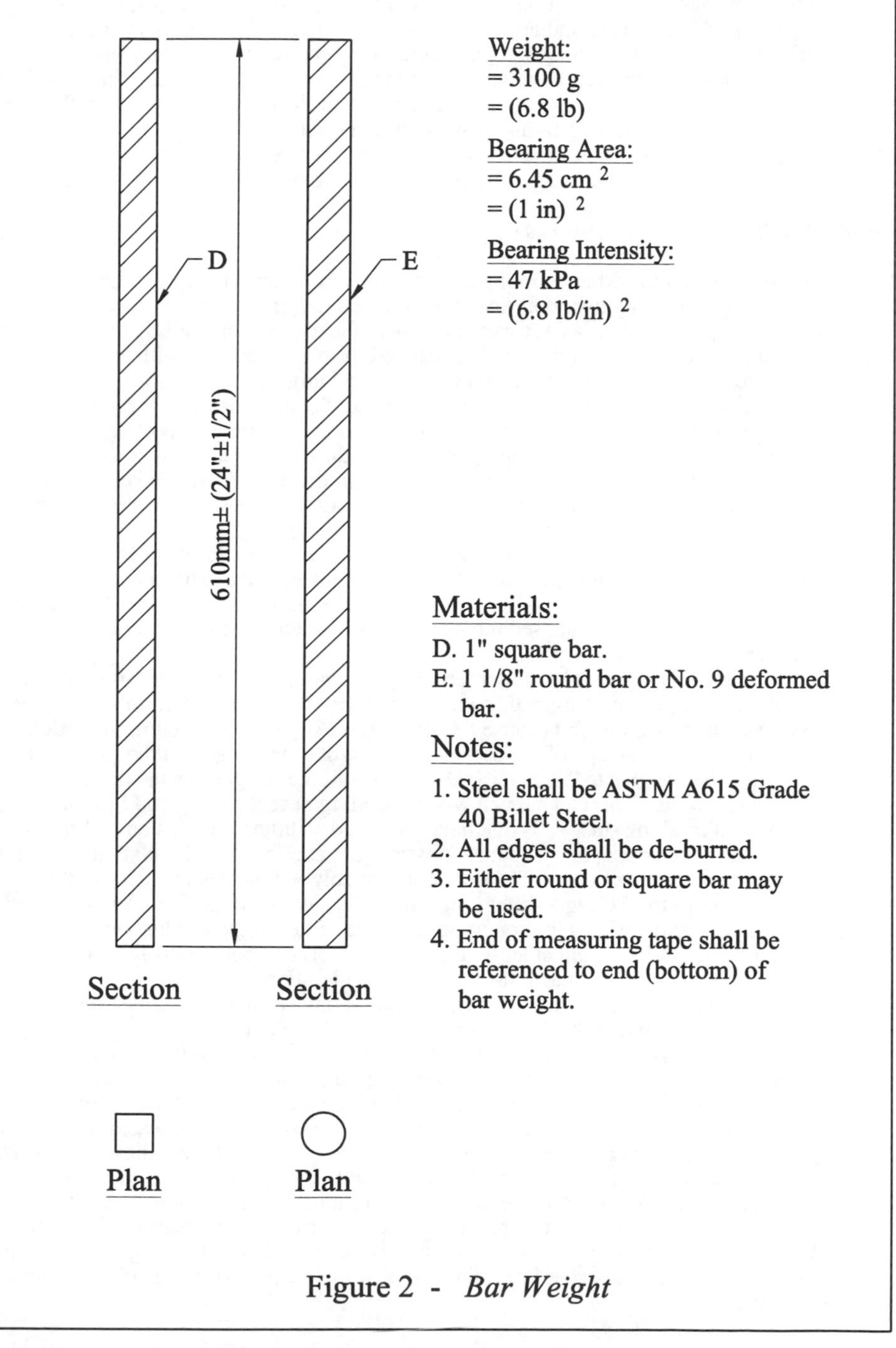

Figure 2 - *Bar Weight*

- Sediments are suspended in a thixotropic fluid of varying gel strength.
- Sediments vary in composition and size (clay balls, granular particles, etc.) The particle packing density and angularity of the granular sediments also varies.
- Frictional drag on the falling weights, particularly the flat weight, may play a significant role in the depth within the sediment at which the weight is supported.
- Support provided by the underlying hard bottom of the excavation which influences the location of the bearing failure plane within the sediments.

The flat weight has been observed to slowly sink into the bottom sediment, even when suspended under slight tension, suggesting sediment may behave as a viscous fluid.

Depth Measurement Procedures

1. Use normal caution when working adjacent to slurry supported excavations. A man-bridge spanning over the excavation may be required to provide access for use of the weights in some cases, especially where the excavation walls have caved and the center of the excavation cannot be reached from the side wall with an outstretched arm. Use work gloves when performing soundings.
2. Specifications often require the contractor to take depth measurement readings. This is satisfactory for the contractor's quality control, but the inspecting engineer should operate and feel the weight performance during all quality assurance measurements for bottom approval. Proper technique is needed to take repeatable, reliable depth soundings and such functions are normally best performed by the engineer with contractor support.
3. Bar and flat weight depth readings are compared in order to define sediment thickness. These depth measurements should be taken at the same location and time, in immediate sequence. Bar and flat weight readings both need to be taken to document bottom depth and sediment thickness following a cleaning attempt for final bottom approval.
4. Lower the sounding weights under tension to maintain a continuously plumb line. When the weight is carelessly thrown in the slurry, it may not be plumb below the assumed sounding station because the slurry viscosity slows its return to vertical. Allow the graduated tape of the descending weight to feed through the fingers. Do not allow the weights to "free fall." Do not permit the weight to slide down the excavation walls. This can happen when sounding near the steep end face of an excavation or along side walls that have sloughed. Sliding weights on end faces can mislead the completion of panel construction, at corners of excavations and at the leading edge of a trench. The operator probably will not realize the weight is sliding, except for feeling a slight drag. If this is suspected, pull the weight up out of the trench and perform the sounding from the center of the excavation, As an alternative raise the weight at least one meter (3 feet) and bob it slowly to attempt to "catch" the excavation face, defining it for consideration.
5. The weights, particularly the flat weight, should have a slow, controlled descent near the bottom so that they land without impact. The intention is that the flat weight will rest on top of sediment, not penetrate it. Conversely, bar weight soundings must be coaxed to penetrate through the slurry-supported sediment. A drag on the bar weight or a "soft bottom" feel indicates sediments are present. Progressively work the bar weight into the bottom sediment by repeated up and down motion, raising the bar weight about 300 mm (one foot) and allowing it to fall through the sediment as required. Confirm that the weights are not "laying down."
6. Pull the tape taut and read the depth of the weight below the reference elevation. Flat weight readings should be recorded next to corresponding bar weight readings.
7. Provide a description of bottom sediment recovered with the flat weight. Be aware that material recovered could be "filter cake" if the weight drags the side wall during removal.

Project Experience

The two weight system has been used successfully to identify bottom sediments on four projects we have been directly involved with. The amount of sediment detected has varied considerably, in a manner consistent with our understanding of sediment generation discussed above. Our experience with the two weight system has been similar on each of the four projects. Typically, at the start of a project, the two weights reveal some amount of sediment, undetected by the "drag feel" of the bar weight. If the sediment problem became sufficiently troublesome, the contractor was required to refine the bottom cleaning and/or excavation method in order to achieve sediment criteria given in the specifications. Where extensive sediment removal was required, various modifications were attempted to mitigate sediment generation. Modifications to the excavation system which met with some degree of success included: increasing slurry viscosity, dragging the bucket parallel to the trench bottom when cleaning, increasing the number bucket teeth and using short teeth, welding plates between long bucket teeth, and separating the toe of the backfill from the excavation area. On two trench projects backfill placement was held at one location until no additional backfill could be accepted. We believe this practice reduced the quantity of sediment by minimizing mixing of backfill with the active slurry column. Using the two weight system allowed observation of this influence.

Once a proper cleaning and excavation system was developed, the two weight system sediment check became confirmatory, typically showing little sediment. However, it is the opinion of the authors that the two weights should still be used routinely. Using the weights can detect small side wall collapses below the slurry level as observed and documented on one project, can detect unusually thick slurry as occurred on two projects, and in general promotes good trench quality, giving information about the excavation bottom to the inspection staff.

Notes For Field Inspection

1. The precision and repeatability of depth measurements is dependent on the reference used and the care exercised when sounding with the weights. The writers recommend using survey stakes with elevation reference marks about every 3 m (10 feet) along trench and panel wall alignments.
2. The excavation equipment used and its operation will influence the bottom roughness and therefore the repeatability of bottom soundings. Holding the bucket teeth parallel to the trench bottom while the backhoe tracks backwards often improves sediment retrieval and bottom cleaning, and enhances sounding repeatability. This method should typically be employed during the last few feet of excavation advance and for final bottom cleaning.
3. A sufficiently viscous slurry will prevent excessive sedimentation; conversely, thin slurry will increase sediment development. Adding water, chemicals, or thin slurry to trench slurry must be done with caution, as this procedure may generate large amounts of sediment along the entire trench alignment in areas where backhoe excavation equipment cannot reach for sediment removal.
4. In our experience, mixing fresh slurry at a 6% bentonite content to a viscosity of about 48 Marsh funnel seconds provides a workable slurry with good particle holding capacity. Fresh slurry pumped to excavations should be hydrated, well blended and have consistent viscosity. In a long and deep trench slurry will stratify by density and viscosity. Heavy and viscous slurry will fall to the lower portion of the trench.
5. Sediment thickness will increase with time as large and heavy particles fall from suspension. Soundings for sediment thickness measurement should therefore be performed immediately prior to allowing placement of backfill over the excavation bottom.

6. Overly viscous slurry is capable of suspending coarse sand and gravel particles or soil clods and may build up at the bottom of the slurry column. Chemical attack can both thin and thicken slurry. Slowed weight penetration rates and poor reproducibility of flat weight soundings are indicators of a thickened slurry condition. In cases of extreme slurry thickening the flat weight may rest on the surface of the submerged slurry. Thick slurry can be sampled with the use of the flat weight.
7. Effective removal of sediment from the bottom of a slurry filled trench using a backhoe is bucket dependent. Bucket shapes, teeth length, teeth spacing and the use of side cutters, ripper teeth, etc. can influence the efficiency of material removal and opportunity for sediment generation. Typically, buckets with long and few teeth cannot retrieve sediment, but instead create sediment. Short teeth and flat cutting edges promote sediment retention for removal. Holes cut into the back of the buckets to drain slurry increase slurry sand content and can generate large amounts of sediment. These observations were made by repeated use of the flat weight under various conditions.
8. Overfilled backhoe buckets can drop debris as they are raised through the slurry column. This can be particularly worrisome when the bucket is operated above the backfill slope when raised. Consider use of an end stop pipe to maintain a rigid separation between the excavation and backfill ends of trench excavations.

Instrumentation to Monitor Landfills

Steve B. Taylor,[1] Chris C. White,[2] and Ron D. Barker[1]

A Comparison of Portable and Permanent Landfill Liner Leak Detection Systems

---

REFERENCE: Taylor, S. B., White, C. C., and Barker, R. D., "**A Comparison of Portable and Permanent Landfill Liner Leak Detection Systems,**" *Field Instrumentation for Soil and Rock, ASTM STP 1358*, G. N. Durham and W. A. Marr, Eds., American Society for Testing and Materials, West Conshohocken, PA, 1999.

ABSTRACT: Monitoring of the integrity of electrically non-conductive geomembrane liners installed at waste sites using electrical geophysical techniques has been carried out for a number of years using above-liner leak location surveys and, more recently, below-liner monitoring systems. We compare the theoretical response of both types of survey to a hole in a liner and then compare with measurements made in the field. The theoretical leak response indicates that above-liner surveys are sensitive to leaks over a greater area, though both responses result in comparable leak detectability. However, field data suggest that in practice, measurements made on a sparse grid below the liner have the greater sensitivity to certain leaks. This may be due to the differing leak geometries and background conditions present above and below the liner.

The results indicate that a sparse below-liner monitoring grid, with its long-term monitoring capabilities, combined with above-liner surveys to pinpoint leaks accurately offer a successful approach to ensuring liner integrity throughout the lifetime of a lined waste site.

KEYWORDS: leak detection, geomembrane liners, landfill monitoring, geophysics

---

Introduction

Plastic lined impoundments are frequently used for the containment of waste, with the aim of reducing the possibility of environmental damage due to escape of contaminants into the

[1]Researcher and Senior Lecturer respectively, School of Earth Sciences, University of Birmingham, Edgbaston, Birmingham, B15 2TT, UK.

[2]Technical Director, Technos Ltd., Walford Manor, Baschurch, Shrewsbury, Shropshire, SY4 2HH, UK.

surrounding land. Monitoring of the integrity of these liners is essential for the early detection of leaks from lined waste sites, in order to reduce the extent of pollution of the groundwater and the financial costs of remediation.

To this end, electrical geophysical techniques have been employed in the detection and location of leaks in electrically non-conductive geomembrane liners installed in waste impoundments, for a number of years. This monitoring normally takes the form of a single Geomembrane Leak Location Survey (GLLS) carried out above the liner on a dense half to two metre grid using portable equipment. Such surveys are regularly employed in the UK as the final part of a Construction Quality Assurance (CQA) plan to ensure the integrity of the liner prior to waste disposal. A theoretical analysis of the above-liner electrical leak location method has been presented by Parra (1988) and Parra and Owen (1988), where the response of a gradient array to a leak in a lined, liquid filled impoundment is assessed. Adopting this array type, Darilek et al. (1989), reported on electrical leak location surveys conducted at a number of lined sites in the United States where an average of 26 leaks per hectare were detected. The first GLLS to be conducted in the UK was in February 1993 (Laine and Moseley 1995) at Craigmore Landfill Site near Belfast, Northern Ireland. Ten holes were located in a 1.4 ha site with 0.3 m of soil cover above the geomembrane liner.

More recently, permanently installed monitoring systems, comprising a grid of electrodes located beneath the geomembrane liner, have been developed offering the possibility of long term monitoring of the liner's integrity. Frangos (1994) presents a system comprising electrodes on an 8 x 10 m grid located beneath a lined waste-disposal pond. The system proved successful in the location of six holes in the liner.

This paper compares the theoretical responses to leaks for both above and below-liner survey methodologies. Field data collected using a permanent below-liner monitoring system (White and Barker 1997) are discussed with data using an above-liner GLLS also conducted at the same site, to offer direct comparison of the capabilities and limitations of each survey type.

In some cases, leaks detected with the permanent below-liner system, which can be nominally located to within a 20 m grid square, have proved difficult to locate easily with a GLLS. Knowledge of the presence of such holes provided by the permanently installed system has required that the holes be located with a GLLS, but it has only been through perseverance that all holes have eventually been located and repaired. This experience suggests that some of these holes would have gone undetected with an above-liner GLLS alone, and appears to indicate that the permanent below-liner system does provide a greater sensitivity in practice.

The Methodology of Leak Detection and Location

The method of leak location in lined impoundments relies on the electrically insulating properties of the geomembrane liner onto which the waste is deposited. These liners, constructed of high density polyethylene (HDPE) or polyvinyl chloride (PVC) have typical resistivities of between 109 Ωm and 1014 Ωm. A current source located inside the lined impoundment and a current sink located outside the impoundment set up a potential difference across the geomembrane liner. The resulting distribution of potential is sampled on a regular grid either within the impoundment, for manual GLLSs, or below the liner in the case of permanent monitoring systems. In the ideal case where the site is electrically isolated, i.e. where there are no paths for current to flow over the edges of the liner and there are no holes in the liner, current flow is impeded by the high resistivity of the liner material and the resulting potential distribution is relatively uniform (White and Barker 1997).

If holes are present in the liner, conductive liquids passing through these holes will form conducting paths along which current may flow. These conductive paths offer the route of least resistance to current flow and a high current density will be present in the region of the leak. The highest potential values from the resulting distribution of electrical potential are therefore located in the vicinity of the leak. As the dimensions of the hole are normally small compared with all other dimensions of the impoundment, the leak effectively behaves as a point source of current.

Parra (1988) determined the above-liner electrical response of a leak in a lined impoundment. He characterised the impoundment using a three layer model comprising an above-liner layer of material with resistivity ρw and thickness h, a thin highly resistive liner of resistivity ρl and thickness t, and a semi-infinite half-space beneath the liner of resistivity ρs (Figure 1). The formulation combines the response of this three layer model with the response of a leak acting as a point source of current located in the thin highly resistive liner layer.

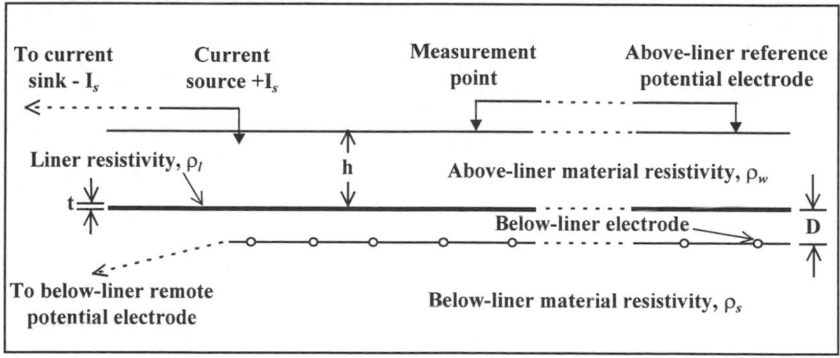

Figure 1 - Schematic representation of impoundment indicating electrode arrangement both above and below the liner.

Extending this to measurements made beneath the liner, the contribution to the below-liner potential distribution due to the current flowing in the material above the liner is negligible due to the presence of the highly resistive liner. The resulting distribution of potential, ϕ, below the liner is therefore dominated by the leak response which approximates to the response of a point current source, Is, located at the position of the hole on the surface of a semi-infinite half-space of resistivity ρs, simply

$$\phi = \frac{I_s \, \rho_s}{2\pi \, r} \tag{1}$$

where r is the distance of the measurement point from the leak.

Parametric Study

Introduction

Most above-liner leak location surveys conducted in lined impoundments use the gradient array configuration (Figure 2a). The response of such an array to a leak is bipolar with the leak located at the amplitude midpoint of the resulting anomaly. One of the problems associated with sampling the electrical potential using this type of electrode array is that the potential difference measurements tend to be very small. This requires accurate equipment and stable electrodes if these small potential differences are to be accurately recorded. A second more significant problem is the fact that the gradient tends to decrease to background levels within a very short distance of the leak source. Consequently if the thickness of material covering the liner is too great, or the grid on which measurements are made is too sparse, there is a likelihood that leaks will not be detected at all. Indeed, the parametric study with such an array conducted by Parra and Owen (1988), using the model described in Parra (1988), concludes that practical leak detectability occurs only within 0.1 m of the leak with the dipole (P1 and P2 in Figure 2a) close to the liner.

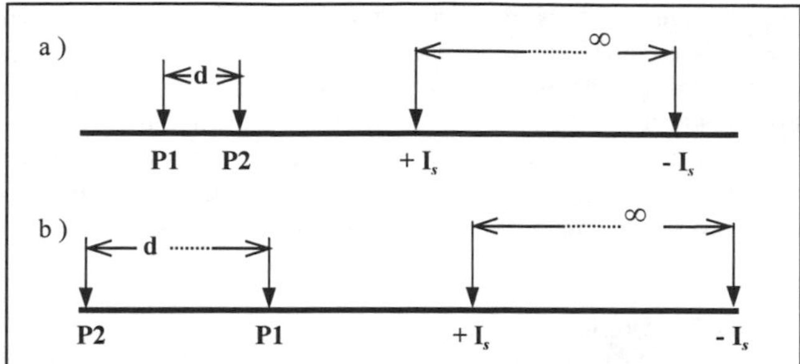

Figure 2 - Electrode array configurations, with - I_S, the current sink, located outside the impoundment, and the current source, +I_S, located inside but distant from the potential electrodes, P1 and P2. a) Gradient array, with moving potential electrode dipole, P1 and P2, with separation, d, of typically 1 - 2 m. b) Pole-Pole array where only P1 moves and the separation, d, of P1 and P2 is maximised.

Parra and Owen's study was directed specifically at liquid containment sites where the distance at which measurements are made above the liner could be minimised. In lined solid waste sites the distance to the liner is set by the thickness of the drainage layer which is normally found to lie between be 0.3 m and 1.0 m. This automatically reduces the level of detectability possible using above-liner leak detection methods.

The Pole-Pole Configuration

An alternative electrode arrangement to the gradient array is the pole-pole configuration of electrodes (Figure 2b). With the potential distribution in the impoundment set up in the same way as for the gradient array case, this configuration samples the electrical potential distribution using a single moving electrode and compares the value to that measured at a distant stationary electrode also in the impoundment. The potential difference measurements made with this configuration are therefore much larger than those measured with the gradient array. Secondly, the position of the leak is coincident with the maximum amplitude of the anomaly curve associated with it and not at the amplitude mid-point of the anomaly as with the gradient array. This makes identification of leaks and their location during a survey more straight forward and does not necessarily require the plotting of the data to determine the location of any leaks.

Using the equations presented in Parra (1988) and equation 1 above, a theoretical comparison of the above-liner and below-liner responses using the pole-pole array can be made. In this comparative study, an impoundment of 4 ha with a hole of radius 0.4 mm is chosen. From Parra (1988), the amount of current passing through this hole in an impoundment of this size is approximately 75 % of the current injected above the liner. The current source electrode is located on the surface of the above-liner material of resistivity ρw =30 Ωm, forty metres from the start of the profile line. For the data calculated above the liner, the reference potential electrode is located forty metres along, and offset forty metres from the profile line. For the below-liner data the reference potential electrode is located two-hundred metres along the profile line. The liner is characterised by a resistivity, ρl = 1014 Ωm and thickness, t = 2.5 mm, with the material beneath the liner characterised by a resistivity, ρs = 80 Ωm. The above and below-liner resistivities are chosen as they are representative of the site discussed in the case study.

Comparison of Profile Shapes

Figure 3 shows a comparison of profiles over a leak located ten metres along the profile line where the thickness of material above the liner is h = 0.5 m. The data are displayed as normalised potential difference, that is the potential difference divided by the source current, I_S. Note that for a current source, I_S, above the liner and return current beneath the liner, potentials measured in the vicinity of a leak lead to a relative low above the liner and a relative high below the liner. The data are calculated at the same distance from the leak both above and below the liner. It is clear that the response to a given leak and impoundment conditions is greater when measured above the liner than below. It also appears from the shape of the curves that the above-liner response to the leak is laterally more extensive than the response below the liner, with a larger gradient present in the below-liner response than in the above-liner response. This evidence suggests that any given leak should be easier to detect with an above-liner survey than with a below-liner survey.

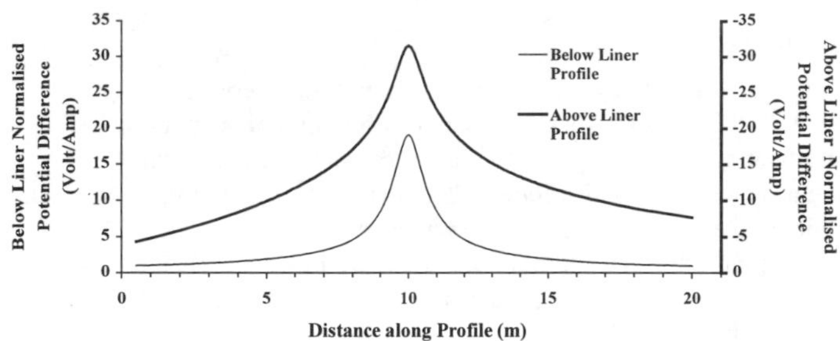

Figure 3 - Comparison of above and below-liner leak profile shapes calculated at the same distance from a leak located 10 m along the profile.

Effect of Distance to Liner and Material Resistivity

Figure 4 shows the variation of the maximum normalised potential difference against logarithmic resistivity calculated across a leak for varying material resistivity and distance from the liner. The plot displays data calculated on the surface of the above-liner material with varying thickness and resistivity, with the resistivity beneath the liner kept at $\rho s = 80\ \Omega m$. Also displayed are data calculated at different depths below the liner for differing below-liner resistivity values, with the above-liner material resistivity kept at $\rho w = 30\ \Omega m$. Due to the electrical properties of waste material inside the impoundment it will in general have a lower resistivity than the ground beneath the liner, and the different ranges over which the data have been collected reflect this.

Both sets of data indicate that the nearer to the liner a survey is conducted, the greater the maximum leak response and hence the better the likelihood of leak detection. As the distance from the liner increases, the maximum leak response decreases, suggesting a limit on the distance from the liner that a survey can be conducted for successful leak detection.

A greater maximum leak response is also observed for measurements made in more resistive materials. This highlights the potential problem of very small leak anomaly responses when conducting surveys over highly conductive waste or drainage materials. Indeed comparison of the two data sets presented in Figure 4 indicates that for below-liner materials that are more than five times as resistive as above-liner materials, the maximum normalised potential difference response to a leak will be greater below the liner than above.

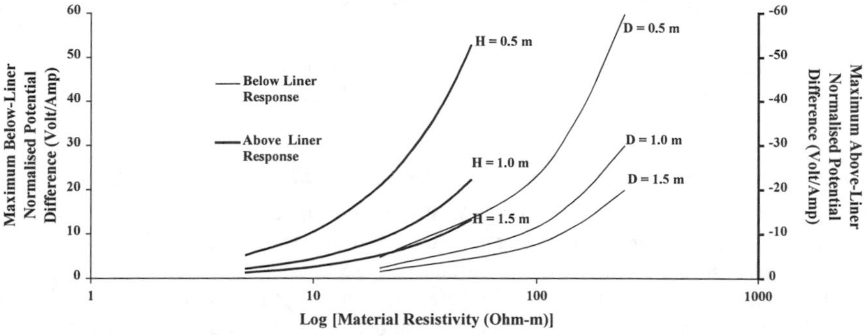

Figure 4 - Variation of maximum normalised potential difference over a leak with above and below-liner material resistivity and distance from the liner.

Variation of Profile with Lateral Offset

The range of profiles presented in Figures 5a and 5b are calculated across a single leak at a distance of 0.5 m from the liner. The figures show how the theoretical leak response varies with offset distance from the leak centre both above and below the liner. As expected, the smaller the offset from the leak, the greater the maximum leak response and the greater the likelihood of detection. For the data collected above the liner, it is clear that for offsets of 0.25 m or less, there is little variation (less than 5 % from that at zero offset) in the observed maximum leak response. This suggests a minimum line separation for practical leak location surveys of approximately 0.5 m with no extra useful information being gained from having survey lines any closer together. Similarly, at greater distances from the liner, the minimum offset increases, with less than 5 % variation in the maximum leak response observed at offsets of 0.45 m and 0.65 m for distances above the liner of 1.0 m and 1.5 m respectively. At offset distances of approximately 2 m or greater, the peak of the anomaly becomes much broader and less well defined in both above and below liner cases.

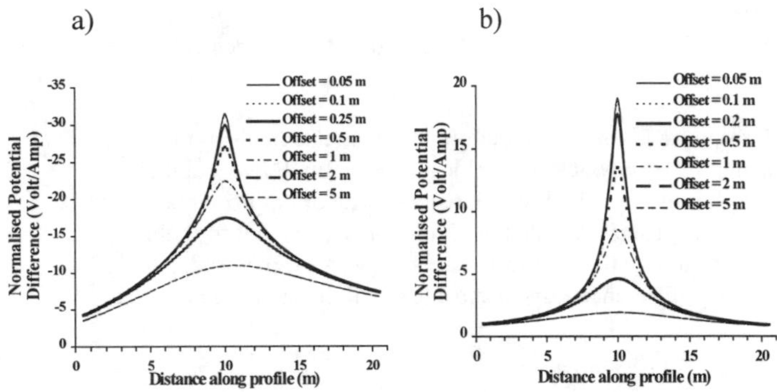

Figure 5 - Variation of leak profile with offset from the centre of a leak, calculated at the same distance, 0.5 m, a) above, and b) below the liner.

Variation of Fractional Current Conducted Through Leaks

The most important parameter in determining the ability of a leak detection survey to detect a specific leak is the amount of current conducted by that leak. The profiles displayed in Figures 6a are calculated on the surface of a 0.5 m thick above-liner drainage layer across a single leak conducting different fractions of the source current. Figure 6b shows profiles calculated at a depth of 0.5 m beneath the liner across the same leak. As the fractional current flowing through the leak is reduced, so is the maximum normalised potential difference response. As the fraction of current flowing through the leak drops below approximately $0.1I_S$, the above-liner leak anomaly becomes indistinguishable from the background potential. Below the liner the peak anomaly amplitude also reduces sharply, though the anomaly shape is still observable even at very small fractional currents.

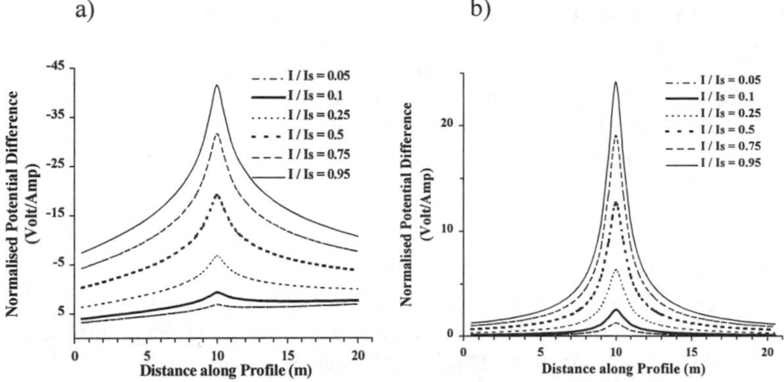

Figure 6 - Variation of profile shape with fraction of current conducted through a hole in the liner calculated at the same distance (0.5 m) a) above and b) below the liner.

Variation of Below-liner Response with Grid Size

It is impractical to have a grid of electrodes located beneath the liner sampling the potential distribution with the spatial frequency achievable with above-liner surveys. It is therefore important to assess the limits of detectability of leaks using as sparse a sampling grid as possible. The plots in Figure 7 show contoured below-liner data calculated 0.5 m

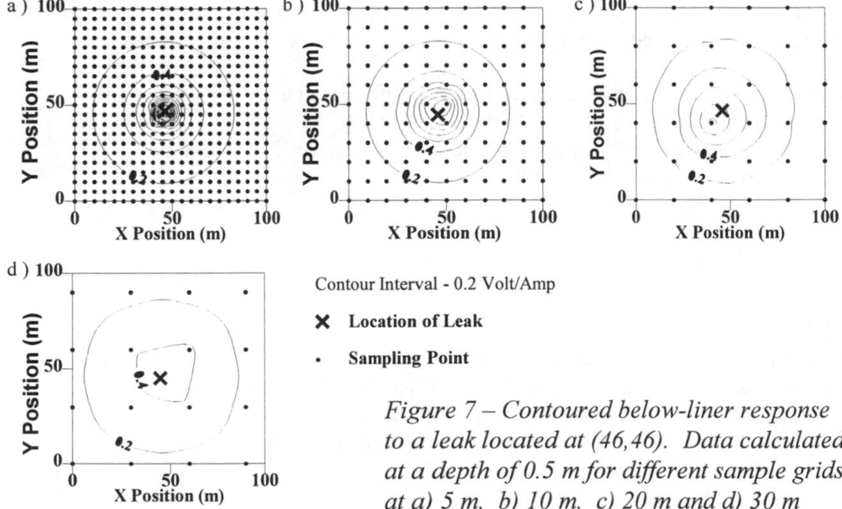

Figure 7 – Contoured below-liner response to a leak located at (46,46). Data calculated at a depth of 0.5 m for different sample grids at a) 5 m, b) 10 m, c) 20 m and d) 30 m

beneath the liner for a leak located at (46,46) and sampled on a 5, 10, 20 and 30 m grid. The concentric pattern of contours indicate the presence of a leak, though the location of the leak becomes less clear with the increasing size of the sampling grid. The presence of a leak is most clearly seen in the data sampled on 5 m and 10 m grids and is still apparent, though the maximum amplitude is smaller, in the data sampled on a 20 m grid. At larger grid sizes, the anomaly is still apparent, though less identifiable, and the location of the leak is less well defined.

Anticipated response implied by the theoretical analysis

The results of the parametric study indicate that for the given set of impoundment characteristics, the above-liner leak response produces a leak anomaly that is larger than the leak response produced below the liner. This holds for below-liner materials that have resistivity values up to five times those of the material above the liner. For ratios larger than this, the maximum leak response below the liner is larger than that above.

The largest anomaly responses are produced closest to the leak and therefore to achieve the greatest level of leak detection, measurements should be made as near to the liner as possible, echoing the findings of Parra and Owen (1988). Similarly, the further laterally from the centre of a leak, the smaller the leak response. This is important in determining the line separation for GLLS surveys. The results show that there is a minimum practical line separation of 0.5 m for an above-liner material thickness of 0.5 m, increasing to 0.9 m and 1.3 m for thicknesses of 1.0 m and 1.5 m respectively. At these offsets, the difference between the maximum leak anomaly and that observed at zero offset is less than 5 %, with smaller line separations offering little, or no improvement on leak detectability. With data collected on a sparse grid beneath the liner, it is still possible to detect a leak even with a grid spacing as large as 30 m, though at this spacing the maximum anomaly amplitude is small and the location of the leak is poorly defined.

By varying the amount of current conducted by a leak, the maximum anomaly response is observed to change, becoming smaller for smaller conducted currents. To achieve the greatest level of leak detectability in any leak location survey, the highest possible current should flow through any leaks in the liner.

Comparison of Field Data

Introduction

The examples presented below compare data collected above the liner during a GLLS with that collected using a below-liner monitoring grid of electrodes at a landfill site in the UK. The site, Sandy Lane Landfill Site, operated by Cleanaway Ltd., is the first fully operational landfill site in the UK to have a below-liner geophysical monitoring system installed.

Brief Description of the Below-liner Monitoring System

The data presented in the following section were collected during evaluation of the integrity of the geomembrane liner in Phase 1 of the Sandy Lane Landfill Site, (White and Barker 1997). The present extent of the landfill covers an area of approximately 5.5 ha and is located in the base of an old sandstone quarry. The liner construction comprises a 2.5 mm thick HDPE liner underlain by a 0.3 m thick layer of bentonite enriched sand (BES). Directly beneath the BES layer are located monitoring electrodes on a 20 m grid, with alternate east-west lines having electrodes at 10 m separation. Presently, the grid comprises 242 below-liner electrodes, all of which are connected to an electronically controlled switchbox installed within the landfill weighbridge building, enabling any of the electrodes to be selected for current conduction or potential measurement.

Data are collected using the pole-pole array with return current and reference potential electrodes located over 200 m away from the electrode grid. A constant current supply injects a 20 mA current through a stainless steel plate located in the 0.5 m thick sand drainage layer at the base of the landfill. Potential differences are recorded between each of the grid electrodes and the reference potential electrode.

To date three cells have been completed within the landfill with the geophysical monitoring system installed during each construction phase. Its use as an additional CQA survey has revealed 3 holes in Cell 1, 19 holes in Cell 2 and 6 holes in Cell 3. This averages 5 holes per hectare, compared to 26 cited by Darilek et al. (1989). Giroud and Peggs (1990) relate the number of holes per hectare to the effectiveness of the CQA programme. The number of holes located in Cells 1 and 3 relate to good CQA whilst Cell 2 relates to average CQA. However, this needs to be put in the context that only four of the holes detected could be easily seen with the naked eye. In both Cells 2 and 3, the holes were concentrated in distinct areas; Cell 2 along a berm mid way up the landfill slope where the upper and lower halves of the liner were welded together; and Cell 3 where construction was most intense in the sump area.

Real Data Examples

Example 1 — Figure 8a shows a typical example of contoured data collected using the below-liner system when there is a single hole present in the liner. Displayed on the plot are the locations of the electrodes in the monitoring grid and overlaid on the data is an outline map of Phase 1 of the landfill. Towards the centre of the plot there is a single sharp positive peak anomaly which causes a departure from the otherwise smooth distribution of normalised potential difference. This peak is indicative of the presence of a leak in this area.

With the data collected using the below-liner grid of electrodes, it is possible to estimate the location of any leak to within a 20 m grid square. To improve on this location accuracy it is possible to model the data to produce a best-fit estimate of the leak location. This can normally improve the location accuracy to a 1 to 2 square metre area, though multiple leaks can reduce this accuracy. To locate the source of the leak precisely it is still necessary to conduct a GLLS over this localised area. The results of the GLLS conducted over the estimated position of the leak is shown in Figure 8b.

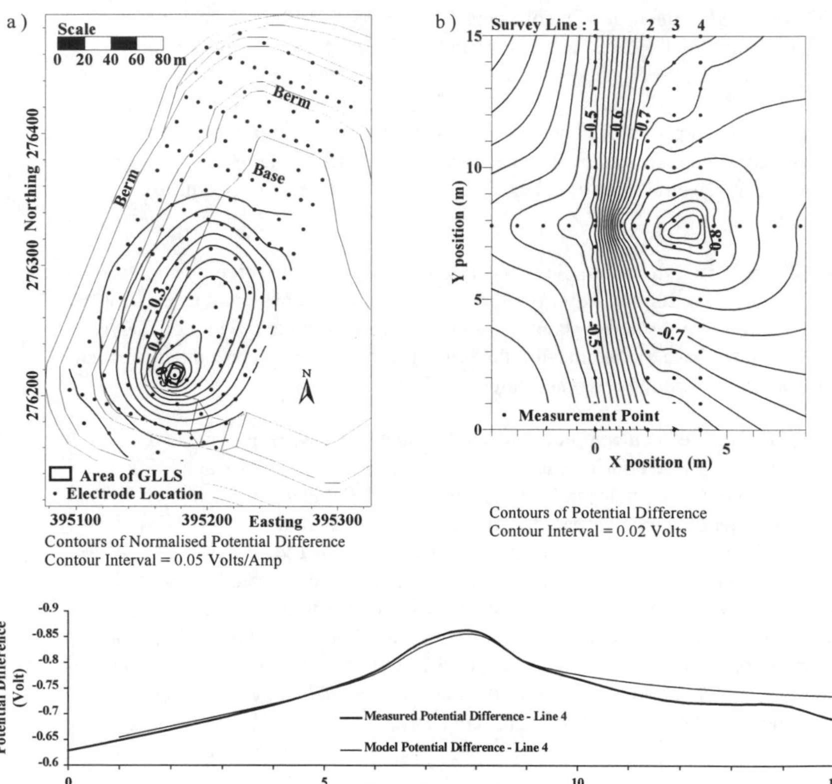

Figure 8 - a) Data collected using the below-liner monitoring grid of electrodes. Data indicate the presence of a single leak. b) Above-liner survey results over the leak clearly show its location. c) Comparison of model fit to the data measured along survey line 4.

The survey was conducted on the surface of the 0.5 m thick drainage layer of sand present in the base of the landfill on a 1 m grid using a constant 250 V supply. The data clearly show a single localised sharp peak anomaly indicating the location of the leak. Also observed is a sharp drop in potential between survey Line 1 and Line 2. This is caused by the presence of a gravel drain running parallel to, and located between these survey lines. The leak was found to be two adjacent 5 mm knife cuts in the basal liner. Using the survey parameters and resistivity values measured at the site of 30 Ωm for the above-liner sand and 80 Ωm for the sandstone below the liner, the model of Parra (1988) was used to reproduce the observed data and a comparison is shown in Figure 8c. This example shows that both above and below-liner methods are sensitive to very small leaks and that the model is able to reproduce the measured data with a high degree of accuracy.

Example 2 - Figure 9a shows another example of data collected using the below-liner system. Located in the north of the site are a number of leak anomalies. Concentrating on the largest anomaly, anomaly A (Figure 9b) this anomaly is observed to have a maximum amplitude of 0.32 Volt/Amp (V/A).

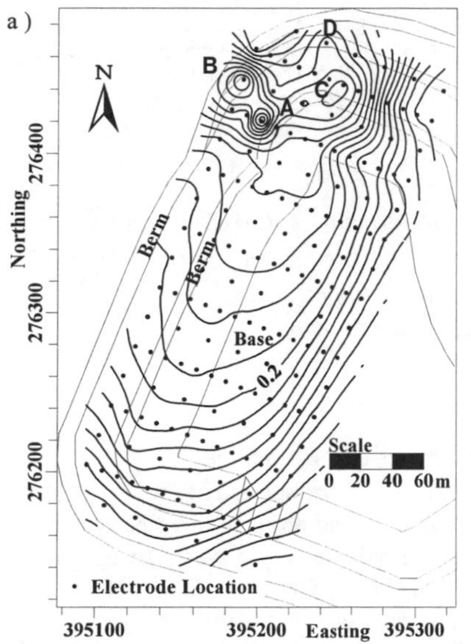

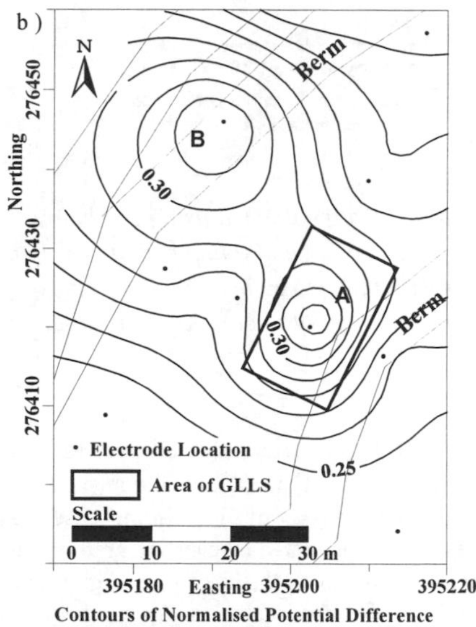

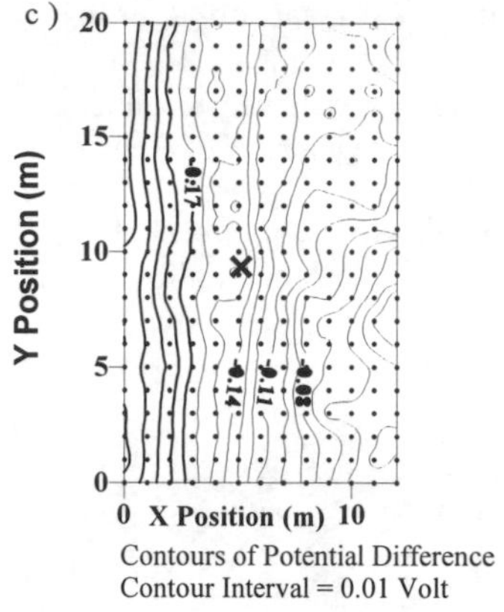

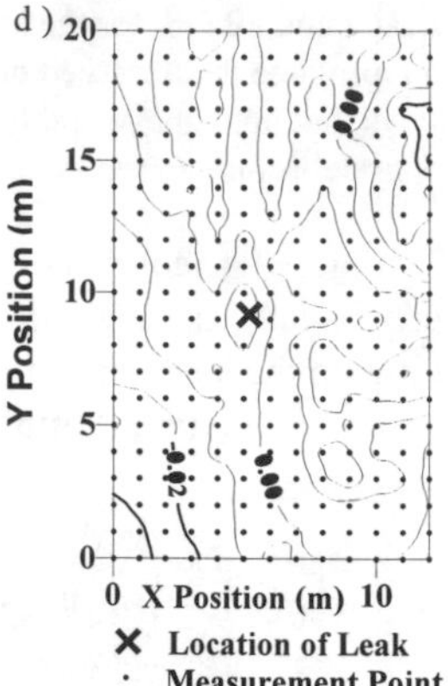

Figure 9 - a) Data collected using the below-liner monitoring grid of electrodes indicating the presence of a number of leaks. b) Expanded view of the Northwest corner of data in Figure 9a showing the area of the above-liner survey. c) Above-liner survey data before and d) after background removal.

The results of the GLLS conducted over the estimated position of the leak is shown in Figure 9c. Once again, the survey was conducted on the surface of the 0.5 m thick drainage layer of sand. In this case the data set shows no obvious sharp peak anomalies and is dominated by a strong gradient running approximately east-west. Even with a best fit estimate to this background gradient removed (Figure 9d), the data do not clearly indicate the presence of a sharp anomaly of the kind predicted in theory or of the type observed with the data collected using the below-liner grid of electrodes. Knowledge gained by the below-liner system that a hole was present necessitated locating the other holes first, in order to create a more identifiable anomaly at this precise location.

Further GLLSs were conducted to locate other leaks observed to be present in the data shown in Figure 9a. These surveys did establish clearly identifiable anomalies similar to the anomaly presented in Example 1. All the holes associated with these anomalies were repaired, the above-liner survey repeated over the remaining leak and the hole, a pin-hole sized split on a weld joint, was finally located.

Discussion

Though most above-liner GLLSs have proved successful in the swift and accurate location of leaks detected with the below-liner system, there have been a number of occasions, similar to the one presented in Example 2, where leak location, although always successful, has proved difficult and time consuming. These results highlight the fact that above-liner GLLSs may not be sensitive enough to detect leaks in certain situations and there is a possibility when conducting a GLLS that some leaks may go undetected.

Example 1 has shown that the model fit to certain leak data is very accurate. However, other findings (Example 2) also indicate a disparity between the results of some above-liner surveys and the theoretical response, which may be due to the assumptions made in the development of the theory. Primarily, the observed maximum amplitude response over a leak is often much less than that predicted in theory for a hole and impoundment of given size. The model predicts the percentage of current flow through a circular hole of given diameter. However, in reality holes are very rarely circular and tend to be very thin knife cuts, pin holes or splits on welds. In such circumstances, the level of connectivity between the material inside and outside the impoundment may result in less current being able to flow through the hole than predicted in theory.

As the amount of current available to flow through any hole is such an important parameter, it is important that the lined impoundment be electrically isolated when leak location surveys are performed. If the material inside the impoundment is not adequately electrically isolated from the material outside the impoundment, current will flow out of the impoundment at the places of inadequate isolation. This may be caused by sand or soil lying over the edge of the liner, or due to conduction along water filled pipes, metal guide ropes, or earth wires leading out of the impoundment. Any low resistance pathways such as these will significantly reduce the current densities in the vicinity of leaks and as a consequence reduce the sensitivity of any leak location survey. A GLLS conducted in these circumstances is likely to produce a false negative.

Masking of holes can occur if there are a number of leaks present in the liner, as the current passing through each hole will be reduced and, moreover, current will preferentially flow through the leaks offering the least electrical resistance. Hence as a consequence, this can preclude detection of smaller leaks in the presence of much larger ones. A further reason for the departure of the measured response from that predicted in theory is the assumption that the impoundment can be approximated by a model consisting of three infinite homogeneous horizontal layers. In the base of a landfill when no waste is present, this model is fairly representative. However, on the slopes of the landfill, or when waste is present, these assumptions are less valid due to the presence of topography and heterogeneity in the waste complicating the potential distribution.

The above comments apply equally well to below-liner surveys as they do to above-liner surveys. However, there are still important differences in the results of the two types of surveys which should be considered.

With the below-liner survey it is possible to locate the remote current and potential electrodes at a significant distance from the measuring grid. This is advantageous as it removes the strong background gradient that is observed to be present in the above-liner pole-pole data. Location of the potential reference electrode well away from the grid also precludes the random chance that it is located near to a leak and therefore in the region of a steep gradient. This is a possibility with the above-liner survey method, though this can be negated with good field practice. A further advantage of the below-liner system is that the current flows through a larger volume of ground than in the above-liner case and the resulting potential distribution is less affected by localised heterogeneities in the ground. Though prior to landfilling the drainage layer above the liner should be relatively homogeneous and the potential distribution relatively uniform, once landfilling has begun, the heterogeneity of the waste can result in significant localised variation in the potential. Also, as has been shown in example 1, the presence of gravel drains, gravelled sump areas and bunds[3] may complicate normally uninterrupted current flow and lead to complicated gradients.

Another factor that may explain why the results of below-liner surveys are more sensitive to certain leaks than above-liner surveys is the leak geometry. During construction, the external side of the geomembrane liner is laid directly in contact with the secondary BES liner or engineered base, with little overlap and no repair taking place on this side of the liner. All flaps, secondary cuts, welds, patches and repairs are located on the upper side of the liner, onto which the drainage cover is laid. In some circumstances, such as T-weld junctions or overlapping patches, a very complicated and tortuous path for current flow may be set up. A small hole present in the basal liner underneath a weld flap, for example, will allow current to flow from above to below the liner. When 'observed' from below the liner, the current flows into the material below the liner through this small hole which acts as a point source of current. However, above the liner the hole is not in direct contact with the drainage material, but located beneath an electrically insulating flap. Current is dispersed between the flap and the basal liner and only flows into the above-liner material at the edge of this flap. The area over which the current is dispersed into the above-liner material is now much greater than for the original hole, and the current density is reduced resulting in a much smaller anomaly. When an above-liner survey is conducted over such a leak, the leak may not be detected, or if detected, the broad, small amplitude anomaly may not be recognised as a that due to a leak.

Other benefits afforded by below-liner systems include their ability to detect leaks in regions that would normally not be surveyed using GLLS's. For example, partially constructed or removed bunds. If current can enter the bund and holes are located beneath the bund they may be detected using a below-liner monitoring system.

By far the biggest benefit of below-liner systems is their ability to continue to monitor the integrity of the geomembrane liner throughout the lifetime of the landfill, and to continue monitoring post closure. Though repair of the liner may become impractical and prohibitively expensive for leaks that develop late on in the life of the landfill, such systems can be used to monitor the development of leachate migration in the subsurface.

---

[3] Bund - intercell boundaries usually formed by sand enclosed by a sheet of membrane above the basal membrane and should by inference be electrically isolated

This extra knowledge offers advantages in the implementation of remediation systems as these can be targeted more cost effectively.

Conclusion

A comparison of the theoretical leak response calculated both above and below the liner suggest both offer comparable leak detectability, though the theoretical above-liner leak response produces a laterally more extensive anomaly than the below-liner response and an anomaly that is not always observed in practice. The reason for the departure of the observed above-liner response from that predicted in theory is thought to be due to a smaller than predicted current flowing through the leaks in the liner. Field data collected over a number of leaks both above and below the liner have indicated that the below-liner system is more sensitive to certain leaks. Possible explanations, including the differing leak geometries and background conditions present above and below the liner, have been discussed.

The use of the GLLS for leak detection is a successful and proved method. However, from our experience, small leaks that have been detected with the below-liner monitoring system have sometimes proved difficult to locate with a GLLS and without the below-liner system may have gone undetected. The use of a sparse below-liner monitoring grid, with its long-term monitoring capabilities, combined with above-liner surveys to pin-point leaks accurately offers a successful approach to leak detection and long term monitoring of the liner's integrity.

Acknowledgements

The authors would like to thank Cleanaway Ltd. for allowing us to present the data in this paper and Professor D. Griffiths for his comments and deliberations.

References

Anderson, W. L., 1979, "Numerical Integration of Related Hankel Transforms of Orders 0 and 1 by Adaptive Digital Filtering," Geophysics, Vol. 44, pp. 1287-1305.

Darilek, G. T., Laine, D. L., Parra, J. O., 1989, "The Electrical Leak Location Method for Geomembrane Liners : Development and Applications," Industrial Fabrics Association International Geosynthetics '89 Conference, San Diego, CA, pp. 456-466.

Frangos, W., 1994, "Electrical Detection and Monitoring of Leaks in Lined Waste Disposal Ponds," Proceeds of the Symposium on the Application of Geophysics to Engineering and Environment, 1994, Englefield, USA, Vol. 2, pp. 1073-1082.

Giroud, J. P., and Peggs, I. D., 1990, "Geomembrane Construction Quality Assurance," In : Waste Management Systems: Construction, Regulation and Performance. Geotechnical Special Publication No 26, ASCE, pp. 190-225.

Laine, D. L., Mosley, N. G., 1995, "Leak Location Survey of a Soil Covered Geomembrane at a Landfill Site in the UK," In : Waste Disposal by Landfill, A. A. Balkema, Rotterdam, pp. 151-156.

Parra, J. O., 1988, "Electrical Response of a Leak in a Geomembrane Liner," Geophysics, Vol. 53, pp. 1445-1452.

Parra, J. O., Owen, T. E., 1988, "Model Studies of Electrical Leak Detection Surveys In Geomembrane-Lined Impoundments," Geophysics, Vol. 53, pp. 1453-1458.

White, C. C., Barker, R. D., 1997, "Electrical Leak Detection System for Landfill Liners: A Case History," Ground Water Monitoring & Remediation, Vol. 27, No 3, pp 153-159.

Michael J. Byle,[1] Monica L. McCullough,[2] Rodney Alexander,[3] N.C. Vasuki[4] and James A. Langer[5]

Instrumentation of Dredge Spoil for Landfill Construction

REFERENCE: Byle, M.J., McCullough, M.L. Alexander, R.,Vasuki, N.C., and Langer, J.A., **"Instrumentation of Dredge Spoil for Landfill Construction,"** *Field Instrumentation for Soil and Rock, ASTM STP 1358*, G.N. Durham and W. A. Marr, Eds., American Society for Testing and Materials, West Conshohocken, PA, 1999.

ABSTRACT: The Delaware Solid Waste Authority's Northern Solid Waste Management Center is located outside of Wilmington Delaware at Cherry Island, a former dredge disposal site. Dredge spoils, of very low permeability, range in depths up to 30 m (100 feet) which form a natural "liner" and the foundation for the 140 ha (350-acre) municipal solid waste landfill. The soils beneath the landfill have been extensively instrumented to measure pore pressure, settlement and deflections, using inclinometer casings, standpipe piezometers, vibrating wire piezometers, pneumatic piezometers, settlement plates, liquid settlement gages, total pressure cells and thermistors. The nature of the existing waste and anticipated settlements (up to 6 m (19 feet)) have required some unique installation details. The instrumentation data has been integral in planning the landfilling sequence to maintain perimeter slope stability and has provided key geotechnical parameters needed for operation and construction of the landfill. The performance of the instrumentation and monitoring results are discussed.

KEYWORDS: instrumentation, piezometers, total pressure cells, landfills, inclinometers, thermistors, settlement, municipal solid waste

---

Background

In 1984 the Delaware Solid Waste Authority (DSWA) acquired 140+ ha (350+ acres) of land at the U.S. Army Corps of Engineers (USACE) Cherry Island dredge disposal site from private landowners and the City of Wilmington. DSWA also initiated the development

---

[1]Region 3 Geotechnical Practice Manager, Gannett Fleming, Inc., 650 Park Avenue, King of Prussia, PA 19406.

[2]Project Engineer, Gannett Fleming, Inc., 650 Park Avenue, King of Prussia, PA 19406.

[3]Senior Engineer, Delaware Solid Waste Authority, 1128 S. Bradford Street, Dover, DE 19903.

[4]Chief Executive Officer, Delaware Solid Waste Authority, 1128 S. Bradford Street, Dover, DE 19903.

[5]Vice President, Gannett Fleming, Inc., 207 Senate Avenue, Camp Hill, PA 17011.

of a complex memorandum of understanding between the USACE, Delaware Department of Natural Resources and Environmental Control (DNREC), and DSWA under which USACE agreed to release portions of the dredge disposal site for DSWA to use as the new landfill site for New Castle County.

The Cherry Island Landfill (CIL) site is bordered by I-495 on the west side, the City of Wilmington regional wastewater treatment plant on the north side, the Delaware River on the east side, and the Christina River on the south side. The site offers a suitable location because of the ready access, available land area and lack of nearby sensitive neighbors. However, the site is very challenging from a geotechnical viewpoint because of the depth and nature of existing unconsolidated sediments.

The CIL design was based on five phases of construction starting with Phase I in 1984. The location for Phase I was selected on reasonable access from I-495 and the least depth of unconsolidated sediments. DSWA planned a sequential construction program assuming normal growth rate of solid waste generation. However, the closure of several landfills previously operating in New Jersey and Pennsylvania resulted in a large increase in the commercial and industrial waste being delivered to DSWA landfills. Consequently, the planned phasing construction of the cells had to be accelerated to accommodate the substantial increase in solid waste delivery. This had a significant effect on the performance and design requirements for the latter phases of the CIL.

Site Conditions

The CIL is situated at the confluence of the Delaware and Christina Rivers and comprises approximately 138 ha (342 acres) of land divided into five filling sections, or Phases. A wastewater treatment plant is located to the north and an active dredge spoil disposal area is located to the west. The active landfill areas consist of Phases III, IV and V which are approximately 18, 20, and 24 ha (45, 50, and 60 acres), respectively. All three active phases are being filled concurrently to accommodate required consolidation time between lifts in each phase. Phases I, IA, and II are no longer receiving municipal solid waste. They are used for stockpiles, asbestos disposal, and haul roads. The five phases are located on dredge spoils recently deposited by the USACE.

During the pre-design investigations, the dredge material was so soft that tripod drill rigs on plywood rafts and low contact pressure all terrain vehicle drill rigs were necessary for the geotechnical investigations for each phase. The subsurface profile consisted of three primary horizons: dredge spoils (beneath the landfill interior and in the dikes), dense sands of the Pleistocene Columbia formation and very stiff silts and clays of the cretaceous Potomac Formation.

The dredge spoils typically classify as silt (ML), and elastic silt (MH). Standard penetration test (SPT) N-values range from weight of rods to 4 within the interior dredge spoils and from 2 to 7 blows per foot (bpf) in the dikes. Moisture content of dredge spoil samples is typically greater than 50%. Undrained shear strengths from unconfined compression tests and unconsolidated undrained triaxial tests were typically very low, generally 0.4 to 1.9 kPa (100 to 400 psf). The dredge spoil thickness generally increases from a low of 8 m (25 ft) in the northwest corner of the site to a maximum of 35 m (114 ft) near the confluence of the Christina and Delaware Rivers at the southeast corner of Phase V.

The soils used in the construction of the perimeter dikes consisted of materials excavated from the desiccated crust material removed from the interior dredge spoils. This crust material, having been compacted at a lower moisture content, is typically of higher strength than the hydraulically placed interior materials. The dike soils have gradations and Atterberg limits identical to the interior dredge spoils.

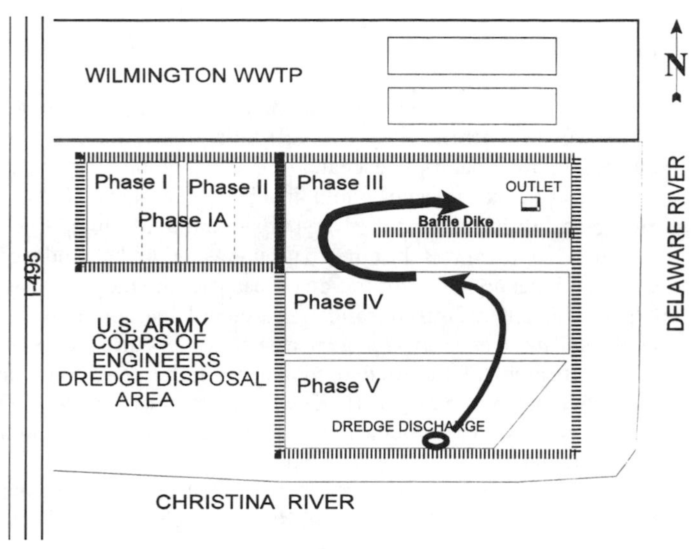

Figure 1 Dredge Placement Pattern in Phases III, IV and V

The particle sizes of the dredged materials vary across the site, since the dredge disposal area was designed to allow the solids to settle out of the slurry as it flowed from the dredge discharge point at one end of the disposal site, around a baffle dike to an outlet weir at the other. Then, after the solids settled out, the water was discharged through a weir structure to the Delaware River. The dredge spoils typically have more than 85% passing the No. 200 sieve and permeability in the range of 1×10^{-6} to 5×10^{-8} cm/sec. The area of Phases I and II consisted of upland dredge disposal cells with reduced thickness of dredge spoil. The dredge spoil was normally to slightly under consolidated at the time of landfill construction. Some natural alluvial deposits underlie the dredge spoil, but are so similar as to be indistinguishable from them.

The Columbia Formation which underlies the dredge spoil consists of dense coastal plain sands and gravels with interbeds of silty sand, silty clay, and clayey silt from 3 to 20 m (10 to 66 ft) thick. SPT values typically range from 35 to 100 bpf. The Columbia Formation is an aquifer of limited quality with local hydraulic gradients toward the southeast confluence of the Christina and Delaware Rivers. The deeper Potomac Formation is a lower confining layer for the Columbia Formation. The soils consist of very stiff silts and clays interbedded with silty sand and gravel.

The hydrogeology of the site is somewhat complex. A water table, perched in the dredge spoil, actually has an upward gradient due to the under-consolidated nature of the dredge spoils. The hydraulic head increases with depth through the near surface dredge spoil to the middle of the dredge material and then decreases toward the bottom where the dredge materials/recent deposits meet the Columbia Formation. There are granular layers within the recent deposits, isolated from the Columbia Formation, that locally distort the hydraulic head.

Landfill Design

The landfill design and operation are controlled by geotechnical issues. The great thickness of low permeability dredge spoils creates an excellent barrier to protect the groundwater from landfill leachate. DNREC granted an exception to the Subtitle D requirements for a composite bottom liner based on a demonstration of equivalent hydraulic performance of the existing dredge spoils to a manufactured liner system. An additional factor in the approval was the upward groundwater gradient induced by consolidation of the dredge spoil that prevents infiltration of leachate into the subsurface. The upward gradient is expected to remain until consolidation is completed, approximately 50 to 70 years. At that time the soils will have an average permeability less than 1×10^{-7} cm/sec.

Settlement of the dredge spoils under the weight of the municipal waste was estimated to be up to 6 m (19 ft). The leachate collection system beneath the landfill has been designed to accommodate this settlement while maintaining adequate slopes for positive drainage.

The low strength of the dredge spoils controls the height and steepness of slopes that can be safely constructed. With a high drained angle of internal friction, on the order of 38°, the dredge materials will be capable of supporting the proposed 3H:1V slopes when fully drained. However, consolidation of the dredge materials must take place for this drained strength to be realized. This consolidation is time dependent and cannot be accelerated, because methods that could be used to improve drainage of the dredge spoils would compromise its effectiveness as a liner.

Therefore, the landfill filling operation requires staged filling among the phases so that the dredge spoils can consolidate under each lift before the next lift is placed. The original designs of each phase projected the expected rate of consolidation and magnitude of settlement. Because of the very low strength of the dredge spoils, it is not possible to construct slopes with high factors of safety. Slope stability analyses were completed using the Modified Bishop Method of Slices which resulted in factors of safety on the order of 1.2 for the design slopes under each successive lift of waste.

Because of the relatively small factor of safety, consequences of failure, very slow rate of drainage, uncertainties associated with predicting time rate of consolidation and strength gain in these materials, and the need to verify soil conditions prior to each successive lift of waste, the design included instrumentation to verify the design assumptions and improve the time rate correlations. This was considered essential to control the rate of filling to maximize the available air space without over stressing the weak foundation soils.

The proposed final landfill height for all phases is 24 m (80 ft). Because of the low dredge spoil strength, geogrids were included at the bottom of the waste in Phases II and III to reinforce the base soils against basal failure. The geogrids were embedded in the sand drainage layer at the top of the dredge spoils. Phases IV and V incorporated vertical strip drains to accelerate consolidation and the dissipation of pore pressures in the perimeter dikes for improved global stability.

Instrumentation Systems Design

The purpose of the instrumentation systems at Cherry Island is to monitor the behavior of the perimeter and the interior soils to determine if they are behaving as required and as expected. The primary instrumentation focus areas for each phase of the landfill are slope stability of the berms and consolidation of the underlying interior dredge spoils. At present the total instrumentation consists of 109 standpipe piezometers, 5 liquid settlement gages, 33 pneumatic piezometers, 5 vibrating wire piezometers, 43 settlement plates, 45 inclinometers, 32 total pressure cells, and 28 thermistors (Table 1). The manufacturers and model numbers of the discussed instrumentation are included in the following sections, where known. A typical cross-section of the landfill is shown in Figure 2.

Table 1

Instrument Type	Instrumentation Summary				
	Phase I	Phase II	Phase III	Phase IV	Phase V
Casagrande Piezometers	14(Perimeter)	17(Perimeter)	24(Perimeter)		
Well-Screen Piezometers				17 (Perimeter) 3 (Interior)	17(Perimeter) 6 (Interior)
Pneumatic Piezometers			15 (Interior)		18 (Interior)
Vibrating Wire Piezometers				5 (Interior)	
Inclinometer	6 (Perimeter)	6 (Perimeter)	12 (Perimeter)	8 (Perimeter)	13 (Perimeter)
Stick-Up Settlement Plate	7 (Interior)	9 (Interior)	9 (Interior)	6 (Interior)	10 (Interior)
Settlement Transducer			0 (Interior) [5 abandoned]		
Total Pressure Cell			10 (Interior)	10 (Interior)	12 (Interior)
Thermistor			8 (Interior)	8 (Interior)	12 (Interior)

Pneumatic Piezometers

Pneumatic piezometers were included in the interior of Phases III and V to verify the direction of groundwater gradients for permit compliance and to measure consolidation related pore pressures. The piezometers were planned at elevations of the upper, middle and lower sections of dredge spoil. The pneumatic piezometers specified are two-tube normally closed piezometers read under no flow conditions with 50 micron stainless steel filters and synthetic rubber diaphragms (Slope Indicator Model 514178 in Phase III and RST Model P-102 in Phase V). The piezometer tips are placed in 0.9 to 1.2 m (3 to 4 ft) of well sand, with a bentonite seal and bentonite grout. Small sand layers were used to keep the monitoring zones small.

The Phase III piezometers were installed using hollow stem augers. The Phase V design used the same type of piezometers but used the direct push method of installation. The piezometers are enclosed in a perforated stainless steel housing that is pushed into the subsoil without a borehole or below the bottom of a casing.

The Phase III design routed the tubing from the boreholes, through the leachate collection sand layer to the remote readout locations on the perimeter berms. In Phase III, the pneumatic tubing consists of two 5 mm (3/16 inch) outer-diameter Nylon 11 tubes inside a polyethylene sheath with fittings of 316 stainless steel. These materials were selected for resistance crushing and chemical degradation. The distance between the instruments and monitoring stations range from approximately 60 to 200 m (200 to 700 feet).

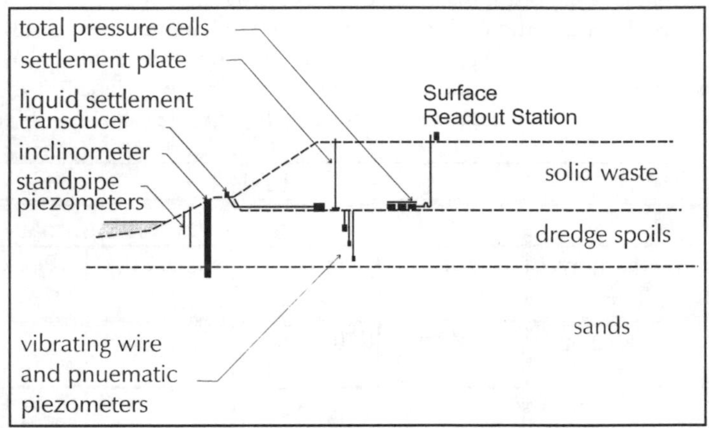

Figure 2 Typical Instrumentation Cross Section

The Phase V design called for high density polyethylene (HDPE) pneumatic tubing routed through a 152 mm (6-inch) diameter HDPE conduit, buried in the leachate collection zone. Flexible rubber hose was used to protect the tubing where sharp bends occurred.

Several pneumatic piezometers were installed after the landfill was in operation to augment the original instruments (Rocktest Model FPC-2). These piezometers were installed through about 6 m (20 ft) of municipal solid waste (MSW) via boreholes drilled through the MSW in the interior of Phase III. The tubing for these instruments could not be trenched through leachate collection zone so these piezometers are read from temporary movable monitoring stations in the landfill interior. The tubing is routed vertically through the waste within telescoping 76 and 102 mm (3 and 4-inch) Schedule 40 galvanized steel casing. A bentonite slurry is placed inside the telescoping casing to provide lubrication and extra protection for the pneumatic tubing.

Vibrating Wire Piezometers

Vibrating wire piezometers were used in the interior of Phase IV. These piezometers include a 50 micron sintered stainless steel filter and a stainless steel body (Irad Gage Model PWS). The vibrating wire piezometers were installed in a 50 mm (2 in) PVC casing with a 1.5 m (5 ft) slotted well screen. The well screen was set in a 1.8 m (6 ft) filter sand pack.

The piezometers cable was routed vertically from the dredge layer to the landfill surface through the MSW inside a steel casing. Additional same diameter steel casing is installed as the MSW height increases. These instruments are monitored from the landfill surface. The frequency and temperature of the instruments are monitored with an IRAD Gage Vibrating wire readout unit.

Standpipe Piezometers

Conventional standpipe piezometers are used in the perimeter berms surrounding all five phases of the landfill. The Phase designer's preferences resulted in two general types of standpipe piezometers. In Phases I, II and III, the piezometers are Casagrande piezometers (from various manufacturers) containing porous tips connected to 19 mm (0.75-inch) PVC riser pipes (Casagrande 1949). The tips were installed in 900 to 1500 mm (three to five feet) of sand and with a bentonite seal and bentonite grout to the surface.

In Phases IV and V, the standpipe piezometers are well-screen type piezometers containing 3 m (10-foot) lengths of 51 mm (2-inch) diameter PVC well screen with a 0.2 mm (0.006-inch) slot size and 51 mm (2 in) PVC casing. These well screens were backfilled with 4.6 m (15 feet) of filter sand and hence have a longer monitoring zone. Phases IV and V also contain interior standpipe piezometers constructed in the same manner. However these interior piezometers include a 400 mm (16 in) HDPE outer protective casing through the MSW. This interior placement was intended to verify initial vertical gradients for the first two lifts of waste only.

Stick-Up Settlement Plates

Each phase of the landfill contains stick-up settlement plates installed during cell construction. The plates in Phase I, II and III are 610 x 610 x 6 mm (24" x 24" x 1/4") structural steel plate with a 19 mm (3/4 inch) threaded steel riser pipe attached to the plate center. The plates were placed on a prepared subgrade of sand constructed on the dredge spoil surface, prior to placement of the sand drainage layer. As additional lifts of MSW are placed, extensions are added to the riser pipes. In Phase I and II, the riser pipe is protected from the MSW with 51 mm (2-inch) outer-diameter threaded steel casing extended as MSW heights increase. In Phase III, the riser pipe is protected by telescoping 51 and 76 mm (2 and 3 inch) steel pipe that is grouted with a bentonite slurry.

In Phases IV and V the plates consist of 1219 x 1219 x 6 mm (48 x 48 x 0.25 inch) structural steel plates with a 19 mm (3/4 inch) steel riser pipe attached to the plate center. The plates are supported on a prepared sand base similar to the other phases. The riser pipes are protected from the MSW with 70 mm (2.75 inch) outer-diameter ABS plastic telescoping inclinometer casing.

Liquid Settlement Transducers

Five settlement plates in Phase III are attached to vertically aligned liquid settlement transducers (LSTs) (Sinco Model 51483). Like the stick-up plates, these plates are supported on prepared sand base placed directly on the dredge spoil surface during cell construction. A transducer was attached to each plate and backfilled with sand. The transducer tubing was routed through the leachate collection zone to readout stations on the perimeter berms. The pneumatic instruments (piezometers and LSTs) were constructed in clusters to minimize trenching and share common readout stations. The tubing and fittings for these instruments are similar to those described for the Phase III pneumatic piezometers. Unplasticized nylon 11 tubing with a polyethylene sheath was selected for low permeability to gas and water (Dunnicliff 1993).

One pair of the transducer tubing for the LSTs contains a deaired glycol solution that is used to provide a fluid head for the measurement. The glycol solution is used to prevent freezing. The remote transducer consists of a pneumatic piezometer, in a liquid filled cell, connected by tubing to a reservoir outside the area of filling. The reservoir of glycol solution is situated at a higher elevation than the transducer, on the perimeter berm. The elevation of the glycol reservoir is surveyed and the elevation of the plate is calculated from the differential head measured by the piezometers. The reservoirs are protected within metal housings but must be refilled regularly because of evaporation. The piezometers in the transducers are three tube, normally closed piezometers which are read under two-tube normally closed conditions like the other pneumatic piezometers (per manufacturer's instructions).

Inclinometers

Inclinometer casing was installed along the perimeter berms of all phases of the landfill to monitor berm stability. Each inclinometer casing was installed to the bottom of the dredge layer, approximately 15 to 34 m (50 to 110 feet), so that a complete profile of the dredge movement could be developed.

In Phases I, II and IV, 70 mm (2.75 inch) O.D. ABS plastic standard inclinometer casing was installed. In Phases III and V, 85 mm (3.34 inch) O.D. ABS plastic casing with telescoping couplings was installed. The telescoping couplings allow for up to 150 mm (6 in) of settlement per coupling (Slope Indicator 1996). Couplings are spaced every 1.5 to 3.0 m (5 to 10 ft) allowing for a minimum shortening of approximately 5% to accommodate consolidation settlement of the berms. The casings in all phases were grouted in place with cement-bentonite grout, except for Phase IV where the casings were backfilled with gravel. Surveys were conducted with standard inclinometer probes.

Total Pressure Cells and Thermistors

Total pressure cells (TPCs) were installed in Phases III, IV and V of the landfill to measure the load applied by the MSW on the dredge spoils. Thermistors were included to monitor temperature variations that could affect the TPC readings (Felio and Bauer 1986).

Retrofit Installation Phases III and IV -- During the TPC design process for Phases III and IV, MSW was already in place, to depths of approximately 3 to 6 m (10 to 20 feet). The installation required excavation through the existing MSW. Four clusters of five TPCs and four thermistors were installed to minimize excavation and provide redundancy. Corrosion resistance was a key factor in the TPC design, since the equipment would be exposed to landfill leachate.

In order to minimize the effect of point loads from the mixed materials in MSW, large instrument bearing areas were preferred. The largest off-the-shelf TPC size was approximately 305 mm (12-inches) diameter. This diameter was considered too small to accurately monitor the pressure because of the potential for nearby interference from large point loads and/or load arching. The concept of clustering multiple TPCs beneath a structural steel bearing plate was considered to spread the load over several TPCs. This option was ruled out because of the corrosive environment of the steel, high cost, and installation difficulty.

In lieu of the steel plate, a geomembrane sandwich was developed. This approach relies on load distribution through two layers of sand and tensile reinforcement by the membranes. The reduced stiffness of this system results in less concentration of landfill loadings than for a steel plate would induce, while providing adequate protection of the instruments.

Five 305 mm (12 in) TPCs and four thermistors were placed in a square pattern in a 3.4 x 3.4 m (11 x 11 ft) area. The area was then covered with two layers of 60 mil HDPE geomembrane and sand. The instrumentation was aligned so that at least 1305 mm (1-foot) spacing was maintained between instruments with at least 305 mm (1-foot) geomembrane extends beyond the outermost instruments. This configuration is illustrated in Figure 3.

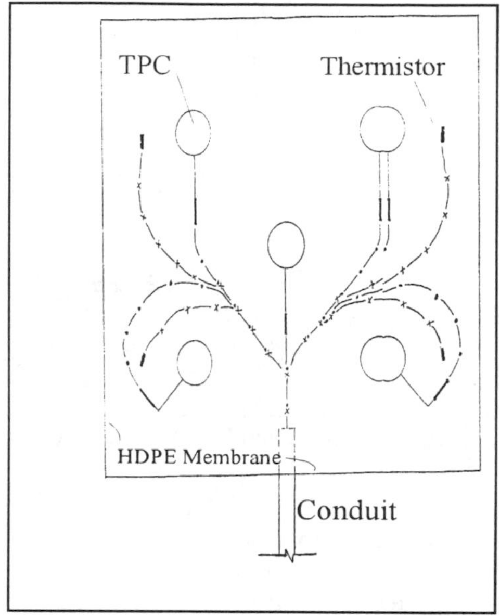

Figure 3 Phase III and IV Typical Total Pressure Cell and Thermistor Layout

A typical cross-section of Phase III and IV TPC installation is included as Figure 4. This illustrates the designed system of alternating layers of compacted sand backfill with two 3.4 x 3.4 m (11 x 11 ft) HDPE sheets. HDPE was the chosen geomembrane type because of its likely resistance to chemical degradation. The geometry of the system was intended to promote even bearing across the 3.4 x 3.4 m (11 x 11 ft) area and to increase protection for the instruments against puncture from objects in the MSW. Areas larger than 3.4 x 3.4 m (11 x 11 ft) were considered and rejected in order to minimize the amount of excavation through existing MSW.

Resistance to chemical degradation was an important factor when ordering the TPCs and thermistors from the manufacturer. The chosen 305 mm (12-inch) diameter cells are normally supplied with stainless steel bodies (RST Model TP-101-P-12). The cells and thermistors were specially ordered with epoxy coating.

A unique protective conduit system was designed to carry the pneumatic tubing and thermistor cable to readout locations. Trenching through the waste to the periphery of the landfill was not considered practical. Instead, tubing was routed vertically to temporary readout stations on the landfill surface. The protective conduit is designed to maximize protection against settlement-induced stresses on the TPC/thermistor tubing system.

The conduit system is shown in Figures 4 and 5. The system was constructed of Schedule 80 PVC for robustness and durability. The system consists of a horizontal PVC

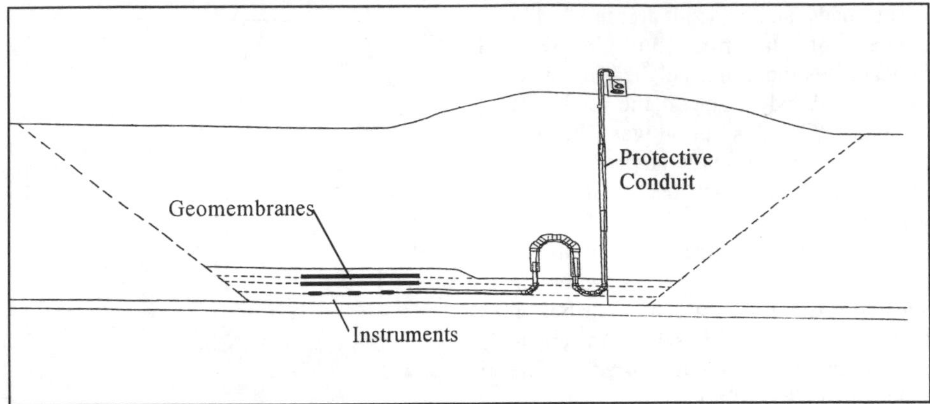

Figure 4 Phase III and IV Total Pressure Cells

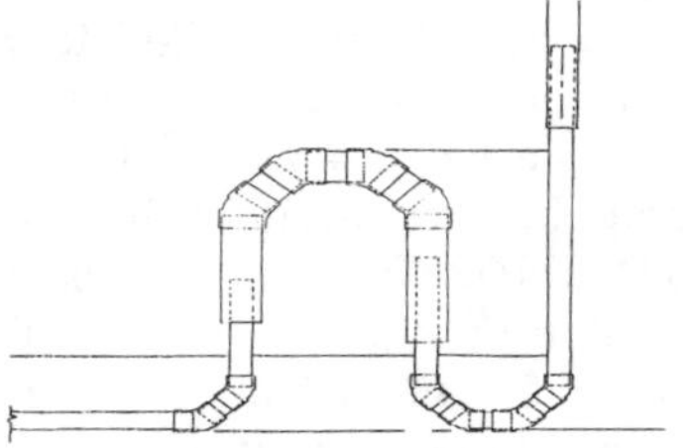

Figure 5 Expansion Assembly

conduit leading from the instrument locations to an Expansion Assembly. The Expansion Assembly isolates the instruments from settlement of the dredge spoils below and compression of MSW above. The Expansion Assembly is similar to a sink trap in configuration, with the center U-shaped section free to slide up or down relative to the instrument side and waste side conduits. The vertical sections consist of telescoping lengths of 127 mm and 152 mm (5-inch and 6-inch) diameter PVC pipe, to the landfill surface. The telescoping design reduces stress on the conduit from MSW settlement. The PVC conduit was filled with a bentonite slurry for extra tubing protection. The entire excavation was backfilled with MSW. The tubing is stored in protective containers which serve as moveable readout stations on the landfill surface. As the MSW height increases additional telescoping sections will be installed.

New Installation - Phase V -- In Phase V, the TPCs and thermistors were installed during cell construction before the first lift of MSW was placed. Again redundancy being an important issue, six sets of two TPCs and two thermistors each were installed. The pneumatic tubing and thermistor cable were routed in HDPE pipe embedded in the leachate collection zone to remote readouts on the perimeter berms.

Two total pressure cells and two thermistors were placed at each interior cluster. The TPCs were embedded in 300 mm (12 in) of sand with 900 mm (1.5 ft) of aggregate placed above the sand. The tubing utilized the same HDPE piping as the pneumatic piezometers to travel to the remote terminal. Twelve total pressure cells (RST Model TP-101-P-12) were installed. The thermistors were also provided by RST. The tubing supplied for the project was a twin 5 mm (3/16") OD HDPE tubing in a polyethylene sheath. Thermistor cables were 22 AWG in a 6.3 mm (0.25") diameter heavy duty, direct burial cable. Six terminal posts were constructed of 100 mm (4") steel box tubing, 3 mm (1/8") wall and zinc plated.

Instrument Installation and Performance

In general, all of the instruments were installed as describe above. Several items of interest were noted. The protective casings around instruments tend to settle; often the instrument cap falls below the top of the riser pipe which prevents closing and locking the cap. The protective casing around the Phase V interior standpipe piezometers consisted of HDPE pipe that tapers so that each additional section of pipe reduces in diameter. The pipe sections were so heavy that wood forms were required prior to landfilling around them.

Five pneumatic piezometers were recently placed in Phase III through existing MSW to depths ranging from 1.5 to 18 m (5 to 60 ft) below the bottom of waste (top of dredge spoil). This installation occurred after two lifts, about 6 m ± (20' ±), of MSW had been placed. This required drilling through existing MSW to reach the dredge layer. Nested installations were used to minimize the number of boreholes required through MSW. The pneumatic tubing for these installations was extended to the landfill surface inside the steel casing backfilled with a bentonite slurry. The installation had to be suspended several times to allow for venting of natural gas in excess of the lower explosive limit. The gas was from decomposition of solid waste and organic matter in the dredge spoil. Continuous monitoring with a combustible gas monitor was required throughout the installation.

Phase V interior pneumatic piezometers are push-in pneumatic piezometers. The tubing was conducted through a flexible rubber hose that turned 90 degrees and inserted into the minimum 150mm (6 in) SDR 15.5 HDPE protective conduit. The conduit was placed in the leachate collection zone one foot above the subgrade. A protective layer rock gravel was placed over the HDPE protective conduit before waste placement. This surface installation required no trenching and was simple to construct, but left ridges of gravel across the landfill floor that became obstacles to construction equipment.

Generally standpipe piezometers, both the porous tip and well screen types, are performing satisfactorily. Sediment accumulates in the well screen piezometers and has to be removed by flushing periodically. Several piezometers in Phase IV have experienced such high pore pressures after waste placement, that they overflowed and required long extensions to reach the hydrostatic level. Some of these require a lift to read and others have been read with a pressure gage attached to the top of the casing. In the winter, the water can freeze in the riser making the piezometers unreadable until a thaw occurs.

Some of the newer standpipe piezometers encountered gas pockets during their installation. These piezometers overflow with effervescent water and cannot be read except with a pressure gage. The gas pressure typically dissipates over time and the water level in the casings stabilize.

Settlement stick-up plates are generally performing well. Any instrument that sticks up through the landfill can be damaged by the waste placement process. Settlement plates were being regularly damaged and required frequent repair until the landfill operator's contract required him to pay for the replacement of instruments damaged by his operations. Prior to waste being placed around the settlement plates, several of the riser pipes snapped at the base and fell over due to wind load.

Some of the inclinometer casings in the early phases have had to be replaced because of excessive deflection. Many of these inclinometers have sustained more than 1 m of horizontal deflection. In general, the larger diameter casings have a longer life and tolerate larger deflections.

The liquid settlement transducers installed in Phase III have failed to provide consistent data. All five LSTs have failed despite repeated purging and de-airing. All of the piezometers in the LSTs became unstable and would not give stabilized readings. The problem appears to be with the instruments rather than the tubing, since the tubing is of the same material and shares a common trench with the pneumatic piezometers which are functioning quite well.

Of the ten total pressure cells installed in Phase IV, one may have been damaged by the installation process and has not shown a stabilized reading. Another one initially indicated a steady decrease in pressure inconsistent with the unchanged loading conditions, but has since recovered with the placement of additional waste. This behavior appears to be the result of load redistribution and demonstrates the need for multiple instruments when measuring pressures under waste with relatively small instruments.

A few of the Phase IV interior standpipe piezometers had to be abandoned due to tilting. They were placed in a triangle to prevent damage by landfill equipment working between the riser pipes. The lack of compaction on the waste between the piezometers resulted in the riser pipes settling inward towards each other to a point where the instruments were not readable.

The thermistors indicate generally rising temperatures under the waste. The temperature measurements appear to be affected by both seasonal changes and biodegradation of the waste.

In Phase V, several standpipe piezometers were located adjacent to the pneumatic piezometers and screened at similar elevations to correlate the two types of instruments. The standpipe piezometer data correlated well with the pneumatic piezometer data. However, potentiometric groundwater gradients consistently differ by approximately 3 m (10 ft) between the standpipes and the pneumatics. This is likely due to the differences in the size of the screened zone between the standpipe and pneumatic piezometers.

In order to preserve the pneumatic tubing in Phases III and V, the first lift of waste material excluded bulky waste with large, heavy, or pointed objects. Extending the interior pneumatic tubing, vibrating wire settlement plates and standpipe riser pipes through the MSW interferes with the normal waste filling operation. Waste tends to settle around interior instruments due to poor compaction creating areas where water ponds around the instrument making monitoring difficult. Compacting around the interior instruments is a delicate task.

Findings

Landfill operations have had a substantial and profound influence over the interpretation of instrumentation data. Early readings from the instruments did not agree with the known waste surcharge on the dredge spoils. Record keeping by the facility manager had to be improved to include information about the area of filling, location and size of stockpiles, tonnage of waste received, tonnages of cover material used, etc. Only with good filling and operations records can variations in the instrumentation data be attributed to real life conditions.

The pore pressure data support the design assumption that strength gain via consolidation due to the applied waste loadings will occur within the dredge spoil stratum. The measured piezometric heads indicate that the excess pore water pressures develop consistent with this consolidation process and dissipate as vertical flow in two directions

(Figure 6). The flow is upward into the landfill's leachate collection system within the upper half of the dredge spoil stratum and downward into the underlying Columbia Formation within the lower half of the dredge spoil stratum thickness. The instrumentation generally indicates that excess pore pressures dissipate more quickly in the lower one quarter of the dredge spoils than in the upper portions.

Pore pressure measurements, while accurate, have proven to be difficult to interpret. Uncertainties regarding the surcharge loading, initial degree of consolidation in the dredge spoils prior to waste placement and determination of the initial hydrostatic water table in the dredge spoils made the estimation of future behavior from pore pressures alone impossible. Pore pressure readings in Phase III after placement of the first lift of waste indicate that pore pressures continued to increase when no additional waste was being placed.

It was essential to measure the load imposed by waste placement in order to gain an understanding of the pore pressure data. The TPCs were installed for this reason. TPCs cannot directly measure the overburden pressure because they are subject to the "inclusion effect". That is, they, being stiffer than the surrounding soil attract loading. Reduction of the TPC data required estimating this effect. This was accomplished by embedding the TPCs in sand of known density and computing the adjustment by taking a reading after the three feet of protective sand cover was in place. Weiler and Kulhawy (1982) recommend such in-soil calibration of load cells. Based on this calibration the measured TPC pressures are 10% greater than the overburden pressure. Temperature corrections are small and calibration curves between TPCs show large scatter. Temperature corrections were based on a statistical mean of many calibrations. The TPC data indicate that the average waste density is 84 pcf. This value represents the total unit weight of the waste layer including cover soil and absorbed moisture at the time of the most recent measurement. This value is increasing with time (Figure 7). A typical thermistor record is included in Figure 8.

Settlements measured from the settlement plates (Figure 9) are consistent with calculated values for the first 3 to 9 m (10 to 30 ft) of waste. The data indicates wider scatter and increasing settlements with additional lifts. This appears to be the result of distortion of the pipe extensions as the waste settles. Though the extensions are protected from downdrag forces by protective casings, it is possible that some downdrag loading is transmitted to the base plate, either by the distorted extensions coming into contact with the protective casing or from the casing being forced down into contact with the base plate.

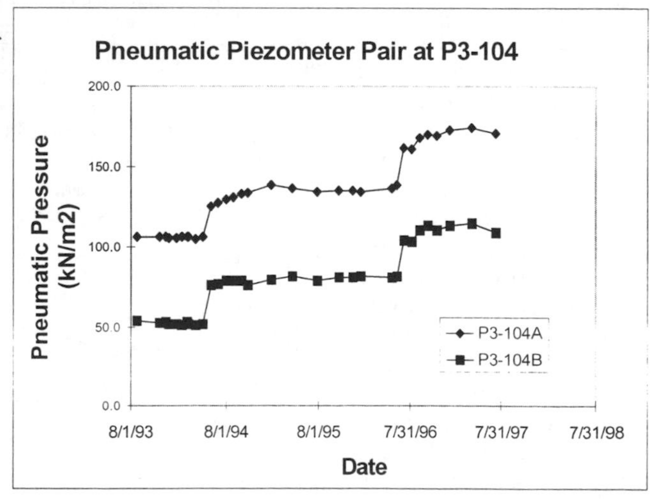

Figure 6

The inclinometers have shown that the perimeter dikes are deflecting outward under the waste loading. The pattern of movement is indicative of general shear with no defined shear plane. The maximum inclinometer deflections have been plotted over time and coincide with expected rates of consolidation for the dredge spoils. This data is consistent with horizontal consolidation within the dikes. Cumulative inclinometer deflections up to 1 m (3 ft) of horizontal movement have been measured without evidence of failure.

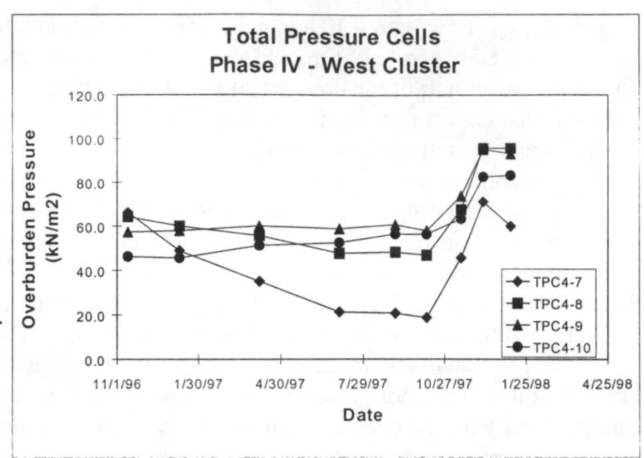

Figure 7

Implications for Design

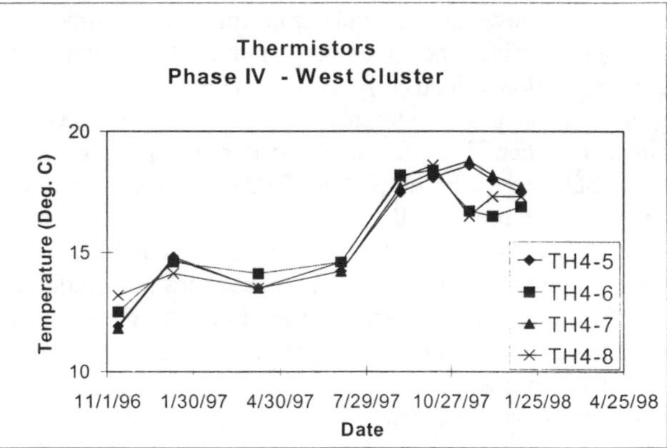

Figure 8

The landfill has been in operation since the opening of the 33-acre Phase I in October 1985. Filling in Phase I was completed in 1987 without incident. The Phase II filling began in 1987. Routine monitoring of the instrumentation in 1990, showed accelerating movement in two inclinometers in Phase II after placement of the first four lifts of waste. Based on these measurements, three additional inclinometers were installed and monitored weekly. The monitoring data from these instruments confirmed that a shear plane was forming in the dredge spoils. A topographic survey of the waste revealed that the

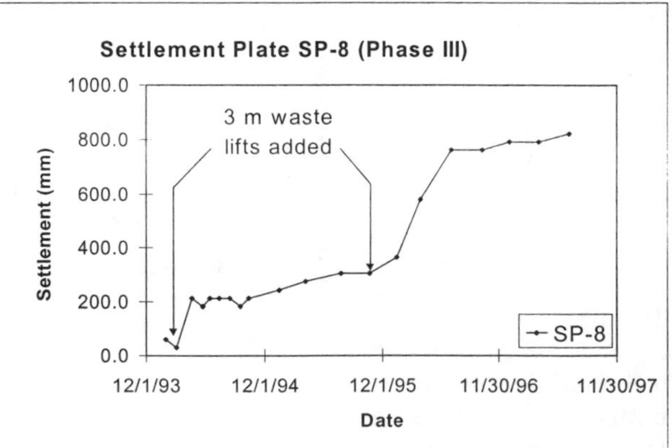

Figure 9

waste had been placed about 3 m (10 ft) higher than the design waste height. The actual slope configuration was analyzed and found to have a safety factor near unity. The slope was quickly stabilized by constructing a toe berm from amended sludge imported from the neighboring wastewater treatment plant. The movement abated and filling was resumed. This failure and analysis confirmed that soil parameters used in the design of Phase II were generally accurate and showed the value of instrumentation monitoring to the safe operation of the facility. Phase II filling topped out without further incident in 1991.

Inclinometer casings in Phases I and II have deflected sufficiently to prevent insertion of the inclinometer. The early installations in these phases used 70 mm (2.75 inch) casings with rigid couplings. The later phases using 85 mm (3.34 inch) casings with telescoping couplings have undergone similar deflections and remained readable. The telescoping couplings can be problematic for taking repeatable readings and care must be used to avoid positioning the inclinometer instrument across a coupling. This problem has been overcome by logging the installation and carefully positioning the inclinometer during readings. All replacements are now being made with the larger diameter casing with telescoping couplings.

Phase III was constructed in 1991 and continues to receive waste. The instrumentation has confirmed that pore pressures caused by the placement of waste are dissipating very slowly. In fact, more slowly than would be expected from the design data. This has resulted in modifying operation plans of the facility to alternate waste filling among the three active phases to permit a longer period of time for pore pressures to dissipate and dredge spoil strengths to increase in the critical slope areas. The second lift of waste to be placed in Phase III was set back 45 meters (150 feet) from the perimeter slope to maintain stability since the excess pore pressure had not dissipated sufficiently under the first lift of waste. Continued monitoring and evaluation indicated that pore pressures had dissipated sufficiently to fill this bench and place a third lift with a setback of 30 m from the slope face.

Fill placement in Phase IV began in 1995. Placement of the first lift of waste was completed in October of 1995. Based on the designer's review, the second lift was placed in Phase IV with a set back of 135 m (450 ft). On-going analysis is required to determine when the setback area can be filled. Phase V received its first lift of waste later in 1996. An evaluation of this phase is also in progress to assess the stability of the second lift to be placed.

Conclusions

We have made the following conclusions based on the instrumentation data:

- Interpretation of pore pressures is closely tied to estimated waste loading.
- Pore pressures were often higher than expected considering the amount of waste placed. This was initially assumed to be a lack of consolidation under the waste, but has been determined to be a result of higher than expected waste density.
- Total pressure cells indicate that the average waste density is 25 to 100 percent higher than estimated in the design.
- The waste load is not a constant for a given lift. The waste density increases as it absorbs moisture from storm water infiltration and as soil cover materials are worked into the waste pore spaces.
- Consolidation is proceeding. Evaluation of consolidation requires periodic

reassessment of the waste loads.

- Liquid settlement transducers have not proven to be durable or reliable. All LSTs have been abandoned.
- Total pressure cells do incur an inclusion effect. This has been adjusted by measurements under known loadings during installation of the instruments in final placement.
- Measured settlements generally agree with expected settlement. The accuracy of settlement measurements is variable because of distortion of the vertical extensions and potential for downdrag on the protective casings.
- Larger diameter inclinometer casings with telescoping couplings remain readable to larger deformations than smaller casings with rigid couplings.

As the site continues to be filled, instruments are regularly monitored to validate the design assumptions. The instruments have proven to be valuable tools in assessing the performance of the subsoils and dredge spoil materials. Periodic instrument monitoring combined with evaluation of slope stability is used to maximize both operational efficiency and to maintain stable slopes. DSWA is planning to install additional instrumentation within the existing cells and Phase V to better define dredge spoil behavior and the loadings from the waste. Filling is expected to continue for the next 15 to 20 years and this instrumentation will play a vital role in the successful operation and performance of the Cherry Island Landfill.

References

Casagrande, A. (1949), "Soil Mechanics in the Design and Construction of the Logan Airport," *Journal of the Boston Society of Civil Engineers*, Vol. 36, No. 2, pp. 192-221.

Dunnicliff, J., (1993), *Geotechnical Instrumentation for Monitoring Field Performance,* John Wiley & Sons, New York, NY.

Felio, G.Y., and G.E. Bauer (1986), "Factors Affecting the Performance of a Pneumatic Earth Pressure Cell," *Geotechnical Testing Journal ASTM*, Vol. 9, No. 2, June, pp. 102-106.

Slope Indicator (1996), *Geotechnical and Structural Instrumentation,* Slope Indicator Company, Bothel, WA.

Weiler, W.A., and F.H. Kulhawy (1982), "Factors Affecting Stress Cell Measurements in Soil," *Journal of the Geotechnical Engineering Division of ASCE*, Vol. 108, No. GT12, Dec. Pp. 1529-1548.

Horace K. Moo-Young[1], Chris LaPlante[2], Thomas F. Zimmie[3], and Juan Quiroz[4]

FIELD MEASUREMENTS OF FROST PENETRATION INTO A LANDFILL COVER THAT USES A PAPER SLUDGE BARRIER

REFERENCE: Moo-Young, H. K., LaPlante, C., Zimmie, T. F., and Quiroz, J. "**Field Measurements of Frost Penetration into a Landfill Cover that uses a Paper Sludge Barrier,**" *Field Instrumentation for Soil and Rock, ASTM STP 1358*, G. N. Durham and W. A. Marr, Eds., American Society for Testing and Materials, West Conshohocken, PA, 1999.

Frost penetration is a major environmental concern in landfill design. Freezing and thawing cycles may deteriorate the permeability of the liner or cap. In this study, the depth of frost penetration into a landfill cover that uses paper sludge as the impermeable barrier (the Hubbardston landfill in Massachusetts) was measured using a frost measurement system. A thermistor probe measured the temperature at various depths. Although temperature measurements are important, soil resistivity measurements are required to accurately predict the freezing level, since soil resistivity increases greatly upon freezing. A conductivity probe measured the half bridge voltage between conductivity rings and a ground rod. Data were collected in data loggers.

The data collected from 1992-1996 showed that the frost level did not penetrate the paper sludge capping layer. Heavy snow cover throughout the winters decreased the depth of frost penetration by insulating the landfill. The high water content in the sludge also contributed to the lack of freezing.

KEYWORDS: instrumentation, freezing, thawing, permeability

---

[1]Assistant Professor, Department of Civil and Environmental Engineering, Lehigh University, Bethlehem, PA 18015.

[2]Assistant Professor, Department of Civil and Environmental Engineering, Union College, Schenectady, NY.

[3]Professor, Department of Civil Engineering, Rensselaer Polytechnic Institute, Troy, NY 12180.

[4] Graduate Research Assistant, Department of Civil Engineering, Rensselaer Polytechnic Institute, Troy, NY 12180.

INTRODUCTION

A major environmental problem facing the world today is waste disposal. Although several methods of solid waste disposal are available, landfilling is by far the most popular method of disposal. Low permeability hydraulic barriers are utilized to reduce the movement of leachate from the landfill. Although regulatory requirements for the permeability of the hydraulic barrier vary, the most common maximum allowable value is 10^{-7} cm/sec, and compacted clay soils are commonly used as the cover material.

When clay is not locally available, the cost of landfilling is significantly increased. This has sparked interest in the use of unconventional material such as paper sludge to substitute as the barrier protection layer. Paper sludges have been successfully used to cap landfills in Maine, Wisconsin and Massachusetts (Moo-Young and Zimmie, 1997; Moo-Young et al., 1997; Zimmie et al., 1997; Moo-Young and Zimmie, 1996a; Moo-Young, 1995; Zimmie et al., 1995; Moo-Young, 1992).

A major concern in landfill design is the effect of freezing and thawing on the hydraulic conductivity of the hydraulic barrier. Numerous researchers have studied the effects of freezing and thawing on the hydraulic conductivity of compacted clays (Othman et al., (1994) summarizes these works). In general, freeze and thaw cycles can cause an increase in the hydraulic conductivity of compacted clays of one to two orders of magnitude. In studying the effects of freeze/thaw cycles on paper sludge, Moo-Young and Zimmie (1996a, 1996b) obtained a similar increase in hydraulic conductivity to compacted clay. However, unlike compacted clays that show a greater increase in hydraulic conductivity at low effective stresses (two to three orders of magnitude), the paper sludge's hydraulic conductivity increased about one order of magnitude, regardless of the effective stress. The better performance of the paper sludge may be due to the fibers in the sludge and the high compressibility of the sludge. Fibers give the sludge some ability to resist tension. There are a wide variety of paper sludges, and the observation on the role of fibers may be unique to the type of sludge tested (Moo-Young and Zimmie, 1996a).

Although there is a significant amount of data that supports the detrimental effect of freezing and thawing cycles on clay and paper sludge low permeability layers, there is little information available on the amount of frost penetration at landfills. In this study, instrumentation to measure frost penetration was installed in the paper sludge low permeability layer at the Hubbardston landfill in Massachusetts and at the Corinth landfill in Corinth, NY and into the clay cap at the Greenwich landfill in Greenwich, NY and at the Corinth landfill. .

Frost Action in Soils

Most of the literature on frost heave action in soils pertains to highways, foundations, construction in permafrost, and chilled pipelines buried in unfrozen soils. . Frost action in soils has received considerable attention in the literature (Jumikis, 1966, Tsytovich, 1975; Nixon, 1991; Anderson and Morgenstern, 1973). Frost heaves are caused by the freezing of in situ pore water and by the flow of water to the freezing front. Pressure develops in the direction of crystal growth that is determined by the direction of cooling. The frost fringe transports water to the active ice lens from the unfrozen soil under a suction gradient. Pore water in fine grained soils does not necessarily freeze at 0° C. In some clays, as much as 50% of the moisture may remain as a liquid at -2° C (Tsytovich, 1975; Anderson and Morgenstern, 1973). The freezing temperature depends on pore size, water content, applied pressure, and solute concentration.

Extensive research has been conducted on the effects of applied pressure on frost heaves. Applied pressure inhibits frost heaves and affects the freezing temperature and permeability of the soil. When the rate of cooling is near zero, the freezing front attracts water as long as the applied load is less than the actual shutoff pressure in which no water flows to the ice lens (Konrad and Morgenstern, 1982). For all practical purposes, the shutoff pressure in fine grained soils is quite high and would never be exceeded in landfill covers, which typically are subjected to low values of effective stress.

Some major factors controlling frost penetration in soils are the soil thermal conductivity (k), volumetric sensible heat (C), and latent heat (L). Thermal conductivity is the ratio of heat flow through a unit area under a unit gradient. Volumetric sensible heat is the change in thermal energy in a unit volume of soil per unit change in temperature. It is derived from the specific heat, which is the change in thermal energy per unit mass per unit temperature change. Latent heat is the change in thermal energy in a unit volume of soil to freeze and thaw the soil moisture at its melting point without change in temperature. Sludges require more energy loss to freeze pore water than typical soils, due to the high water contents. The latent heat of soil moisture controls frosts penetration.

Soil Resistivity

In fine-grained material, the pore water does not necessarily freeze at 0° C (Konrad and Duquennoi, 1993). Liquid water well below 0° C, in a super cooled condition, moves towards the frozen fringe to form ice lenses. Although temperature measurements are very important, an accurate determination of freezing requires additional measurements, one possibility being soil resistivity. Soil resistivity increases greatly upon freezing. The resistance, R, of a material is defined as the voltage, V, divided by the current, I, and is proportional to resistivity, ρ. Soil Resistivity, ρ, is equal to the resistance,R, multiplied by the length, l, and divided by the cross sectional area, A (where $\rho = Rl/A = Vl/AI$). Electrical conductors have very low resistivity, and insulators have very high resistivity. Water has a lower resistivity than ice. Before the pore water in the soil forms the active ice lens, the resistivity of the soil is low. As ice lenses form in the soil, resistivity increases.

INSTRUMENTATION EQUIPMENT

Conductivity Probe

Geonor Inc. built the frost measurement system utilized in this study. A conductivity probe measures the half bridge voltage between the given conductivity ring and a reference electrode, when a 60 ms pulse is fed to the ring from a 2.5 V DC source through a 10,000 ohm series resistor. The probe has a length of 79.6-cm and a diameter of 2.22-cm. The stainless steel reference electrode has a length of 94.5-cm and a diameter of 2.22-cm. Each conductivity probe has eight conductivity rings that are spaced 7.62-cm apart and are located on the outside of a schedule 40 PVC pipe. The sensors on the conductivity probe are assigned a number in a top-down direction. The top of the probe connects to the data acquisition cable, which relays the conductivity reading to the data acquisition system. For thawed paper sludge, a voltage reading of 0.3 to 0.5 V is typical; while for completely frozen material (zero liquid pore water), a voltage of about 2.5 V is common. The formation of ice lenses and subsequent freezing is indicated by a voltage increase (Geonor, 1990). Figure 1 illustrates the frost measurement system.

In a DC excitation situation, galvanic potential is a common problem. Two techniques are used in this probe to minimize this effect: the rings and reference electrode are made of the same grade and lot of stainless steel; and the excitation pulse width is short, minimizing electromigration and other undesirable effects.

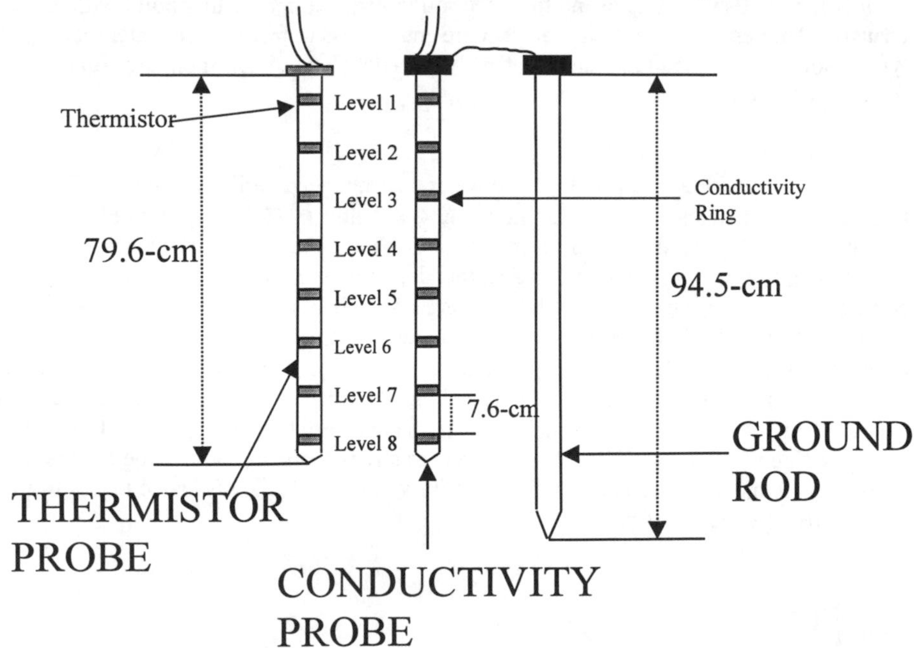

Figure 1 Instrumentation Equipment

Thermistor Probe

Thermistor probes (as shown in Figure 1) measure the temperature at various depths in the soil and consist of eight thermistors spaced 7.62-cm apart. The probe has a length of 79.6-cm. Thermistors are inside a schedule 40 PVC with an outer diameter of 2.22-cm. The thermistor temperature range is between -75° C to 150° C. At the top of the thermistor, a data acquisition cable transmits the temperature readings to a data logger. Correlation between temperature and conductivity readings at the same depth occurs when the thermistors and conductivity rings are installed parallel to each other.

Data Loggers

R.S. Technical Instruments Ltd. (RST) built the data loggers and created the LE8200 Data logger Software. The data logger (station) has three external ports: conductivity, thermistor, and data retrieval. Interface cables from the thermistor and conductivity probes connect to the data logger ports. Microprocessors collect and store the data. A 10,000-ohm series resistor located in the data logger feeds a 60 ms pulse to the conductivity rings.

Each data logger is capable of storing 32,000 bits. Each probe requires 16 bits per reading, hence the data logger can store 1,000 readings overall. The data logger can be programmed to take readings in various modes. Typically, hourly readings are obtained, and thus 42 days of data can be stored.

LANDFILL INSTRUMENTATION

The Hubbardston Landfill, located in Hubbardston, Massachusetts, is a four-acre municipal waste disposal facility that utilized Erving paper mill sludge as the low permeability cover layer. Erving paper mill is conducting studies to measure the performance of their sludge serving as the low permeability barrier in landfill covers.

Two sets of the frost measurement system were installed into the low permeability layer in the Northwest portion of the Hubbardston Landfill in December 1992. The impervious layer is overlain by a 30.5-cm thick top soil layer and by a 15.5-cm thick drainage layer. The drainage layer has a permeability equal to or greater than 1×10^{-3} cm/sec. A frost chisel was used to remove the top 7.6-cm of frozen soil and to outline the site. The remaining soil was excavated, and the site was prepared for installation of the probes. An elevated concrete platform was constructed to place and store the data loggers. The frost measurement systems were placed 3.81-m to the north (Hole #1) and 3.86-m to the west (Hole #2) of the concrete pad.

Two guide holes were drilled into the sludge layer perpendicular to the surface and were spaced 7.62-cm apart on center (the 7.62-cm spacing was suggested by the manufacturer). Electric ground rods were used to puncture the holes; two 1.9-cm diameter water pipes were welded together to provide a guide for the two parallel holes.

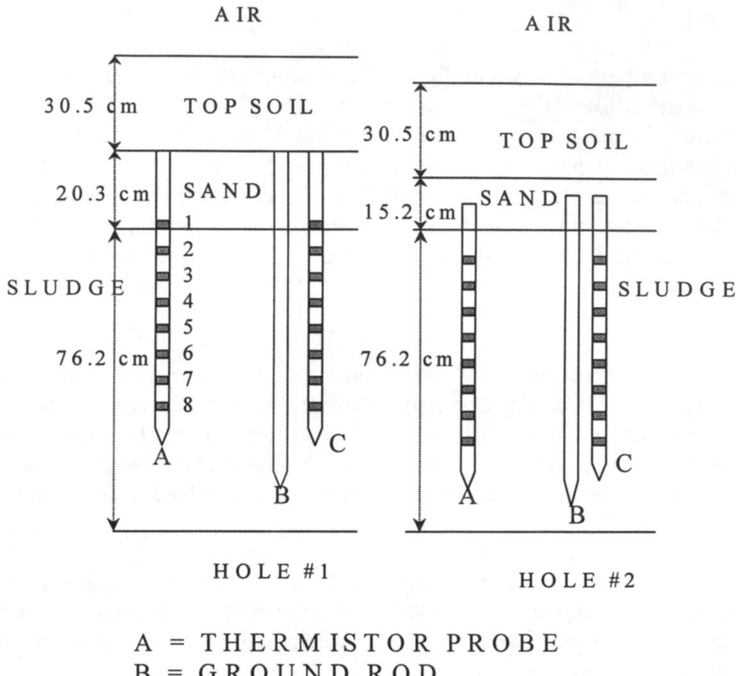

Figure 2 Vertical Profile of the Hubbardston Landfill

The conductivity and ground probes were placed into these holes. The holes were slightly smaller than the probe's diameter to allow a good soil-probe contact. A third hole for the thermistor was punctured perpendicular to the surface.

The probes were pushed by hand into the two holes made in the sludge. Figure 2 shows the vertical profile for the instrumentation equipment into the Hubbardston landfill. In Hole #1, the probes were placed 61-cm into the sludge. The vegetative support layer was 30.5-cm thick, the sand drainage layer was 20.3-cm thick, and the sludge layer was 76.2-cm thick (Figure 2). The top bands of the conductivity and thermistor probes were in the sand layer. In Hole #2, the probes were placed 74-cm into the sludge which allowed all the conductivity bands and thermistors to be contained in the sludge (Figure 2). For Hole #2, the topsoil layer was 30.5-cm thick, the sand drainage layer was 15.2-cm thick, and the sludge layer was 76.2-cm thick.

Two trenches 7.6-cm deep were excavated from the concrete pad to the holes containing the frost measurement system. The thermistor and conductivity data acquisition cables were placed in the trenches and covered with top soil. The data loggers were placed on top of the concrete pad, and the instrumentation cables were connected to the data loggers through a hole in the middle of the concrete pad.

To prevent vandalism and to protect the data loggers from the environment, half of a fifty-five gallon drum was placed over the data loggers and secured to mounts constructed into the concrete pad with a lock. The sand and top soil layers were carefully backfilled into both holes so that the instrumentation equipment was not disturbed (Myers, 1996).

INTERPRETATION OF FROST MEASUREMENT DATA

Although there were two sets of the frost measurement system, Hole #1 will be used to depict the depth of frost penetration into the Hubbardston Landfill, since Hole #2 replicates the data of Hole #1. Figure 3 displays the minimum air temperature at the Hubbardston Landfill from December 1992 to April 1996. The mean air temperature during the four freezing periods, T_s, and the average duration of freezing, t_{avg}, were –5°C and 163 days respectively. The average minimum (T_{omin}) and maximum (T_{omax}) surface temperatures are 2.5°C and 12° C, respectively. In Figure 3, the four periods of freezing (winters) are represented by 0 to 159 days, 348 to 543 days, 746 to 903 days, and 1096 to 1239 days.

Figure 4a displays a three dimensional plot of temperature, time, and voltage for Hole #1 at Level #1, which represents the lower portion of the sand drainage layer (see Figure 2). During the four winters from 1992 to 1996, Figure 4a shows a decrease in temperature below 0°C and an increase in soil voltage in the sand drainage layer. The increase in voltage (i.e. this infers an increase in soil resistivity) indicates that freezing (the formation of ice lenses) occurred in the sand drainage layer during each winter.

Figure 4b displays a three-dimensional plot of temperature, time and voltage for Hole #1 at Level #2, which represents the upper layer of the sludge barrier. Figure 4b shows a decrease in temperature slightly below 0°C during each winter. Since there was no increase in the voltage plane during the four period of freezing, it can be inferred that the sludge layer did not freeze. Levels #3-8 for Hole #1 and Levels #1-8 for Hole #2 follow a similar trend of no freezing to Level #1 for Hole #1 (Moo-Young et al., 1997; Zimmie et al., 1997; Moo-Young, 1995).

ANALYSIS OF FROST PENETRATION DATA

Frost measurement data from the landfill was compared to two conventional thermal analysis tools to determine the applicability of these methods to predicting frost penetration into landfills: Corp of Engineers design charts and Stefan formula (Brown, 1964; Linell et al., 1963; Mitchell, 1993).

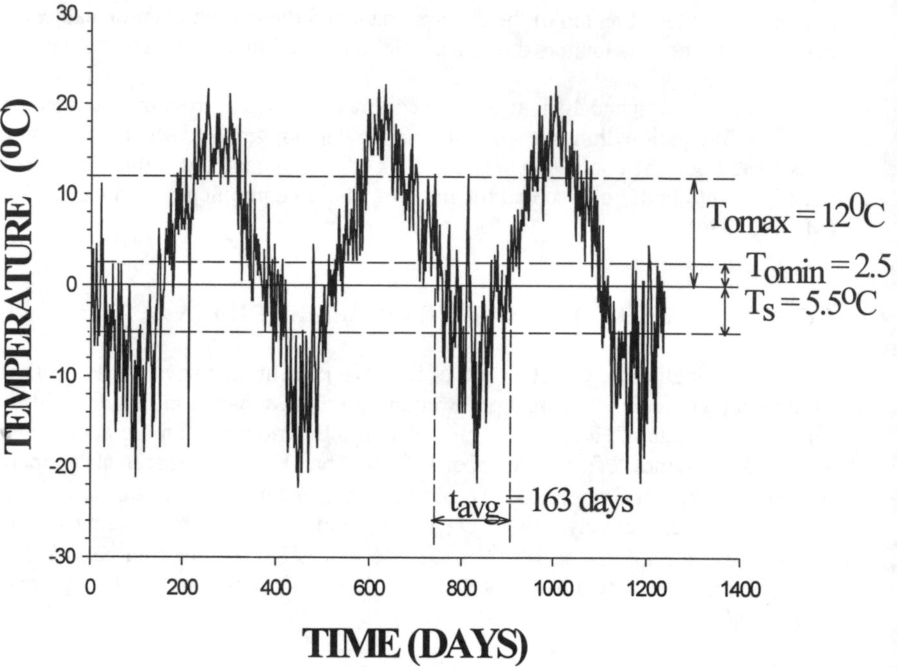

Figure 3 Air Temperature Profile for the Hubbardston Landfill

Estimating Properties

Frost penetration depths in soils depend on numerous factors including weather conditions, ground conditions, and soil conditions. Weather conditions for predicting frost penetration include the freezing index, the mean annual temperature, T_{avg}, and the duration of freezing, t_{avg}. The freezing index, F, is equal to the product of the mean surface temperature during freezing, T_s, and the duration of freezing, t_{avg}. Weather data for the Hubbardston landfill is given in Figures 3.

Frost penetration also depends on the temperature of the ground surface. It should be noted that the ground surface is generally warmer than the air temperature, and the

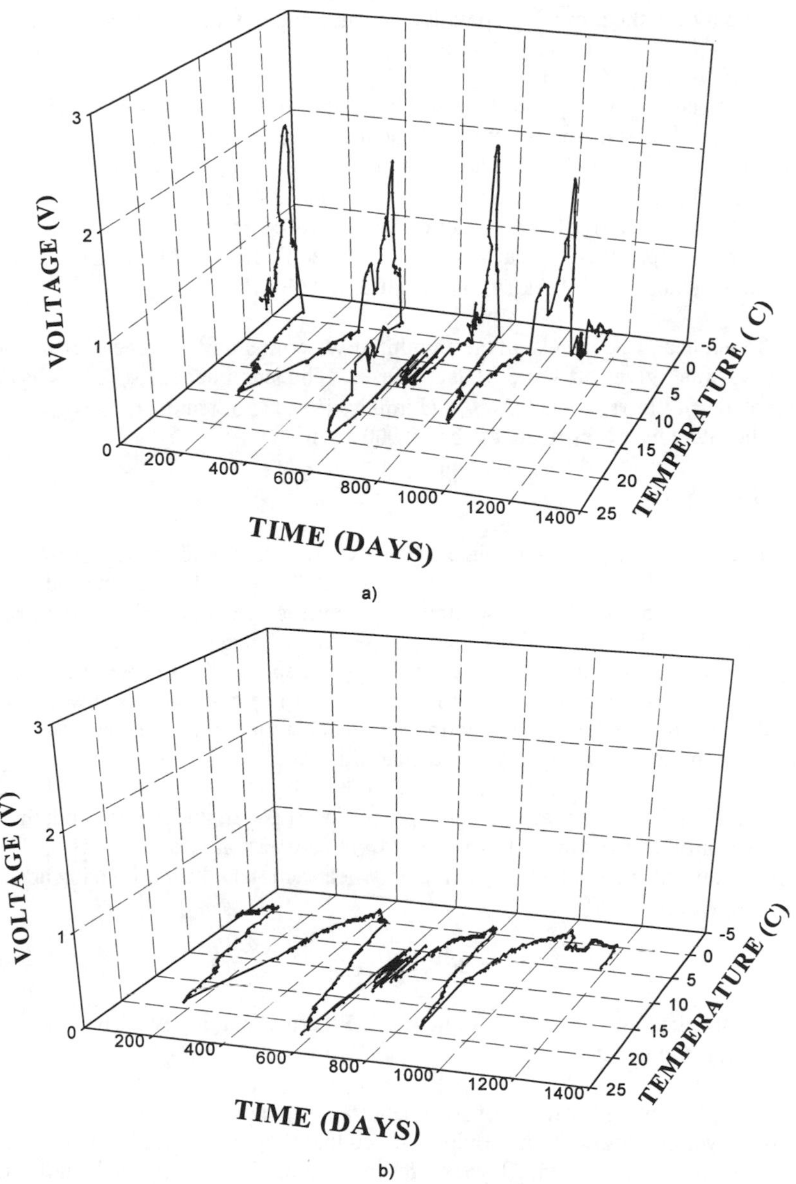

Figure 4a Temperature, Voltage, and Time Profile for Hole #1 at Level #1
4b Temperature, Voltage, and Time Profile for Hole #1 at Level #2

difference between the ground temperature and air temperature is enhanced by snow cover.

Frost penetration depths are controlled by the soils' thermal conductivity, volumetric sensible heat, and latent heat. The thermal conductivity for frozen and unfrozen paper sludge can be estimated from the thermal properties of its constituents. The thermal conductivity for unfrozen paper sludge at 0°C is equal to 0.98 J/m-s°C assuming that paper sludge has an water content of 180% and is composed of 50% clay and 50% organic (paper). Typical values for the specific heat and volumetric sensible heat for paper sludge are 3.1 J/kg°C and 1,630 J/m^3 °C.

Latent heat, L, depends on the amount of water in a unit volume of soil. One kilogram of water gives off 151.3 kJ as it freezes. The latent heat of soil, L (kJ/m^3), can be estimated as follows: $L = 0.80\ w\gamma_{dry}$ (Jumikis, 1966). For paper sludge ($\gamma_{dry} = 530$ kg/m^3), the latent heat is estimated to be 49,900 kJ/m^3.

Thermal Analysis

The Corp of Engineers design chart was utilized to predict frost penetration depths into the landfill covers (Brown, 1964). From Figure 3, the freezing index was predicted for the landfill. Table 1 summarizes the measured depth of frost penetration and compares it to the predicted results from the Corp of Engineers. Table 2 shows that the Corp of Engineers design chart over predicts the depth of frost penetration into the landfill cover. It should be noted that the Corp of Engineers design curve was developed for a well drained granular non-susceptible base coarse material. This method may be applicable to the sand drainage layer in the landfill cover.

Utilizing the thermal and physical properties of paper sludge and sand, the Stefan formula was utilized to estimate the depth of frost penetration into the landfill. This solution assumes that the latent heat is the only heat removed during freezing and is shown in equation 1:

$$Z = (2kF/L)^{1/2} \qquad (1)$$

Where Z is the depth of frost penetration, k is the thermal conductivity, F is the freezing index, and L is the latent heat.

Table 1 displays the depth of frost penetration predicted by the Stefan formula for the barrier layer and for sand. It should be noted that these estimated depth of frost penetration into the paper sludge layer represent freezing into an exposed landfill cover (i.e. no frost protection, vegetation, or drainage layer.). To predict the effects of frost penetration into the drainage and vegetation layers at The Hubbardston Landfill, the Stefan formula was utilized to estimate the depth of frost penetration in sand. For The Hubbardston Landfill, the Stefan formula predicts that frost penetration will be a problem, since the depth of frost penetration predicted for sand in Table 1 is greater than the depths of the vegetation and drainage layers (Figures 2). For the paper sludge barrier

Table 1: Summary of Measured to Predicted Frost Penetration Depths

LANDFILL	YEAR	MEASURED FROST PENETRATION (cm)	STEFAN FORMULA (cm)		CORP OF ENGINEERS (cm)
			Sludge	Sand	
Hole #1	92-96	≤50.8	175	554	140

at the Hubbardston Landfill, the Stefan formula over predicted the depth of frost penetration in comparison to the measured frost penetration. The Stefan formula may have over predicted the depth of freezing, because the removal of the volumetric sensible heats of frozen and unfrozen soil are neglected (Mitchell, 1993).

FACTORS CONTRIBUTING TO THE LACK OF FROST PENETRATION

Four factors may have contributed to the lack of frost penetration in the protected landfill cover at the Hubbardston Landfill: high water content of paper sludge, applied stress created by the drainage layer and vegetation layer, snow cover, and the heat of decomposition in the landfill. The high water content of the paper sludge may have attributed to the lack of freezing. After one year of in situ consolidation, the sludge water content at the Hubbardston Landfill ranged from 100% to 110% (Moo-Young, 1995). For typical sands, the in situ water content is about 10 to 15%, and for clay, the in situ water content typically ranges from 20 to 30%. Moo-Young (1995) showed that the pore water from paper sludge has a freezing point depression of –5°C. Significantly more energy loss is required to freeze the pore water in sludge than to freeze the pore water in a typical sand or clay. The Stefan formula, and Corp of Engineers design chart do not account for the flow of water into or out of the soil during freezing. This may have attributed to the over prediction of frost penetration depth by these methods.

Another factor, which may contribute to the lack of freezing, is the applied stress on the barrier layer. Applied pressure affects the freezing temperature and the permeability of clays (Othman et al., 1994, Konrad and Morgenstern, 1982). For paper sludges, Moo-Young and Zimmie (1996a and 1996b) showed a decrease in hydraulic conductivity with an increase in applied stresses. Since paper sludge consolidates considerably under a low effective stress and since paper sludge is composed of 50% kaolinite, applied stresses may affect the freezing temperature of paper sludge in a similar fashion to clays.

The third factor, which may have reduced the depth of frost penetration, is snow cover at the landfill. Although the Corp of Engineers design chart, and Stefan formula, utilize climatic data, these solutions do not account for the insulation of the soil by snow cover. Figure 5 shows the observed snow cover depth at the Hubbardston Landfill from

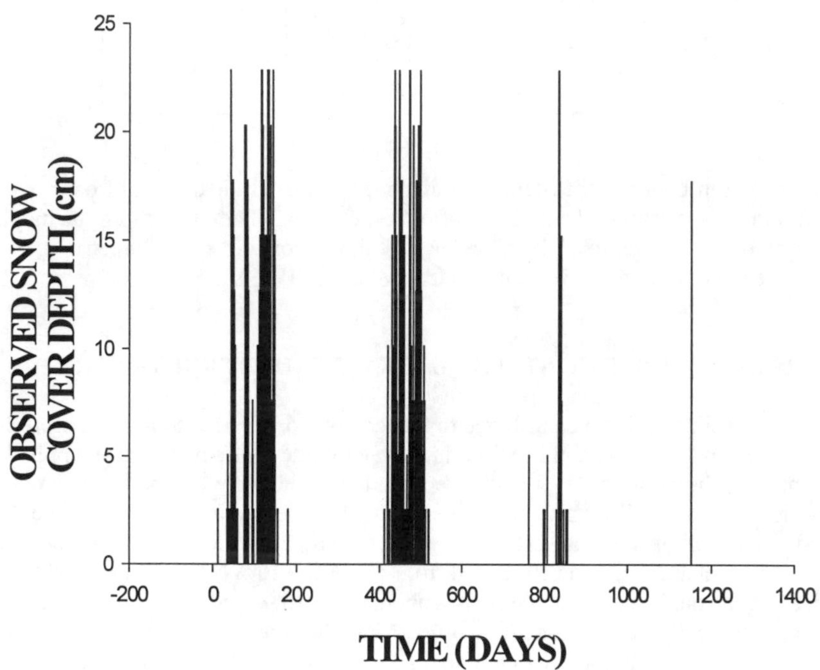

Figure 5 Observed Snow Depth

December 1992 to April 1996. Snow cover throughout the winters of 1992-1993 and 1993-1994 at the Hubbardston Landfill may have insulated the landfill cover. In analyzing the data for Hole #1 at Level #2, although the temperature dropped below 0°C, freezing can not be inferred from the voltage data. During the winters of 1994-1995 and 1995-1996, the observed snow cover depth at the Hubbardston Landfill was less than the previous two winters.

The fourth factor, which can be postulated to reduce the amount of frost penetration into the landfill cover, is the heat of decomposition in the landfill. The heat of decomposition is created by the decomposition of the waste caused by biological activity of the waste mass. The contribution of the heat of decomposition to reducing the frost penetration into the Hubbardston landfill cover was unknown. However, this is an area for future research.

SUMMARY AND CONCLUSION

A frost measurement system was installed into the Hubbardston landfill to measure the depth of frost penetration into clay and paper sludge low permeability barrier layers. A thermistor probe measured the temperature at various depths, and a conductivity probe measured the half bridge voltage between a given conductivity ring and the ground probe.

Data from the landfill were collected from 1992-1996. For the Hubbardston Landfill, there was no frost penetration into the barrier layer. Thus, the paper sludge acted as an effective hydraulic barrier. It should be noted that even though no frost penetrated into barrier layers of the unexposed landfills, future landfill designers should design for maximum frost penetration. Freezing and thawing in the barrier layer has been shown to increase the hydraulic conductivity of paper sludge or clay (Moo-Young and , 1996b and Othman et al., 1994).

The measured frost depths were compared to the two theoretical solutions for predicting the depth of frost penetration: Corp of Engineers design chart and Stefan formula. These solutions over predicted the depth of frost penetration into the unexposed landfill covers.

Three factors may have contributed to the reduction in the depth of frost penetration: high water content, applied stress, and snow cover. Heavy snowfall throughout the winter covered the landfill may have acted as an insulation blanket, which reduced the depth of frost penetration. The high water content of paper sludge may have also contributed to the lack of freezing. Lower temperatures and more energy loss are required to freeze the pore water in paper sludge than to freeze the pore water in a typical sand or clay. This indicates that low temperatures are required to freeze paper sludge. Moreover, the applied stress on the sludge layer that is created by the drainage layer and vegetation layer may have affected the freezing temperature of the paper sludge

ACKNOWLEDGEMENT

The Erving Paper Company, in Erving, MA, the United States Environmental Protection Agency, and the National Science Foundation provided funds for this research. Their generous support and cooperation is greatly appreciated. However, the opinions expressed herein are solely those of the authors. The authors would also like to thank Timothy Myers of the Hydro Group in Bridgewater, New Jersey and Warren Harris and Dr. Carsten Floess of Clough Harbor and Associates in Albany, New York for their assistance in conducting this project.

REFERENCES

Aldrich, H.P., 1956, "Frost Penetration Below Highway Pavements," *Highway Research Board*, Bulletin 135.

Anderson, D.M. and Morgenstern, N.R., 1973, "Physics, Chemistry and Mechanics of Frozen Ground," *Proceedings, 2nd International Conference on Permafrost,* Yakutsk, U.S.S.R., pp. 257-288.

Brown, W.G., 1964, "Difficulties Associated with Predicting Depth of Freeze or Thaw," *Canadian Geotechnical Journal*. Vol. 1, No. 4, pp. 150-160.

Geonor, Inc., 1990, *Frost Measurement System*, Geonor, Inc., Milford.

Jumikis, A.J., 1966, *Thermal Soil Mechanics*, Rutgers University Press, New Brunswick.

Konrad, J.M. and Morgenstern, N.R., 1982, "Effects of Applied Pressure on Freezing Soils," *Canadian Geotechnical Journal*, Vol. 19, pp. 494-505.

Linell, K.A., 1963, "Corp of Engineers Design in Areas of Seasonal Frost," Highway Research Record, No. 33, pp. 77-128.

Mitchell, J.K., 1993, *Fundamentals of Soil Behavior*. John Wiley and Sons, New York.

Moo-Young., *1992, Evaluation of the Geotechnical Properties of a Paper Mill Sludge for Use in Landfill Covers,* Master of Science Thesis, Rensselaer Polytechnic Institute, Troy.

Moo-Young, H.K., 1995, *Evaluation of Paper Mill Sludges for Use as Landfill Covers*, Ph.D. Thesis, Rensselaer Polytechnic Institute, Troy.

Moo-Young, H.K. and Zimmie, T.F., 1996a, "Geotechnical Properties of Paper of Mill Sludges for Use in Landfill Covers," *Journal of Geotechnical and GeoEnvironmental Engineering*. Vol. 122, No. 10, pp. 768-775.

Moo-Young, H.K. and Zimmie, T.F., 1996b, "Effects of Freezing and Thawing on the Hydraulic Conductivity of Paper Mill Sludges Used as Landfill Covers," *Canadian Geotechnical Journal*, Vol. 33, pp. 783-792.

Moo-Young, H.K. and Zimmie, T.F., 1997, "Utilizing a Paper Sludge Barrier Layer in a Municipal Landfill Cover in New York," Testing Soils Mixed with Waste or Recycled Materials, ASTM STP 1275, M.A. Wasemiller and K. Hoddinott, Eds., American Society of Testing and Materials, West Conshohocken, pp. 125-140.

Moo-Young, H.K., Zimmie, T.F., and Myers, T.J., 1997, "Environmental Monitoring of the Depth of Frost Penetration into Landfill Covers," *ISCORD,* H. Zubeck, C. Wollard, D. White, and T.S. Vinson, (Eds.), Fairbanks, AK., American Society of Civil Engineering, Reston, pp. 403-406.

Myers, T.J., 1996, *Landfill Cover Instrumentation to Determine Depth of Frost Penetration*, M.S. Thesis, Rensselaer Polytechnic Institute, Troy.

Nixon. J.F., 1991, "Discrete Ice Lens Theory for Frost Heave Beneath Pipelines," *Canadian Geotechnical Journal.* Vol. 29, pp. 487-497.

Othman, M., Benson, C.H., Chamberlain, E.J., and Zimmie T.F., 1994, "Laboratory Testing to Evaluate Changes in Hydraulic Conductivity of Compacted Clays Caused by Freeze-Thaw: State of the Art," *Hydraulic Conductivity and Waste Contaminant Transport in Soils.* ASTM STP 1142, D.E. Daniel and S.J. Trautwein. American Society for Testing and Materials, West Conshohocken, pp. 227-254.

Tsytovich, N.A., 1975, *The Mechanism of Frozen Ground.* McGraw-Hill. Washington , D.C.

Zimmie, T.F., Moo-Young, H.K. and LaPlante, K., 1995, "The Use of Waste Paper Sludge for Landfill Cover Material," *Green 93--Waste Disposal by Landfill*, R.W. Sarsby, Ed., A.A. Balkema, Rotterdam, pp. 487-495.

Zimmie, T.F., LaPlante, C.M., and Quiroz, J.D., 1997, "Evaluation of Frost Penetration in Landfill Cover Systems," *ISCORD,* H. Zubeck, C. Wollard, D. White, and T.S. Vinson, (Eds.), Fairbanks, AK., American Society of Civil Engineering, Reston, pp. 407-410.

Thomas G. Thomann,[1] Majed A. Khoury,[1] Jack L. Rosenfarb,[1] and Richard A. Napolitano[2]

Stability Monitoring System for the Fresh Kills Landfill in New York City

REFERENCE: Thomann, T.G., Khoury, M.A., Rosenfarb, J.L., and Napolitano, R.A., **"Stability Monitoring System for the Fresh Kills Landfill in New York City,"** *Field Instrumentation for Soil and Rock, ASTM STP 1358,* G.N. Durham and W.A. Marr, Eds., American Society for Testing and Materials, West Conshohocken, PA, 1999.

ABSTRACT: The Fresh Kills Landfill, located in Staten Island, New York, serves as the repository of all municipal solid waste from the five boroughs of New York City. Because of the existence of compressible soils under most of the filling areas and the urban environment surrounding the landfill, considerable importance is being placed on the relationship between filling operations and the stability of the landfill. As a result of this concern and to address Order on Consent requirements, a program of geotechnical site characterizations, stability analyses, and design and implementation of a geotechnical instrumentation program was undertaken. Geotechnical instruments have been installed within the refuse fill and foundation soils to monitor both the magnitude and rate of change of pore pressure, lateral and vertical movements, and temperature. This paper presents an overview of the subsurface conditions, the overall instrumentation plan for assessing the landfill stability, a description of the various instruments, the performance of these instruments to date, an overview of the collected measurements, and a description of how these measurements are used to monitor the stability.

KEYWORDS: geotechnical instrumentation, slope stability, landfill, lateral deformations, settlement, refuse fill

---

Introduction

The Fresh Kills Landfill, located in Staten Island, a borough of the City of New York (see Figure 1), serves as the repository of all municipal solid waste from the five boroughs of New York City. The landfill is operated by the New York City Department of Sanitation (NYCDOS) and has received as much as approximately 115,650 kN (13,000 tons) of municipal refuse per day. Construction of the landfill began in 1948 and now encompasses

[1] Senior Project Engineer, Senior Principal, and Senior Project Engineer, respectively, URS Greiner Woodward Clyde, 363 Seventh Ave., New York, New York.
[2] Senior Project Manager, New York City Department of Sanitation, New York, New York.

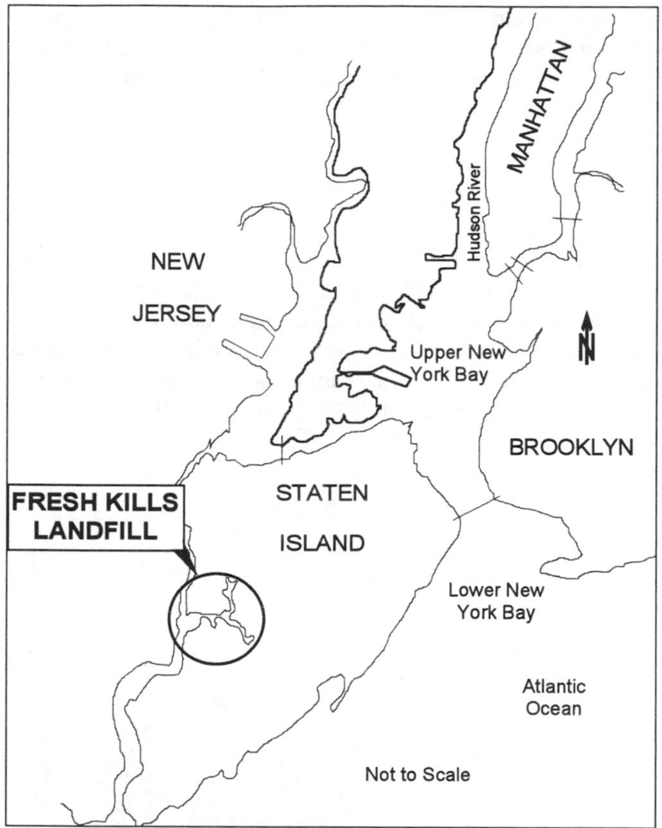

Figure 1 - ***General Site Location Plan***

more than 9.7 sq. km. (2,400 acres) of former marshland. The landfill is comprised of four ections (see Figure 2) identified as the Fresh Kills Bargefill (Sections 1/9 and 6/7), the Victory Boulevard Truckfill (Section 3/4) and the Muldoon Avenue Truckfill (Section 2/8). Sections 2/8 and 3/4 are no longer receiving refuse fill and have undergone final closure. Sections 1/9 and 6/7 are currently receiving refuse fill. Originally, the landfill operations were to continue into the year 2015 to reach a maximum height of about 152 m (500 ft). However, present plans are to close the landfill by the year 2002.

Considerable importance is being placed on the relationship between filling operations and the safety and stability of the landfill because of 1) the planned size of the landfill (Section 1/9 when completed will be about 76 m (250 ft) high based on the most recent plan); 2) the existence of compressible soils under most filling areas, and 3) the presence of an urban environment in the vicinity of the landfill. As a result of these concerns, the NYCDOS has undertaken a program of geotechnical site characterizations, stability and deformation analyses, design and installation of an automated monitoring instrumentation system, and development of an operations and maintenance manual for this system. This program was also designed to fulfill a portion of an agreement between the NYCDOS and the New York State Department of Environmental Conservation (NYSDEC).

Because of the complex stratigraphy at the site and the uncertainties regarding the short-term and long-term behavior of the refuse and foundation soils (in terms of their field behavior versus laboratory behavior), an observational methodology has been adopted (Peck, 1969). This entails installing a monitoring system to obtain measurements of key parameters (e.g, pore pressure and deformation) during landfilling to ascertain the performance of the landfill, and then to modify, if necessary, landfilling operations (e.g., location of filling, rate of filling, schedule of filling) to maintain the satisfactory performance of the landfill. Similar approaches have been adopted for other projects involving earthen embankments and landfills (Handfelt, et. al. 1987, Withiam, et. al. 1995, Oweis, et. al. 1985, Duplanic 1990)

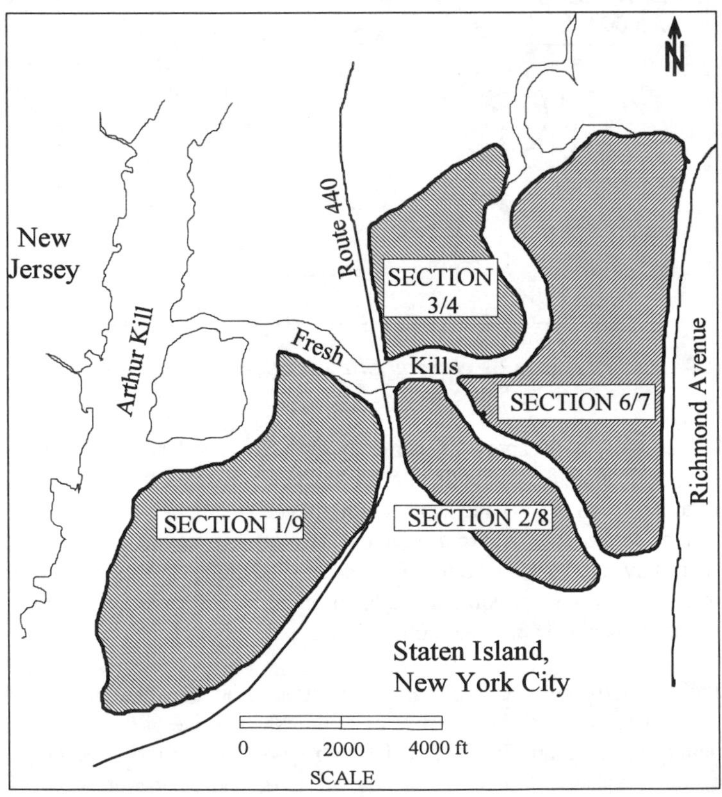

Figure 2 - *Landfill Section Plan*

This paper presents an overview of the subsurface conditions, the overall instrumentation plan for assessing the landfill stability, a description of the various instruments, and the performance of these instruments to date. Summaries of the data, collected to date, are also presented and discussed in the context of their use in conjunction with on-going stability assessment of the landfill.

Site Characterization

Fresh Kills Landfill is located in a very complex lithology due to the past existence of two and, perhaps, three known glacial end-moraines found south of the site. As a result, the soil stratigraphy under the landfill is extremely variable in layer thickness, extent, and material properties. The strata of interest to the evaluation of the stability of the landfill are generally in the following sequence from the ground surface:

- Refuse Fill
- Recent Silt and Clay, Q_{rc}
- Recent Sand, Q_{rs}
- Glacial Sand, Q_{gs}
- Glacio-lacustrine Clay, Q_{gl}
- Glacial Till, Q_{gt}
- Cretaceous sequence of Clays and Sands, K_c and K_s

Not all strata are present throughout the landfill complex. For example, the Q_{rc} deposits are absent from historic upland areas (prior to the construction of the landfill) which are generally located in the eastern portions of Sections 1/9 and 6/7, the southern portion of Section 2/8, and the northeast portion of Section 3/4.

Monitoring System Approach

Considerable importance has been placed on providing adequate instrumentation coverage for the complex and diverse stratigraphy at a reasonable cost. The results from preliminary slope stability analyses indicated that the landfill stability is primarily dependent on the presence, consolidation, and strength behavior of the Q_{rc} and the Q_{gl}. The results from these analyses and geotechnical site characterization studies were then used to select representative cross-sections, or profiles, and locations for installation of instrumentation. The criteria used for selecting representative profiles included the presence and characteristics of the Qrc and Qgl layers and the planned height and side slopes of the landfill.

Along a profile, the instruments are typically located in groups referred to as clusters. In general, the cluster locations were successfully selected so that the instruments would be positioned within a potential failure surface zone. Consideration was also given to locating the clusters near an existing or future bench so as to minimize the amount of lateral trenching across the landfill (to protect instrument signal cables) and to facilitate installation, manual readings, and future maintenance. Typically, three to four instrument clusters are installed within a profile (see Figure 3). One cluster is generally located near the toe, one near the mid-point of the existing slope, and one or two clusters are located inward of the present landfill crest, so that they will be below the edge of the crest of the future landfill stages.

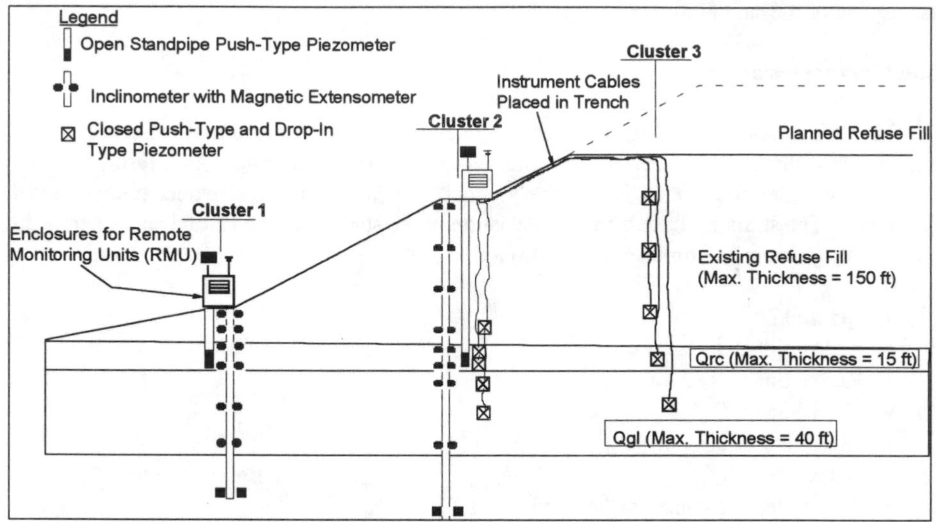

Figure 3 - ***Typical Profile Instrumentation***

In general, the distribution of instrument types within the clusters at each profile is designed to measure the following:

- Pore pressures in the cohesive foundations soils (Q_{rc} and Q_{gl});
- Pore pressures in the sand layer(s) (Q_{rs} and Q_{gs});
- Pore pressures in the refuse fill;
- Vertical deformation (settlement or heave) and horizontal deformation of the refuse fill and the foundation soils; and,
- Temperature within the refuse fill.

The instruments were designed and installed in a phased approach to follow the progress and location of refuse filling. This phased approach also spread the cost over several years and allowed for the refinement of instrument designs and installations based on experience gained during previous phases. A total of three (3) instrument installation phases were performed between 1992 and 1997 resulting in a total of sixteen (16) instrument profiles. The majority (13) of these profiles have been located in Sections 1/9 and 6/7 since these sections continue to receive refuse fill.

Instrument Types

To date, over 260 geotechnical instruments have been installed within the refuse fill and foundation soils to monitor both the magnitude and rate of change of pore pressure, lateral and vertical movements, and temperature. These instruments and their associated components (e.g., electrical cables, inclinometer casing, readout locations, etc.) had to be designed to survive on-going landfilling and closure activities, and environmental hazards at and below the landfill surface. These hazards include temperatures within the refuse fill in excess of 65°C (150°F), temperatures within the foundation soils near the refuse fill in excess of 32°C (90°F), lateral and vertical movements within the refuse fill on the order of tens of centimeters and meters, respectively, and the harsh chemistry of the landfill leachate. An Automated Data Acquisition System (ADAS) was also installed because of the large number of instruments to be installed, difficult access conditions, large areal coverage, and the necessity to make timely changes in landfill operations in case of instability concerns.

Piezometers

The following six types of piezometers were selected for making pore pressure measurements (Dunnicliff 1988):

- Closed push-type vibrating wire (VW) piezometers;
- Dual closed push-type VW piezometers;
- Closed drop-in type VW piezometers;
- Closed drop-in type vibrating strip (VS) piezometers;
- Open standpipe push-type piezometers.

The closed push-type VW piezometer refers to a single pressure sensor installed at the end of a 1.5 m (5 ft) long rod with special couplings and attachments. The dual version of this piezometer contains two VW sensors along a maximum 4.6 m (15 ft) long rod. The spacing of the transducers along the rod is either 1.5 m (5 ft), 2.1 m (7 ft), or 3.0 m (10 ft). The purpose of this version is to measure pore pressure at two different elevations within the same borehole. This arrangement eliminates the need for an additional boring and results in reduced installation costs.

The closed drop-in type VW piezometer and VS piezometer refer to a pressure sensor that is placed in a small sand-filled burlap bag and lowered into the borehole to a predetermined location. These types of piezometers are only installed in the upper portions of the refuse fill.

The body of each sensor used in the VW and VS piezometers is made of stainless steel and contains a thermistor (temperature sensor). The filter stone in front of the pressure-sensitive diaphragm is made from sintered stainless steel having a maximum opening size of 50 microns and is therefore considered a low air entry filter.

The open standpipe push-type piezometer consists of a porous tip attached to a polypropylene riser pipe. The porous tip consists of sintered bronze manganese with a maximum opening size of 50 microns (i.e., low air entry) and is designed to be pushed from within a borehole to the desired location. The riser pipe consists of Schedule 80 polypropylene pipe. To protect this pipe from lateral and vertical deformations within the refuse fill, a Schedule 40 steel pipe is installed around the polypropylene pipe. To allow for automated measurements, a VW pressure sensor is installed within the riser pipe. One of the primary advantages of an open standpipe piezometer is that it allows for the retrieval and replacement, if necessary, of the sensor and/or permits the water level to be read manually.

Inclinometers

Measurement of horizontal deformations are made in inclinometer casings that are installed in boreholes (Dunnicliff 1988). These casings are constructed of self-aligning pultruded fiberglass. Measurement of the angular distortions (i.e., tilt) along the length of the casing is by means of a portable inclinometer probe. The probe is manually lowered and raised in the casing and the readings are taken at 0.61 m (2 ft) intervals. The readings are recorded in a portable computer and then transferred to a desktop computer for further data analysis.

The refuse fill generally undergoes large vertical deformations relative to the foundation soils. To accommodate this, telescoping coupling sections are provided in 1.5 m (5 ft) lengths in the refuse fill in order to minimize the development of "down drag" forces on the inclinometer casing.

Magnetic Extensometers

Vertical deformation measurements within the refuse fill and foundation soils are made using a series of magnetic extensometers installed at pre-determined locations along the outside of each inclinometer casing (Dunnicliff 1988). Two types of magnetic extensometers were used for this project: datum ring magnets and spider magnets. Datum ring magnets are fixed to the bottom of the inclinometer casing and provide a reference point. Each spider magnet is fitted with six leaf springs that are extended by remote means from the ground surface during installation and serve to anchor the magnet into the surrounding soil or refuse fill. The spider magnets are not fixed to the casing; therefore, they move with the soil or refuse fill as it compresses. The location of each spider magnet relative to the datum magnet (and hence the vertical movement of the surrounding refuse fill or soil) is determined by using a magnetic reed switch probe which is manually lowered into the casing.

Temperature Probes

Vibrating wire temperature probes are used to obtain temperature measurements at varying depths within the refuse fill (Dunnicliff 1988). These measurements were useful in designing instruments within the refuse fill. In addition to the probes, thermistors are included within each vibrating wire and vibrating strip piezometer sensor. These provide additional temperature data and are also used in applying temperature correction factors to VW piezometer readings. The vibrating wire probes and the thermistors generally provide similar temperatures.

Automated Data Acquisition System (ADAS)

The ADAS system is used to remotely obtain data from the piezometers and the temperature probes. The ADAS system allows for nearly real-time readings of the instruments, if needed. The ADAS system also reduces the potential for human input errors and the time needed for manual monitoring and data entry.

The ADAS consists of two main components: a set of individual Remote Monitoring Units (RMUs) and a Central Network Monitor (CNM). Typically, for a given profile, one RMU is installed to collect data from the instrument cluster installed at or near the toe, while another RMU collects data from the clusters installed at the edge of the working crest or at an intermediate bench. Each RMU consists of a Measurement and Control Unit (MCU), an antenna for transmitting data and receiving instructions from the CNM, and a solar panel to provide power. The MCU is a microprocessor-controlled data acquisition unit for taking instrument measurements. The CNM consists of a personal computer, a Network Repeater Unit (NRU) and an omni-directional radio antenna. The NRU provides the interface between the computer and the RMUs.

Instrument Performance

The first phase of instrument installations was completed in 1992; therefore, there is over 5 years of instrument performance that can be analyzed to determine the survivability of the instruments within the landfill and the foundation soils. This information was used to develop a basis for estimating future instrument replacement costs.

Piezometers

The open standpipe piezometers have been very durable to date. Although failures of this type of instrument have been rare, many of these piezometers are located at the toe of the slopes where the vertical and horizontal deformations are relatively small. In addition, the piezometers installed within the landfill were generally located in areas where refuse filling operations were located relatively far away from these piezometers.

Out of a total of 139 closed vibrating wire piezometers installed in landfill Sections 1/9 and 6/7, 104 piezometers of these type are currently functioning properly. A statistical analysis was performed to determine an Annual Percent Failure (APF) rate for these piezometers. The APF is defined as the ratio of the number of properly functioning instruments existing at the beginning of a one year period to the number of instruments not functioning at the end of the same one year period. The APF can be calculated for any time interval; one month intervals were chosen for this statistical analysis. The frequency distribution of the APF for the closed vibrating wire piezometers is shown in Figure 4. The relatively high APF rates of 50%/yr and more were encountered after the first phase of instrument installations when the total number of instruments was relatively small. Therefore, a small change in the number of non-functioning instruments would result in a large change to the APF rate. More recent APF rates

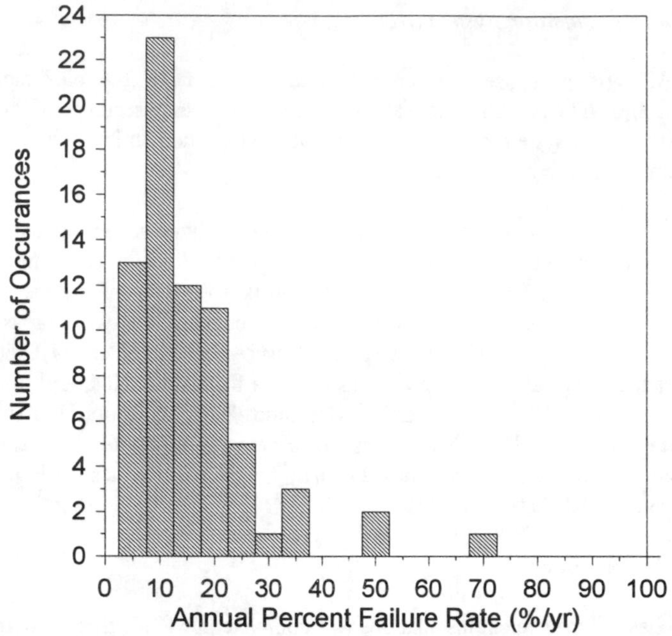

Figure 4 - ***APF Rate for Closed Piezometers***

range from about 10%/yr to 15%/yr. A value of 14%/yr was selected for estimating the number of piezometers that may need to be replaced on a yearly basis.

Inclinometers

In general, inclinometers are installed at the toe of the landfill and within the landfill. Due to the relatively small vertical and horizontal deformations occurring at the toe of the landfill, no failures of the inclinometers at these locations have occurred to date. Failure is defined as not being able to advance the inclinometer probe to the bottom of the casing. A total of eighteen (18) inclinometer casings have been installed within the landfill of Sections 1/9 and 6/7. The number of inclinometers is too small to perform meaningful statistical analyses; therefore, an Annual Number of Failures (ANF) is calculated. In general, the ANF has increased over time because of the greater number of inclinometer casings installed. Based on the more recent years of data and the current number of inclinometer casings, it is estimated that approximately four inclinometers located within the landfill will need to be replaced annually.

Magnetic Extensometers

Since all the vertical deformation measurements are referenced to the datum ring magnets near the bottom of the casing, it is critical that the magnetic reed switch probe be advanced to these magnets located near the bottom of the inclinometer casing. In approximately 20% of the inclinometer casings located within the landfill, it is no longer possible to advance the probe to the datum ring magnets. Therefore, vertical deformation measurements in these casings can no longer be obtained.

Temperature Probes

A relatively few number of temperature probes have been installed within the refuse fill of the four landfill sections. Out of a total of 17 probes installed, only 4 are currently functioning. Approximately 50% of these probes stopped functioning or indicated large temperature variations after approximately two years. As indicated previously, these probes have been primarily used to confirm that the other instruments were properly designed to resist the high temperatures within the refuse fill. They have not been replaced with the same frequency as the other instruments. The reason for the relatively large failure rate of these instruments is not known.

Instrument Measurements

Pore Pressures

The pore pressures within the refuse fill typically increase when additional refuse fill is placed above them and then typically decrease with time. In addition, there appears to be some correlation between the pore pressure and the refuse fill thickness, as shown in Figure 5. The four piezometers within the box of the figure have high pore pressures when compared to pore pressures from other piezometers. At two of these piezometers, placement of approximately

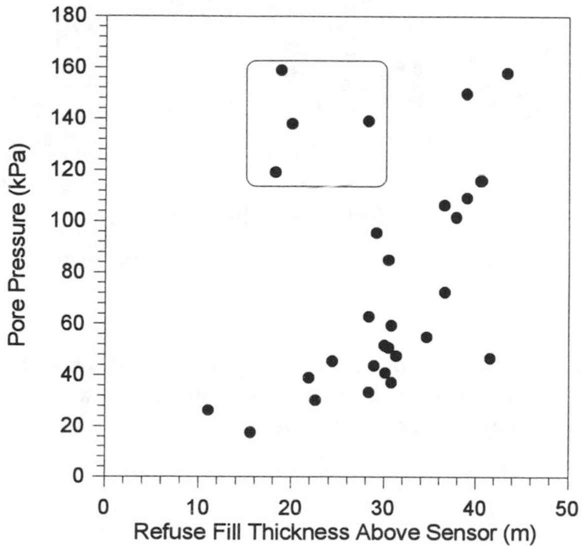

Figure 5 - *Pore Pressure within Refuse Fill*

18.3 m (60 ft) of refuse fill in the vicinity of these piezometers resulted in pore pressure increases on the order of 9.1 m (30 ft) to 12.2 (40 ft), with the greatest increase occurring in the piezometer located closest to the area of refuse fill placement. Approximately 6 months after stopping refuse filling, the pore pressures have dissipated only slightly. The other two piezometers have had relatively high pore pressures since their installation and the pore pressures from these piezometers may be measuring gas pressures since these piezometers may be above the phreatic surface.

The pore pressure within the recent silt and clay (Q_{rc}) stratum of Section 1/9 generally increases with increasing refuse fill thickness, as shown in Figure 6. Also shown on this figure are the pore pressures from those piezometers installed in the strata above (i.e., refuse fill) and below (i.e., recent sand, Q_{rs}) the Q_{rc}. The relatively high pore pressures within the Q_{rc} indicate that this stratum is continuing to consolidate.

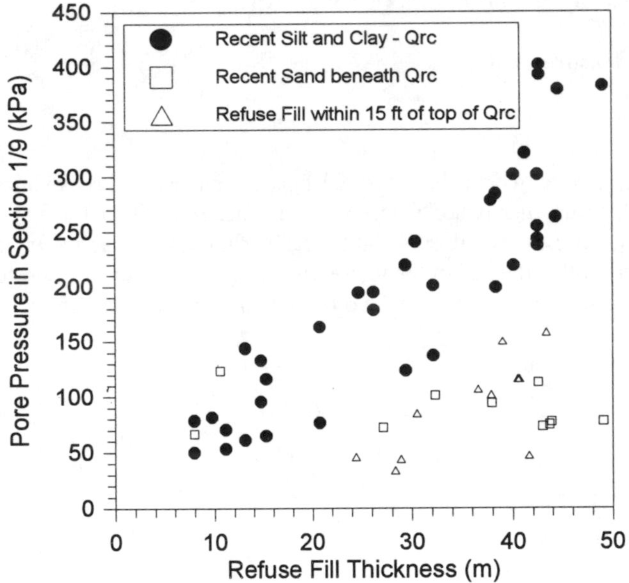

Figure 6 - ***Pore Pressures within Section 1/9***

The pore pressures within the overconsolidated Glaciolacustrine Clay (Q_{gl}) strata have remained relatively low when compared to the Q_{rc} stratum and show very little increase with increasing refuse fill thickness. Similar trends have been observed from piezometers installed in other strata.

Horizontal Deformations

The inclinometers indicate the horizontal deformations that have taken place since their installation. In all cases, the inclinometer casings were installed after the placement of refuse fill. As a result, the measured horizontal deformations only represent a small portion of the total horizontal deformations that have taken place since landfilling began. In addition, since the inclinometer casings were not all installed at the same time, the total horizontal deformations at some inclinometer locations may be greater than others only because they were installed earlier.

The horizontal deformation rate at each inclinometer casing location is shown in Figure 7. The rate is calculated at the location where the greatest horizontal deformations have occurred, which is always within the refuse fill. There are many factors that influence the horizontal deformation rate and not all of these factors are reflected in Figure 7; however, this figure is intended to provide information that may be useful for designing instruments within landfills or other compressible soils. The results indicate that for refuse fill thicknesses less than approximately 18.3 m (60 ft), the horizontal deformations are relatively small. However, for refuse fill thicknesses greater than approximately 30.5 m (100 ft), there is a wide range in rates with some inclinometers having relatively constant deformation rates for almost three years. The higher deformation rates at greater refuse fill thicknesses may be due to greater deformations occurring in the more recently placed refuse fill.

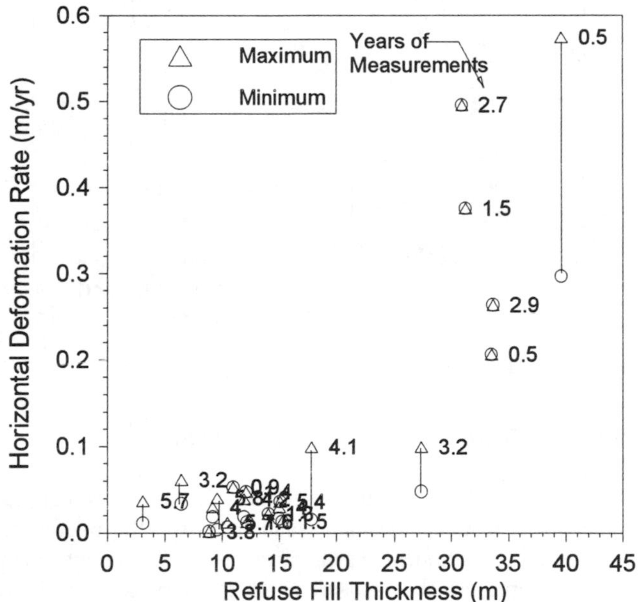

Figure 7 - *Horizontal Deformation Rate*

Vertical Deformations

The vertical deformation rate at each inclinometer location is shown in Figure 8. The rate is calculated using the measurements from the magnet located closest to the landfill surface. Almost all of the vertical deformations occur within the refuse fill. As expected, there is a good correlation between the vertical deformation rate and the refuse fill thickness. As with the horizontal deformation rates, there are many factors that influence the vertical deformation rate and not all of these factors are represented in Figure 8; however, this figure is intended to provide information that may be useful in the design of instruments within landfills or other highly compressible materials.

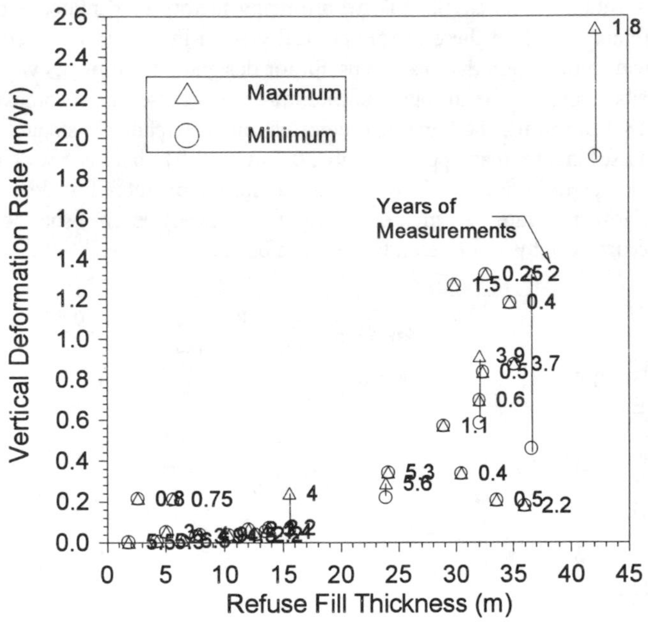

Figure 8 - ***Vertical Deformation Rate***

Use of Instrumentation Measurements for Monitoring Stability

Documents and procedures have been developed for this project to aid in the interpretation of the instrument measurements and to assess the overall stability of the landfill side slopes. One key document that is produced is a monthly monitoring system report. This document presents the measurements in graphical form and is useful in assessing overall spatial and temporal trends in the data. However, due to the changes in the configuration of the landfill side slopes due to active refuse filling and the large volume of data collected, it is difficult to rapidly assess the stability condition of the instrumented profiles. Therefore, a unique computer program has been developed to perform preliminary analyses based on a portion of the key instrument measurements. The input to this program is data from the piezometers and inclinometers. The primary output is an action level for each profile, which is determined based on the factor of safety against slope instability, the maximum shear strain rates within the foundation soils, and the changes in these parameters over time. Associated with each action level are contingency measures that range from maintaining normal operating procedures to stopping refuse fill placement and moving operations to a contingency area. The action levels provide a relatively straightforward and frequent method of informing the client so that grading plans can be developed that take into consideration stability concerns.

Two other key documents include the Slope Stability Monitoring System Operation & Maintenance Manual and the Slope Stability Monitoring System Report. The O&M manual is updated on a yearly basis and includes detailed information concerning the entire monitoring system, information regarding the identification of action levels and contingency measures, the preparation of monitoring system reports, and general guidelines for the evaluation of measurements. The slope stability report is updated on a yearly basis and includes detailed information regarding the performance of the instruments, detailed interpretation of the measurements, and detailed slope stability analyses.

Conclusions

A geotechnical instrumentation system has been designed and implemented to monitor the stability of the Fresh Kills Landfill in New York City. To date, over 260 geotechnical instruments have been installed within the refuse fill and foundation soils for the primary purpose of monitoring both the magnitude and rate of change of pore pressure, and the lateral and vertical deformations. The conclusions regarding the monitoring system are as follows:

- The instruments, which include closed vibrating wire piezometers, open standpipe piezometers, inclinometers, magnetic extensometers, and temperature probes, have been successfully designed and installed to accommodate on-going landfilling and closure activities, high temperatures, large movements, and the landfill leachate.
- Based on the collection of over 5 years of instrument performance data from 139 vibrating wire piezometers, the failure rate ranges from about 10%/yr to 15%/yr.
- Based on the past performance and the current number of inclinometers within the landfill, it is estimated that approximately 4 inclinometer casing will need to be replaced annually.
- The horizontal and vertical deformation rates presented herein may be useful in the design of instruments within landfills or other highly compressible materials.
- Action levels and associated contingency levels have been successfully used to monitor the stability of the landfill side slopes and to provide information to the client for use in the development of grading plans.

Acknowledgments

The design, installation, and monitoring of the instrumentation system for the Fresh Kills Landfill started in 1989 and is still on-going. URS Greiner Woodward Clyde has performed all aspects of this program under contract with the New York City Department of Sanitation. The authors wish to acknowledge the contribution to this work by their colleagues: Melvin Esrig, Min Fan, Drina Ferreira, Aaron Goldberg, Harry Horn, and Niels Jensen. Also, Charles Ladd of the Massachusetts Institute of Technology and John Dunnicliff, Consultant, have provided valuable discussions as members of the Technical Review Board. Finally, the authors want to express their appreciation to Mr. Phillip Gleason, NYCDOS Director of Landfill Engineering, for his support and guidance on this program. His continued interest has been instrumental to the overall success of the project.

References

Duplanic, N., 1990, "Landfill Deformation Monitoring and Stability Analysis," *Geotechnics of Waste Fill, ASTM STP 1070*, Landva and Knowles, Eds., pp. 303-239.

Handfelt, L.D., Koutsoftas, D.C., Foott, R., 1987, "Instrumentation for Test Fill in Hong Kong," *Journal of Geotechnical Engineering*, Vol. 103, No. 2, pp. 127-146.

Oweis, I.S., Mills, W. T., Leung, A., Scarino, J., 1985, "Stability of Sanitary Landfills", *Geotechnical Aspects of Waste Management*, American Society of Civil Engineers Metropolitan New York Section, pp. 1-30.

Peck, R. B., 1969, "Advantages and Limitations of the Observational Method in Applied Soil Mechanics : 9th Rankine Lecture," *Geotechnique*, Vol. 19, No. 2, pp.171-187.

Withiam, J.L., Tarvin, P.A., Bushell, T.D., Snow, R.E., Germann, H.W., 1995, "Prediction and Performance of Municipal Landfill Slope", *Geoenvironment 2000,* ASCE Special Publication No. 46, pp. 1005-1019.

Instrumentation for Monitoring Settlement and Stability

Vincent Silvestri[1]

Field Measurement of Water Contents and Densities by Nuclear Methods in Clay Deposits

---

Reference: Silvestri, V., "**Field Measurement of Water Contents and Densities by Nuclear Methods in Clay Deposits**", *Field Instrumentation for Soil and Rock, ASTM STP 1358*, G. N. Durham and W. A. Marr, Eds., American Society for Testing and Materials, West Conshohocken, PA, 1999.

Abstract : In recent years natural water content and density determinations have been carried out in the stiff clay deposits on Montreal's Island by means of nuclear depth probes. The probes are lowered down aluminum access tubes that are permanently inserted into the ground, thus providing a means of measuring the volumetric water content and the density of the soils surrounding each tube, at any convenient time and depth. Access tubes have an external diameter of 50.9 to 57 mm and a wall thickness of 1.3 to 1.9 mm, and are sealed at the bottom to prevent entry of water. The tubes are pushed into holes of the same diameter which have been reamed out of the soil by a mechanical auger. Tubes have been inserted to a maximum depth of 6 m. Readings in each access hole are taken at 100-200 mm depth intervals along the length of the tube. Values of volumetric water contents and densities are determined by using field calibration curves. The paper gives the results obtained on several projects, reports on the precision of the predicted values and presents typical problems which were encountered during installation of the access tubes and the long-term (up to 8 years) monitoring programs. It is also shown that factory calibration curves furnished with the nuclear probes cannot be used in high-water content clay soils. It is further indicated that even though the water content calibration curves show a high degree of correlation, the neutron scattering technique will not effectively detect changes in soil water content for periods ranging from a few days to about one week. Finally, it is shown that voids around the access tubes adversely affect the nuclear gage readings.

KEYWORDS: water content, density, field, nuclear depth gages, long-term program, precision, clay

Introduction

The use of subsurface gamma and neutron probes for the measurement of soil density and water content, which originated nearly five decades ago (Belcher and Cuyckendall 1950, Gardner and Kirkland 1952), has become common practice in geotechnical engineering. However, despite the progress in understanding the physical phenomena involved and the improvements in equipment design, there is still much apprehension pertaining to the accuracy in the use of the equipment (Tan and Fwa 1991). Reasons commonly invoked for uncertainties associated with gage accuracy are soil

---

[1] Professor, Department of civil, geological and mining engineering, Ecole Polytechnique, P.O. Box 6079, Station Centre-ville, Montreal (Quebec) Canada H3C 3A7.

composition and porosity effects, and adequate methods and procedures for relevant calibration (Tan and Fwa 1991, Ruygrok 1988).

This paper focuses on the use of nuclear density and water content depth gages in the stiff clay deposits on Montreal's Island (QC). It reports on the accuracy of the field-measured values, and presents problems encountered during the installation of the access tubes and the subsequent monitoring programs.

The use of nuclear measurements was found necessary because the clay deposits had to be monitored for several years in order to determine the relationship between soil shrinkage and evapotranspiration. The study was triggered by the severe summer drought of 1983 in the Province of Quebec which affected surficial clay deposits in urban areas and caused substantial structural damage to a large number of buildings founded on these clays. Damage resulted from excessive differential foundation settlement.

Materials and Methods

Nuclear Probes

Two standard Campbell Pacific Nuclear Corp. (CPN) water content/density and water content gages, models 501DR and 503DR, respectively, were used in several stiff clay deposits on Montreal's Island. Each gage was supplied with a 10 m long cable and movable cable stop. The cable was marked at 100 mm intervals. The bottom of the gage contains an oversize hole to allow inserting an adapter ring with a diameter to match the type of access tube being used. With the gage sitting on top of the access tube, the probe is lowered to the depth of measurement and the cable is clamped. Upon retraction of the probe into the shield inside the gage, the probe locks automatically in place for transport. The probes were lowered in thin aluminum access tubes, which had been permanently inserted into the ground. Two sizes of aluminum tubes were employed: a) 50.9 mm outside diameter with 1.33 mm thick wall, and b) 57 mm outside diameter with 1.9 mm thick wall. The tubes were pushed into pre-drilled boreholes of essentially the same diameter. Soil samples were recovered in these boreholes by means of either split spoons or shelby tubes. While split spoon samples were used only for identification and water content determinations, shelby tubes samples served for water content and bulk density determinations.

The maximum length of the access tubes used in these projects was 6.1 m. The upper ends of the access tubes were protected with vandal-proof caps; their lower ends were sealed with flat plates welded to the tubes.

Field Calibrations

The calibration equations for density and water content content, respectively, take the form (Morris and Williams 1990):

$$R_d = F \text{ (water content, density, dry composition)} \qquad (1)$$

and

$$R_m = G \text{ (water content, density, dry composition)} \qquad (2)$$

where R_d is gage response for density, and R_m is gage response for water content.

Each gage parameter, R_d or R_m in Eqs. 1 and 2, represents the count ratio (CR), that is, the ratio of the field count to the equivalent standard count (SC). The SC is the mean value of a series of counts of equal duration, taken with the gage in a fixed position in a medium whose composition and dimensions are stable (Morris and Williams 1990). The use of ratios rather than experimental counts alone minimizes the effects of electronic drift, source and detector decay, and aging.

Following the manufacturer's recommendations the two probes were recalibrated in the field because the soil characteristics were much different from those of the sand used for the original factory calibration of the equipment. Field counts were taken in the access tubes at the exact positions of the soil samples which had been previously recovered in the boreholes.

Density and Water Content Determinations

Nuclear probe readings in the access holes were usually taken at 100 mm intervals along the length of the aluminum tubes. This interval was increased to 200 mm in the deeper soil layers, and also when there was no evidence of any seasonal variation.

For the water content content determination, a 16s count period was used for all measurements. For density determinations, 64s readings were employed. The volumetric water content, $\theta(\%)$, and the bulk density, $\gamma(kN/m^3)$, of the soils were determined using the aforementioned field calibration curves.

Soils and Monitoring Programs

As the vast majority of foundation settlement problems caused by the drought of 1983 occurred in the eastern part of Montreal, field studies were undertaken in that area with the aim of measuring the variation of soil density and water content, especially during the subsequent summer months. Fortunately, in the period of study from 1988 to 1995, Montreal's region experienced moderate droughts in 1988 and 1991, during which significant decreases in volumetric water contents occurred in the surficial clay deposits, leading to excessive foundation settlements (Silvestri et al. 1992, Silvestri and Tabib 1994).

The soils found in the area at study consist of a superficial layer of fill of 0.3 to 1.5 m in thickness, followed by a 2 to 3 m thick layer of stiff brown clay in which the natural gravimetric water content varies between 35 and 45 %. This silty clay is part of the undisturbed sensitive Champlain Sea clay found below, in which the natural gravimetric water content ranges from 60 to 80 % and whose thickness varies between 3 and 15 m on the various sites studied in the program. Below the clay lies a layer of till of variable thickness in contact with fractured bedrock.

Results and Discussion

Gage Precision and Probe Calibration

First, regarding the precision of the gages, several sets of repeated readings of equal duration were taken on different occasions, at fixed positions in the access tubes. For both the 503DR water content and the 501DR density/water content gages, 16s readings showed that for an average volumetric water content $\overline{\theta}$ (%), 95 % confidence limits on $\overline{\theta}$ were computed to be $\overline{\theta}$ ± 0.7 %. In addition, for 64s readings with the 501DR gage, it is expected that for an average bulk volumetric weight $\overline{\gamma}$ (kN/m^3), 95 % confidence limits are $\overline{\gamma}$ (kN/m^3) ± 0.09 kN/m^3. Using the chi-square distribution for the 95 % confidence limits, the two gages were found to work properly with an acceptable level of variation in the devices' electronics.

The precision of the predicted field values of the two gages was determined using the field techniques previously discussed. Results are reported in Figs. 1 to 3. These figures also present the original factory calibration curves. The observed discrepancy between the factory and the field values shown immediately points out to the need of recalibrating the gages whenever in situ soil properties are different from those of the standard materials used by the manufacturer. Note that in Figs. 1 to 3 are also given the straight-line relationships obtains by means of linear regression analyses, and as well as the corresponding coefficients of determination, R^2's, for the field calibrations of the two gages.

Concerning the precision of the predicted field values of the mean volumetric water content, $\overline{\theta}$ (%), the calibration curve of the 503 DR gage shown in Fig. 1 gives, for example, the following 95 % confidence limits: a) $\overline{\theta}$ = 50.29 % ± 1.16 % for CR = 2.0, and b) $\overline{\theta}$ = 67.55 % ± 1.43 % for CR = 2.5. For the data presented in Fig. 2 obtained by means of the 501DR gage, 95 % confidence limits are: a) $\overline{\theta}$ = 52.86 % ± 1.21 % for CR = 0.65, and b)) $\overline{\theta}$ = 64.19 % ± 1.42 % for CR = 0.8. Comparison of these results shows that the precision of the water content contents predicted by both gages is quite similar.

Finally, concerning the precision given by the 501DR gage for the predicted values of the mean bulk volumetric weight, $\overline{\gamma}$ (kN/m^3), the data of Fig. 3 show, for example, the following 95 % confidence limits: a) $\overline{\gamma}$ = 19.00 kN/m^3 ± 0.50 kN/m^3 for CR = 2.6, and b) $\overline{\gamma}$ = 15.96 kN/m^3 ± 0.27 kN/m^3 for CR = 3.1.

Even though the water content calibration curves shown in the preceding figures indicate a high degree of correlation, as shown by the relatively high values of the coefficient of determination, R^2, some remarks are necessary. Consider, again, the calibration curve for gage 503DR reported in Fig. 1. As just mentioned, the 95 % confidence limits on $\overline{\theta}$ were computed to be equal, for example, to 50.29 % ± 1.16 % for CR = 2.0, using just one neutron–probe reading. For a 100 mm thick layer of saturated clay, the quantity of stored water is thus equal to 100 $\overline{\theta}$ or 50.29 mm ± 1.16 mm. The ± 1.16 mm deviation from the mean is ± 2.31 % which at first glance is not very large. However, for the field sites tested, for which the daily evapotranspiration was calculated

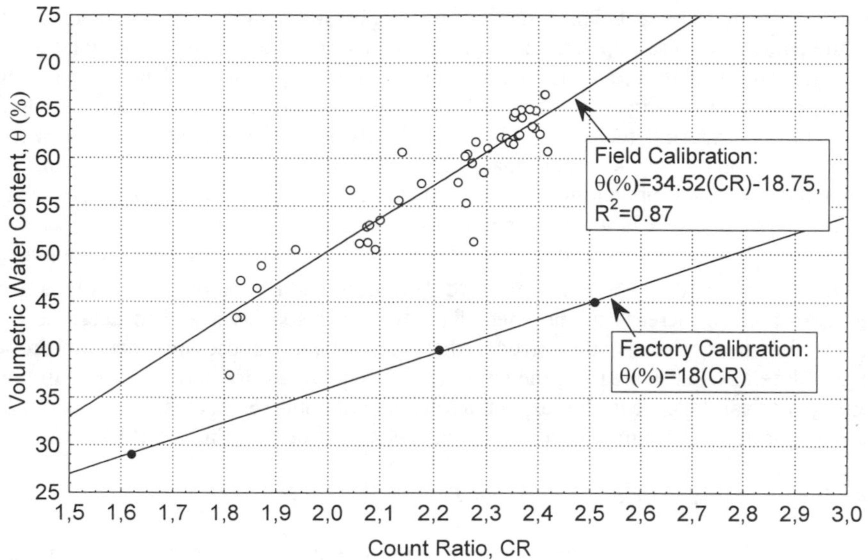

Figure 1 - *Water Content Calibration of 503DR Gage.*

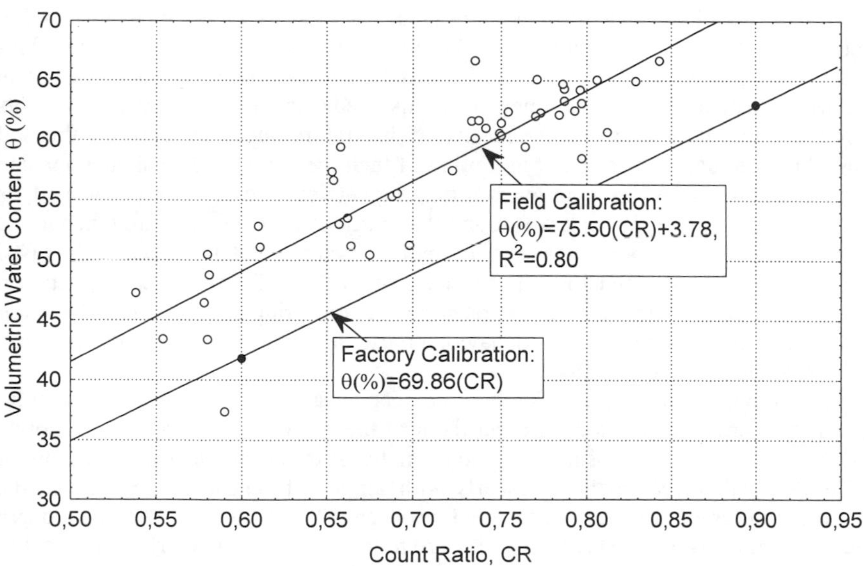

Figure 2 - *Water Content Calibration of 501DR Gage.*

using a simple hydrologic balance model with eight soil layers (Silvestri and Tabib 1994), the cumulative resulting error in millimeters of stored water could be very large, in many cases larger than the daily amount of evapotranspiration. Under operating conditions of one 16s reading per 100 mm depth layer and a total layer thickness of 1 to 2 m, the neutron scattering technique will not effectively detect changes in soil water content for periods ranging from less than a few days to about one week, depending on total root depth and the prevailing evapotranspiration rate. This limitation is primarily due to the expectation of large variances in estimates of soil water over the total soil profile affected by evapotranspiration. In order to decrease the instrument component of the standard error, $\sigma / \sqrt{n}$, where σ = standard deviation, and n = number of readings, the procedure is both to increase the number of readings per soil layer and to decrease the soil layer thickness, that is, the vertical distance between successive neutron-probe readings. Note also that increasing the count time is equivalent to increasing the number of readings per soil layer, e.g., the expected variance of four 16s counts is equivalent to the variance of one 64s count (Haverkamp et al. 1984, Carrijo and Cuenca 1992).

Overboring and Soil Shrinking away from Tubes

In one of the first field studies undertaken in the program, 3 to 4.5 m long access tubes were installed in July 1988, at various distances from a selected number of trees growing in Montreal's Maisonneuve Park. The trees retained were a 22 m high elm, a 20 m high silver maple, a 30 m high cottonwood, and a row of 18.5 m high Lombardy poplars. The investigation aimed at determining the maximum spreading of the drying front caused by water absorption by tree roots.

Due to lack of adequate drilling equipment, some of the access tubes installed at the beginning of the study were inserted in boreholes that were slightly too large. This resulted in two problems. First, due to the fact that the surrounding soil was not in contact with the tube, initial experimental counts predicted rather low water contents compared to the values obtained by means of laboratory measurements on the soil samples which had been retrieved during boring of the access holes. Such a phenomenon is illustrated in Fig. 4 for the depth interval ranging between 1.6 and 3.0 m. Second, as the monitoring program went on, it was found that when count readings were taken in the hours following heavy rain storms, volumetric water content values increased considerably. Such a phenomenon was caused by rain water filling the cavity between the soil and the tube. A similar response occurred in late spring and fall when the water table rose near the surface. Such a behavior is also shown in Fig. 4, for the readings taken, for example, on May 16, 1989.

In addition, because the access tubes had been installed in the summer, an annular cavity began to progressively form around them as the soil continued to dry and slowly moved away from the metallic surface. The diameter of these cavities was maximum at ground surface and decreased almost linearly with depth. At the end of the summer, the maximum thickness of these cavities varied between 20 and 30 mm, and their depth reached approximately 1 m. These cavities were found to form particularly around tubes inserted quite close to the trees, typically at distances of less than 1 H, where H represents the height of the tree.

In order to minimize the formation of such cavities, the access tubes should be installed at the end of the summer, when soil shrinkage caused by evapotranspiration has already attained its maximum and the soil is at its lowest water content content.

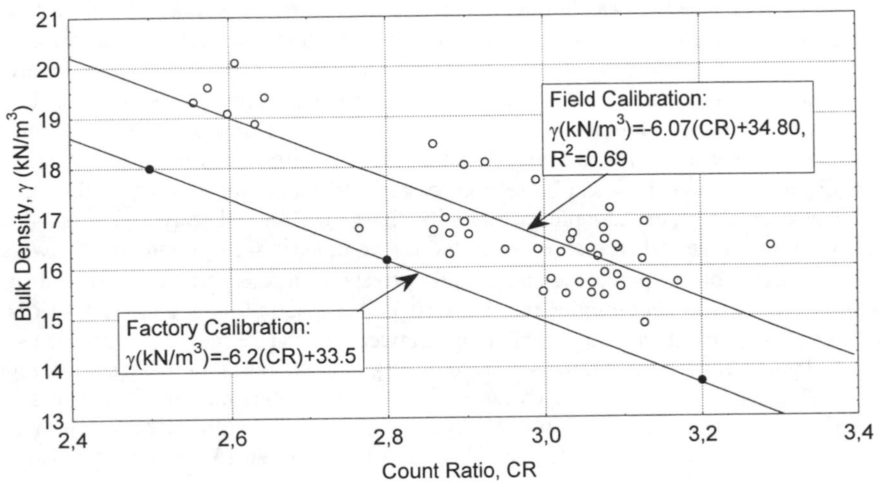

Figure 3 - *Density Calibration of 501DR Gage.*

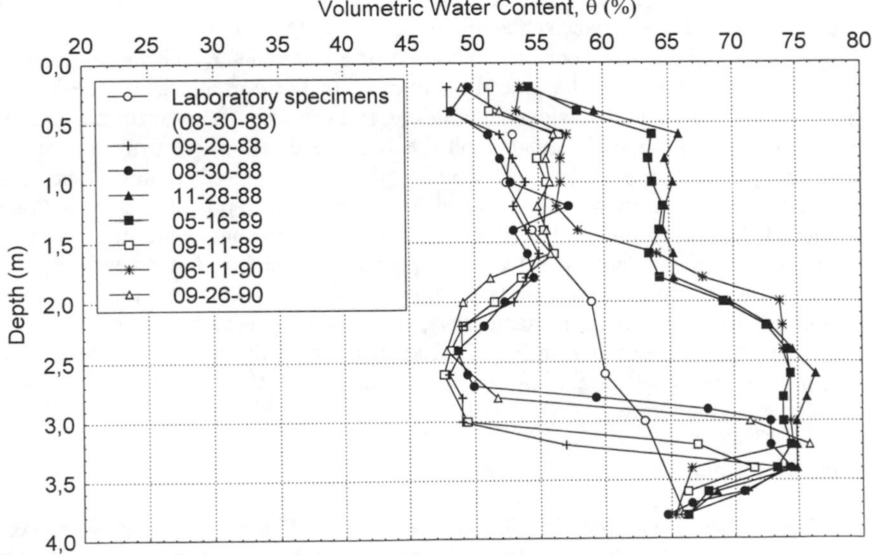

Figure 4 - *Effect of Cavity on Water Contents.*

Water Content Variations

For civil engineering purposes, water content is usually expressed in terms of a gravimetric ratio w (%), that is, the ratio between the mass of water and that of the dry soil. Such an approach is easily implemented in the laboratory as masses can be directly determined. However, when long-term monitoring programs are carried out in the field, it becomes impractical to recover soil specimens and to repetitively determine gravimetric water contents. As a consequence, the volumetric water content θ (%), which is equal to the volume of water divided by the total volume of the soil, is preferred over the gravimetric ratio w, because it can be directly determined using a nuclear gage. In addition, variations in θ can be related to soil settlements for as long as the material remains saturated. For example, in the stiff clays of Montreal, Silvestri et al. (1992) reported that, as the soils remained saturated for volumetric water contents as low as 40 %, the decrease of one percentage point in θ corresponded to a vertical settlement of 1/4 to 1/3 (%) . This finding implies that for a 10 % decrease in θ in a 1 m thick soil layer, the vertical settlement would probably range between 25 and 33 mm.

Typical field water contents obtained using the two gages are presented in Figs. 5 and 6. The data shown in Fig. 5 correspond to water contents measured in access holes drilled on the unwatered lawn around a residential building. One access hole, that is, TA2-1, is located on the front yard, at a distance of 3 m from a 13.7 m high red ash. The other, that is, TA2-3, is situated on the back yard, far away from any tree. The profiles shown for each access holes were obtained when the soil deposit was in its driest and wettest states during the year 1991. Examination of the two profiles for tube TA2-1 indicates that evapotranspiration proceeded to a maximum depth of 4.2 m. As for the soil conditions in the back yard, the two profiles obtained for tube TA2-3 show almost no climatic effect in the soil layers studied. The large variation in the water contents around tube TA2-1 resulted in a surface settlement of approximately 85 mm.

Figure 6 reports water contents measured on a watered site adjacent to the one for which the data is shown in the preceding figure. In the present case, tube TA1-1 is located on the front yard of a residential building, at a distance of 3.4 m from a 13 m high red ash. As for TA1-3, it is situated on the back yard, far away from any tree. A comparison between the driest and wettest profiles measured around tube TA1-1 indicates that evapotranspiration proceeded to a maximum depth of 2.2 m, in spite of continuous irrigation by means of a trickle system. As for the back yard, the two profiles show that the maximum depth affected by evapotranspiration does not exceed 1 m.

In order to compare the results obtained by the two gages, readings were taken with both of them on almost the same days, in the same access holes and at the same depths. The results are shown in Fig. 7. Examination of the data indicates an almost perfect correlation. The straight line relationship and the coefficient of determination R^2 are also given in the figure.

Volumetric Weights

Data obtained with the 501DR gage are shown in Fig. 8. Readings were taken in access hole TA2-1 of Fig. 5. Comparison of the data given in these two figures indicates

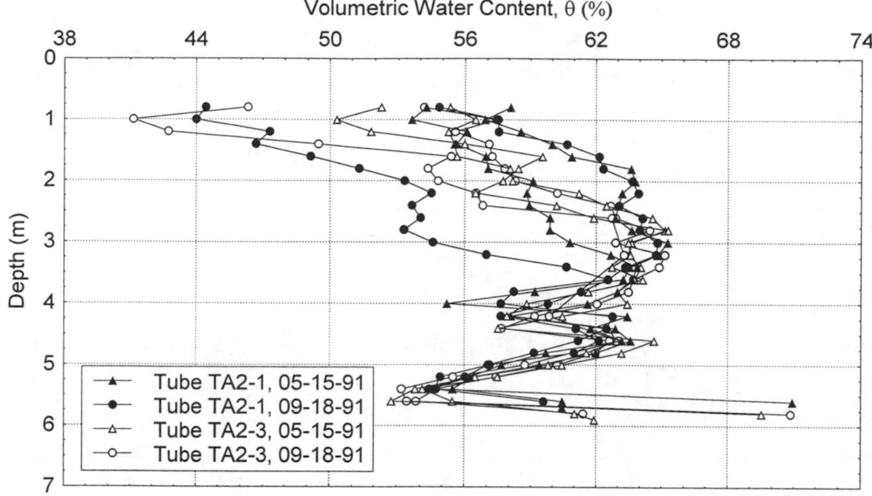

Figure 5 - *Water Content Profiles around Tubes TA2-1 and TA2-3.*

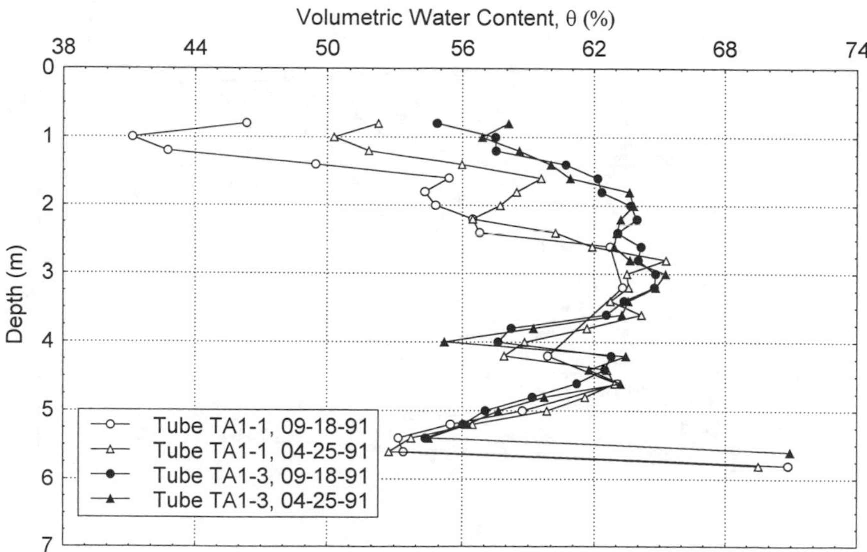

Figure 6 - *Water Content Profiles around Tubes TA1-1 and TA1-3.*

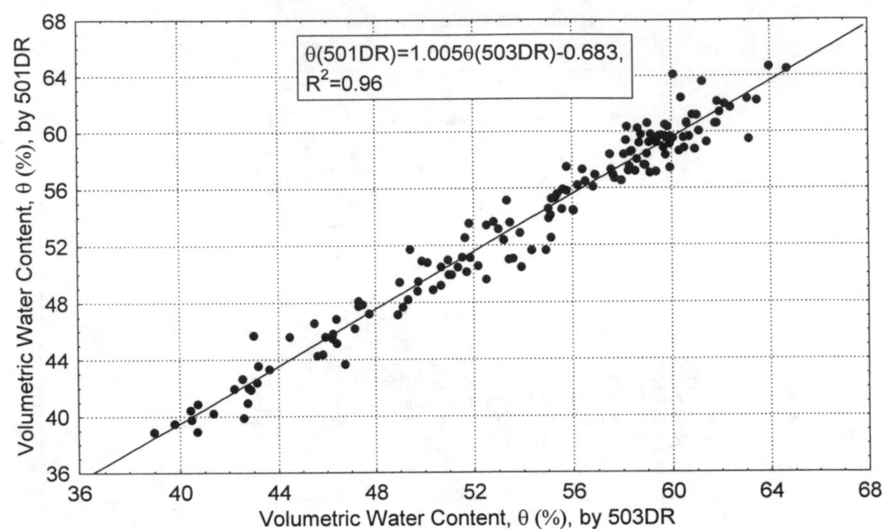

Figure 7 - *Comparison between Water Contents Given by Gages 501DR and 503DR.*

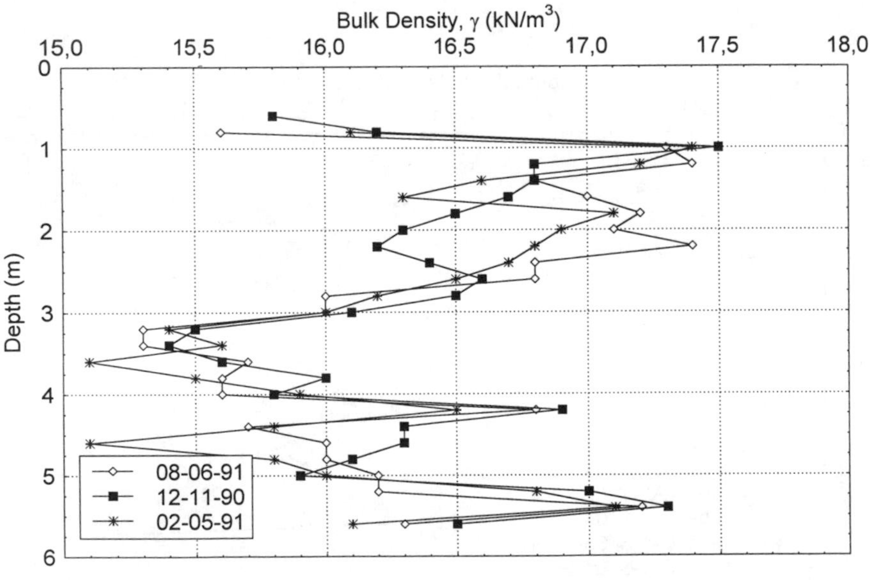

Figure 8 - *Volumetric Weight Profiles around Tube TA2-1.*

that, as expected, whenever the volumetric water contents increase, the bulk volumetric weights decrease, and vice versa.

In addition, to appreciate the effect of the existence of annular cavities around access tubes, measurements of volumetric weights were carried out in the tube for which water contents are reported in Fig. 4. The data are shown in Fig. 9. Analysis of the two relationships given in this figure indicates that in the region affected by the cavity, that is, in the depth interval between 2 and 3.4 m, the bulk unit weight is unrealistic low, i.e., around 14.4 kN/m^3, for an unrealistic high average volumetric water content of 76 % when the probable value of θ was determined to be in the range from 55 to 65 %.

Finally, using both the field density and water content values determined by means of the 501DR gage, Fig. 10 presents the bulk volumetric weight as a function of the corresponding value of θ. In order to better appreciate the results, theoretical relationships given by $\gamma = \gamma_w [\rho_s(1-\theta/100)/\rho_w + \theta/100]$ are also indicated for solid densities ρ_s of 2.6, 2.7 and 2.8 g/cm^3, assuming the soil to be saturated, with γ_w = 9.81 kN/m^3 and ρ_w = 1g/cm^3. Please note that the solid density ρ_s was also determined in the laboratory on twenty nine soil specimens and was found to have a mean value of 2.722 g/cm^3 and a standard deviation of 0.029 g/cm^3. The field data shown in Fig. 10 indicate that most of the values correspond to an average solid density of about 2.7 g/cm^3.

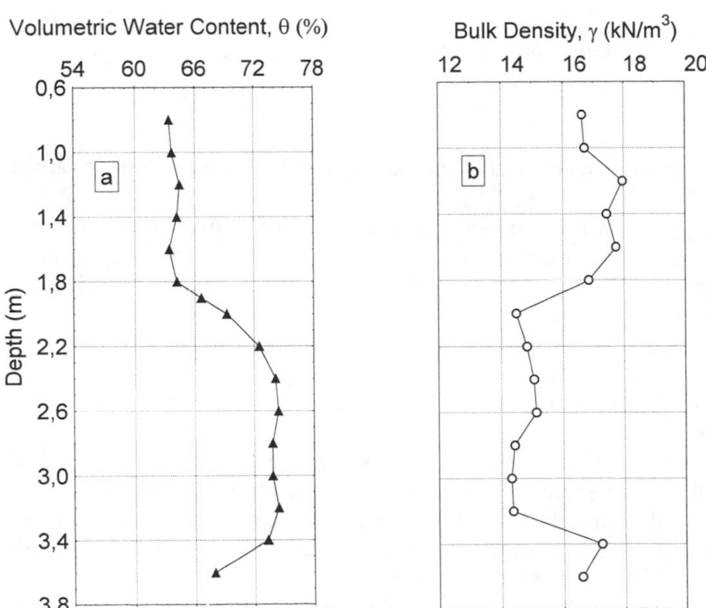

Figure 9 - *Effect of Cavity on Volumetric Weight and Water Content Profiles.*

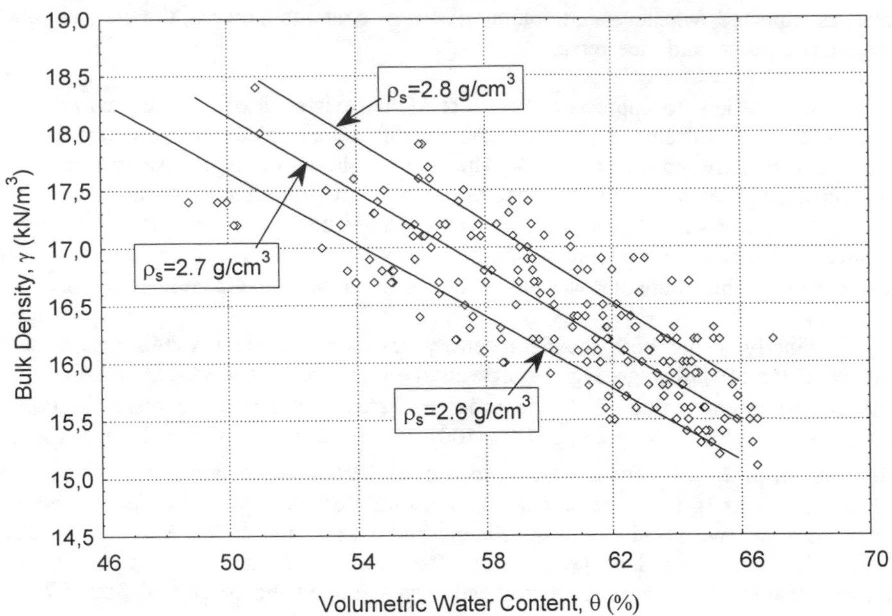

Figure 10 - *Field Volumetric Weight – Water Content Relationship Given by Gage 501DR.*

Conclusion

On the basis of the contents of this paper, the following conclusions are drawn:

i) Nuclear gages can be to effectively useful for long-term monitoring programs of water contents and densities of clay soils subject to evapotranspiration. The two nuclear gages used in this investigation were found to perform adequately. The precisions of the two gages were quite similar.

ii) Nuclear gages cannot effectively measure water content variations over short periods of time in clay deposits. It is estimated that minimum time intervals required to detect meaningful changes of water contents in clays are of the order of one week.

iii) The installation method of access tubes is of prime importance. Methods that result in a void space between the access tube's outside diameter and the wall of the borehole will adversely impact readings made with the nuclear gage. Possible formation of cavities around access tubes may be minimized by installing them when the soil is at its driest point.

iv) Water content and density profiles obtained by means of nuclear gages give a detailed picture of the soil state as function of time.

Acknowledgment

The author expresses his gratitude to NSERC of Canada for the financial support received in the course of this study.

Rerefences

Belcher, D.J., and Cuyckendall, T.R., 1950, "The Measurement of Soil Moisture and Density by Neutron and Gamma Ray Scattering", Civil Aeronautics Administration (USA), Technical Development and Evaluation Center, Washington, Technical Development Report 127, 1950.

Carrijo, O.A., and Cuenca, R.H., 1992, "Precision of Evapotranspiration Estimates Using Neutron Probe", *Journal of Irrigation and Drainage Engineering*, ASCE, Vol. 118, No. 2, pp. 943-953.

Tan, S.-W., and Fwa, T.-F., 1991, "Influence of Voids on Density Measurements of Granular Materials Using Gamma Radiation Techniques", *Geotechnical Testing Journal*, Vol. 14, No. 3, pp. 257-265.

Haverkamp, R., Vauclin, M., and Vachaud, G., 1984, "Error Analysis in Estimations of Soil Water Content from Neutron Probe Measurements: 1. Local Standpoint", *Soil Science*, Vol. 13, No. 2, pp. 78-90.

Morris, P.H., and Williams, D.J., 1990, "Generalized Calibration of a Nuclear Moisture/Density Depth Gauge", *Geotechnical Testing Journal*, Vol. 13, No. 1, pp. 24-35.

Ruygrok, P.A., 1988, "Evaluation of the Gamma and Neutron Radiation Scattering and Transmission Methods for Soil Density and Moisture Determination", *Geotechical Testing Journal*, Vol. 11, No. 1, pp. 3-19.

Silvestri, V., Soulié, M., Lafleur, J., Sarkis, G., and Bekkouche, N., 1992, "Foundation Problems in Champlain Clays During Droughts. II. Case Histories", *Canadian Geotechnical Journal*, Vol. 29, No. 2, pp. 169-187.

Silvestri, V., and Tabib, C., 1994, "Settlement of Building Foundations on Clay Soils Caused by Evapotranspiration", *Proceedings, ASCE Conference on Vertical and Horizontal Deformations of Foundations and Embankments (Settlement '94)*, College Station, Texas, Vol. 2, pp. 1494-1504.

James B. Russell,[1] Christopher A. Lawrence,[2] Michael W. Paddock,[3] and Douglas Ross[4]

Monitoring Preload Performance

Reference: Russell, J. B., Lawrence, C. A., Paddock, M. W., Ross, D., "**Monitoring Preload Performance**," *Field Instrumentation for Soil and Rock, ASTM STP 1358*, G. N. Durham and W. A. Marr, Eds., American Society for Testing and Materials, West Conshohocken, PA, 1999.

Abstract: Reconstruction of State Highway 29 (STH 29) in north central Wisconsin was complicated by two separate deposits of compressible peat and organic silt within the highway right-of-way. The deposits ranged from about 3 to 11 meters in thickness over about 900 meters of the reconstruction alignment. Excavation and removal of the peat and organic silt was evaluated, but was determined to be cost prohibitive. Preloading was determined to be a cost effective and environmentally friendly solution, and was implemented at both deposits to induce the expected consolidation and long-term secondary compression of the final highway embankment and pavement sections. This paper presents the design, instrumentation, and rheological modeling procedures used to complete the project.

Keywords: marsh, peat, organic silt, highway, settlement, vertical drains, preload monitoring, rheological modeling

---

In 1997, the Wisconsin Department of Transportation completed the reconstruction of State Highway 29 (STH 29) between the cities of Hatley and Wittenberg, Wisconsin. The reconstruction was part of the overall reconstruction of STH 29 between the cities of Green Bay and Wausau, Wisconsin, and was undertaken to convert the existing and heavily traveled 2-lane highway (1-lane each way) to a 4-lane highway (2-lanes each way separated by a median). Within Marathon County, the 2 existing lanes became the new westbound lanes, and a new highway embankment was constructed to carry 2 new eastbound lanes.

[1]Managing Engineer, Pare Engineering Corporation, 8 Blackstone Valley Place, Lincoln, RI 02865

[2]Senior Project Manager, CH2M HILL, 1700 Market Street, Suite 1600, Philadelphia, PA 19103-3916

[3]Senior Project Manager, CH2M HILL, 411 East Wisconsin Avenue, Suite 1600, Milwaukee, WI 53211

[4]Project Manager, Wisconsin Department of Transportation, 2811 8th Street, Wisconsin Rapids, WI 54495

Two separate marsh deposits of peat and organic silt adjacent to the existing highway complicated construction of the new highway embankment. At the first deposit (Marsh A), there was about 3 to 4.5 meters of peat underlain by up to 4.5 meters of organic silt. At the second deposit (Marsh B), there was about 3 to 6 meters of peat underlain by up to 6 meters of organic silt. Typical sections at Marsh A and B are presented in Figures 1 and 2, respectively.

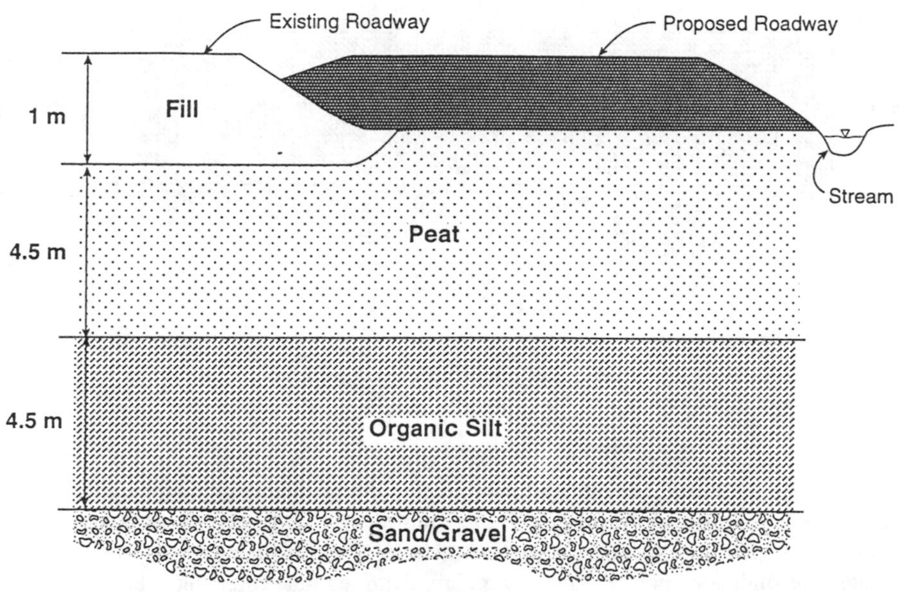

FIG. 1-Typical Section Marsh A

Excavation and removal is the Wisconsin Department of Transportation's preferred alternative when constructing new highway embankments through marsh. However, at Marsh A, there was a Class A trout stream immediately south of the proposed highway embankment. Excavation and removal of the peat and organic silt in this area would have required a braced sheet pile wall to allow the peat and organic silt to be removed without damaging the trout stream. The braced sheet pile wall and depth of the marsh (up to 9 meters) made excavation and removal cost prohibitive in this area. At Marsh B, the depth of the marsh (up to 12 meters) and height of the existing embankment adjacent to the marsh (Figure 2) also discouraged an excavation and removal alternative since the existing roadway would need to be braced during removal.

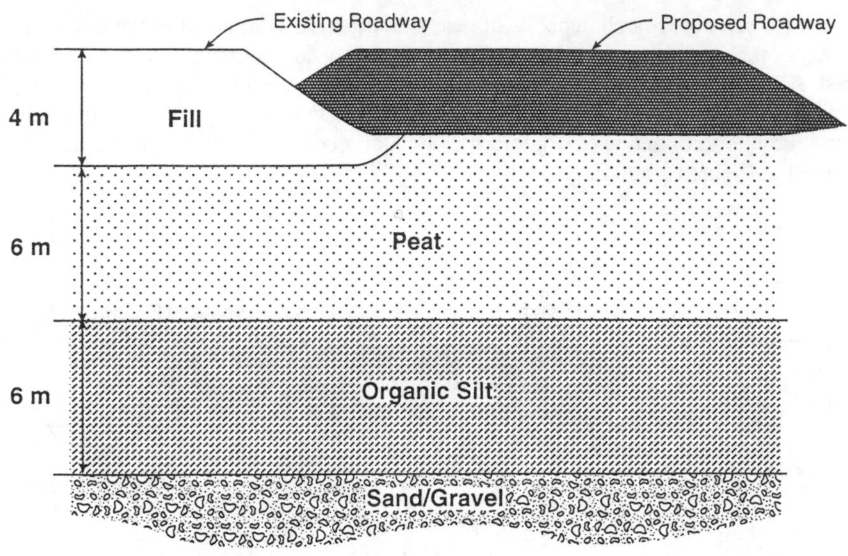

FIG. 2-*Typical Section Marsh B*

Based on the above, preloading of the marshes was evaluated and determined to be a cost effective and environmentally friendly solution. Preloading began at Marsh A in the fall of 1995 and at Marsh B in the spring of 1996. After inducing the expected immediate, consolidation, and long-term secondary compression settlements, the preloads were stripped to the pavement subgrade elevation and the pavement sections installed in the spring of 1997. This paper presents the design, instrumentation, and rheological modeling procedures used to complete the project.

Soils Information

Geotechnical investigations were performed at Marsh A and B to determine the depth and stratigraphy of the marshes, and geotechnical parameters for design. The geotechnical investigations revealed 3 main strata at each marsh, as summarized in Table 1.

TABLE 1-*Generalized Soil Conditions at Marsh A and B*

Stratum	Type of Material	Thickness (m)	Average Moist Unit Weight (kg/m^3)	Average Undrained Shear Strength, Cu (kN/m^2)(1)	Estimated Angle of Internal Friction (∅) (2)
1	Peat	3-6	1,073	14.4	28
2	Organic Silt	0-4.5	1,602	9.6	20
3	Sand/Gravel	3-9	1,922	...	35

Footnotes:

(1) Undrained shear strength, Cu, based on vane shear test results.

(2) Total stress angle of internal friction based on CU triaxial test results.

Stratum 1 is a brown to black fiberous peat. Stratum 2 is a black to grey spongy organic silt with traces of animal shells. Stratum 3 is a well-graded sand and gravel underlain by bedrock. The groundwater table ranged from about 0.3 to 1 meter below existing grade.

Consolidation tests were performed on undisturbed Shelby tube samples collected from the peat and organic silt. The consolidation tests revealed that the peat and organic silt are normally consolidated and highly compressible, as summarized in Table 2.

TABLE 2-*Typical Consolidation Parameters at Marsh A and B*

Stratum	Type of Material	Laboratory Compression Index, Cc	Estimated Secondary Compression Index, C_α	Laboratory Coefficient of Consolidation, C_v (m^2/s) (1)
1	Peat	1.15	0.06	1.9×10^{-6}
2	Organic Silt	0.65	0.03	2.8×10^{-7}

Footnote:

(1) Coefficient of Consolidation, C_v, determined using Cassagrande's Logarithm of Time Fitting Method.

Preload Design

The elevation difference between the existing pavement surface, and the existing ground surface at the proposed eastbound embankment ranged from about 1 to 4 meters (Figures 1 and 2). The finished pavement surface at the proposed eastbound embankment would be at approximately the same elevation as the pavement along the existing westbound embankment.

One-dimensional settlement analyses were performed using the laboratory consolidation data to determine the preload thickness required to induce the expected immediate, consolidation, and long-term secondary compression under the proposed highway embankment and pavement section service loads. Based on the settlement analyses, a 3.5 to 4-meter-thick preload embankment would be required at Marsh A, and a 6 to 6.5-meter-thick preload embankment would be required at Marsh B.

It was estimated that about 1 meter of the preload embankment would be stripped at the completion of preloading for construction of the pavement sections. The initial preload design parameters are presented in Table 3.

TABLE 3-*Initial Preload Design Parameters*

Marsh	Elevation Difference Between Existing and Finish Grade (m)	Maximum Thickness of Peat and Organic Silt (m)	Estimated Total Preload Thickness (m)	Estimated Settlement (m) (1)	Estimated Thickness to be Stripped at Completion (m)
A	1	9	3.5-4	1-1.5	1
B	3	12	6-6.5	1.5-2	1

Footnote:
(1) Total of immediate, consolidation, and long-term secondary over 30-years.

Embankment stability analyses were performed with the UTEXAS computer program using undrained strength parameters (i.e., $\varnothing = 0$). The analyses indicated that the preload embankments would need to be constructed in stages to minimize the chances of a general bearing capacity or slope stability failure. Based on the analyses, the initial preload stage thicknesses were limited to 0.6 meters, and subsequent stages were limited to 1 to 1.2 meters. A total of 5 preload stages were planned to construct the preload embankment at Marsh A, and 6 to 7 preload stages were planned to construct the preload embankment at Marsh B.

Time rates of settlement analyses were performed using the laboratory coefficients of consolidation and assuming double drainage would occur in the field. These analyses indicated that it would take between 60 and 90 days for each preload stage to undergo its full immediate and majority (i.e., 90 percent) of consolidation settlement. Therefore, it was determined that up to 15 months would be required to construct the Marsh A preload embankment (5 preload stages x 3 months/stage) and that up to 21 months would be required to construct the Marsh B preload embankment (7 preload stages x 3 months/stage). Paving was scheduled for the Spring of 1997. The project schedule allowed the Marsh A preload to be constructed within this time period, however, vertical drains were needed at Marsh B to accelerate the consolidation process and allow preloading at this location to be completed before the spring of 1997.

Site Preparation

Prior to field instrumentation and vertical drain installation, brush and vegetation within the limits of preloading were cleared and removed. The sod and grass root mat were not removed because the root mat helped to provide a stable surface. Holes and depressions were filled with granular fill and a woven geotextile fabric was placed over the entire preload area. The fabric was placed perpendicular to the roadway and anchored in the slopes of the existing highway embankment. The fabric was overlapped a minimum of 0.6 meters at the joints. After placing the fabric, a 0.15 to 0.30 meter thick granular fill working surface layer was placed to allow field instrumentation and vertical drains to be installed.

The vertical drains were installed at Marsh B on a triangular grid pattern at 1.8 meters on center. The drains were installed before the field instrumentation was installed to help minimize the chances of damaging the instrument cables. Vertical drains (AMERIDRAIN 407) were pushed about 0.6 to 0.9 meters into Stratum 3 and cut off about 15 cm above the granular fill working surface layer to allow double drainage to occur. The Barron and Kjellman (1992) formula was used to estimate the acceleration of pore pressure dissipation. The analyses indicated that the drains would induce the majority of consolidation settlement within about 2 to 3 weeks for each preload stage, about 5 times faster than without the drains.

Field Instrumentation

Field instrumentation was installed at Marsh A and B and included settlement plates, a liquid settlement gauge, and vibrating wire piezometers, as summarized in Table 4. Settlement plates were generally installed about every 30 meters along the preload embankment. The plates were used to track the actual preload settlement and a schematic of the settlement plates used is presented in Figure 3.

A liquid settlement gauge (GEOKON Model 4600) was installed at Marsh A to serve as a back-up instrument in case the settlement plates were damaged, and to corroborate the settlement plate data. Vibrating wire piezometers (GEOKON Model 4500AL) were installed in the center of the consolidating layers to monitor pore pressure dissipation and help determine when consolidation settlement for each stage was complete and if the next preload stage could be placed.

TABLE 4-*Field Instrumentation Summary*

Marsh	Preload Length (m)	Settlement Plates	Liquid Settlement Gauge	Vibrating Wire Piezometers
A	700	21	1	7
B	200	6	...	2

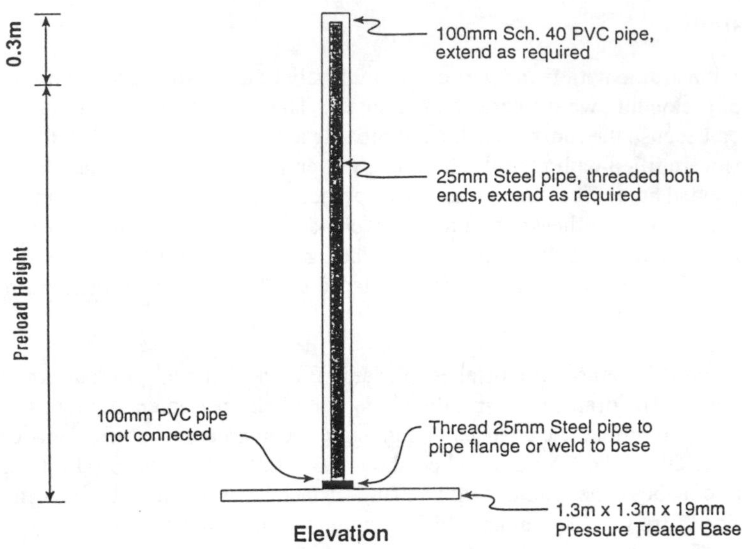

FIG. 3-*Settlement Plate Schematic*

The vibrating wire piezometers and liquid settlement gauge at Marsh A were connected to a central control box located outside the preload area. The control box allowed up to 10 separate instruments to be read from one location. At Marsh B, the instrument cables were connected to wood posts outside the preload limits to obtain readings.

Preload Construction and Monitoring

Marsh A preload construction began in November 1995 and Marsh B construction began in April 1996. For Marsh A and B, a total of 4 and 5 preload stages were constructed, respectively. The first stages were about 0.6 meters thick and subsequent stages were about 0.9 meters thick. The final stages for Marsh A and B were constructed in April 1996 and September 1996, respectively.

The preload embankments were constructed using granular fill from a nearby cut section of the highway alignment and the fill was hauled to the preload areas in 8 m^3 dump trucks. The trucks were able to drive directly on the preload areas and the fill was placed in 0.3 meter lifts and compacted using tracked equipment.

Settlement and pore pressure readings were taken prior to and after placing each preload stage. After each stage was placed, readings were taken at 1 week intervals for the first month and at 2 week intervals for the following months. In general, subsequent preload stages were not placed until about 65 to 90 percent of the excess pore pressures had dissipated. A typical settlement and pore pressure dissipation curve from an adjacent settlement plate and vibrating wire piezometer is presented in Figure 4.

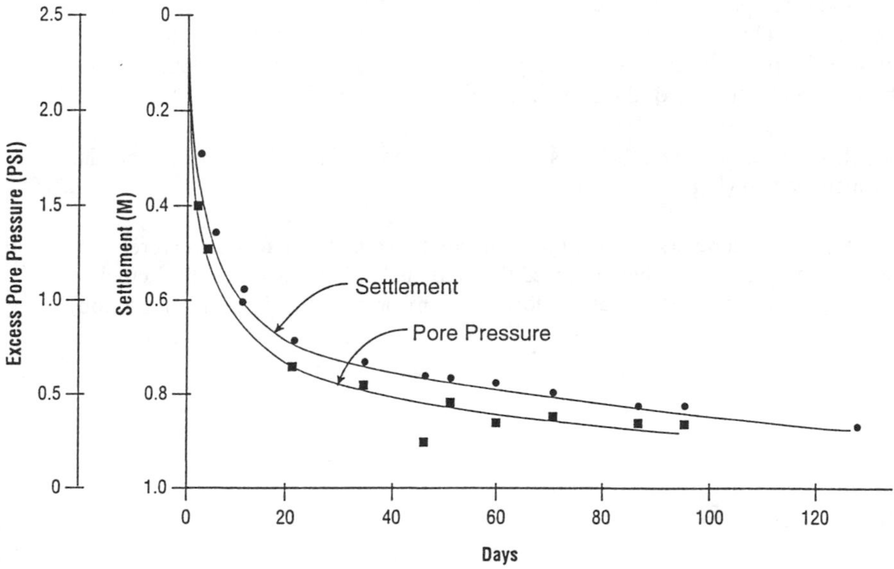

FIG.4-Typical Settlement and Pore Pressure Dissipation Curves

Rheological Modeling

The initial preload design parameters presented in Table 3 were estimated using conventional one-dimensional settlement analyses and laboratory consolidation data. The conventional settlement analyses performed, developed for inorganic silts and clays, was considered approximate. To determine when preloading was completed, the Gibson and Lo Rheological model, as further developed by Edil, was used to generate field settlement curves and predict settlements under the expected STH 29 service loads (Gibson and Lo 1961; Edil 1982). The predicted settlements were compared to actual field settlements to determine when preloading was completed (i.e., when the anticipated immediate, consolidation, and long-term secondary compression was induced). Input parameters for the model were developed using the field settlement data at the Marsh A and B preloads.

The following steps were followed to estimate the rheological parameters for each preload:

(1) Settlement plate data was used to develop vertical strain versus time plots and determine when primary consolidation appears to end and secondary compression begins (Figure 5).

(2) The logarithm of strain rate versus time was plotted for data obtained after primary consolidation had occurred A best fit straight line was obtained through the data points (Figure 6) and the rheological parameters "b" and "λ" were calculated using the slope and intercept of the best fit straight line (Figure 7).

(3) The rheological parameter "a" was determined from "b" and "λ" per the procedure outlined by Edil (Figure 7).

Using the above parameters in Equation 1, the predicted and actual strains were compared. The parameters that produced the predicted strains closest to the actual strains were used to predict the service load (i.e., embankment and pavement section) settlements.

$$\varepsilon(t) = \Delta\sigma[a+b(1-e^{-\lambda/b(t)})] \quad (1)$$

Where

ε (t)	=	Strain at time, t
$\Delta\sigma$	=	Applied Load, F/L^2,
a	=	Rheological Parameter,
b	=	Rheological Parameter,
λ	=	Rheological Parameter,
t	=	Time, t

Using Equation 1 and the marsh deposit thicknesses, settlement prediction curves were developed along different sections of each preload (Figure 8). The preload was intended to eliminate the settlements under the service load over a 30-year design life. The predicted settlement at 30 years was compared to the actual field settlements. If the actual settlement was less than the predicted settlement, the preload was left in place or additional fill added. If the actual settlement was greater than the predicted settlement, the preload was stripped down to the pavement subgrate elevations for construction of the pavement sections.

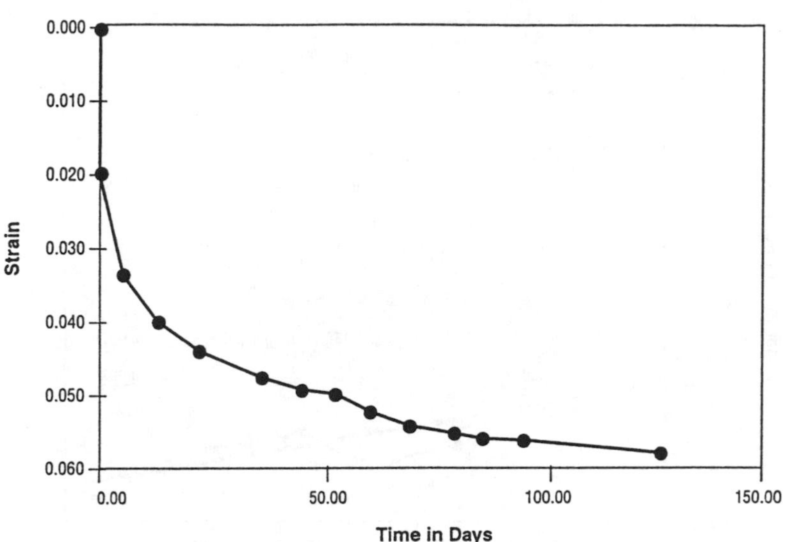

FIG. 5-*Typical Vertical Strain Versus Time Plot*

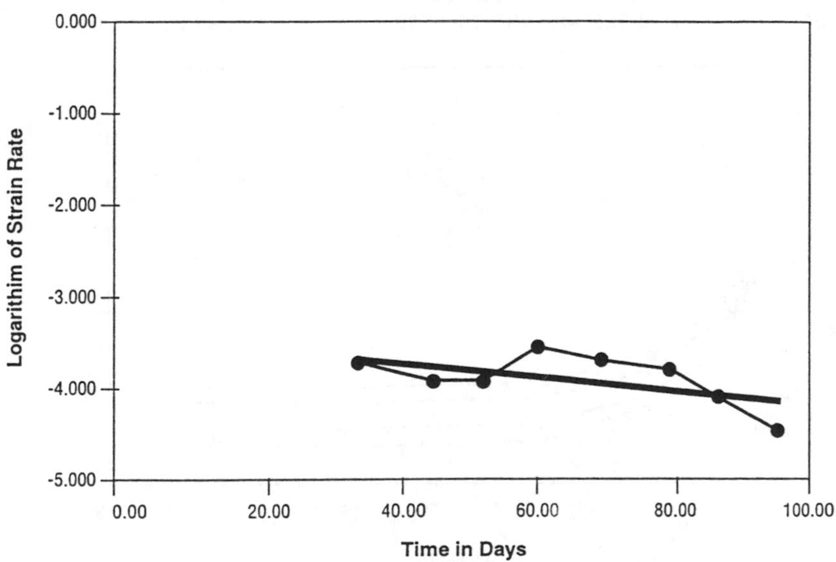

FIG. 6-*Typical Logarithm of Strain Versus Time Plot*

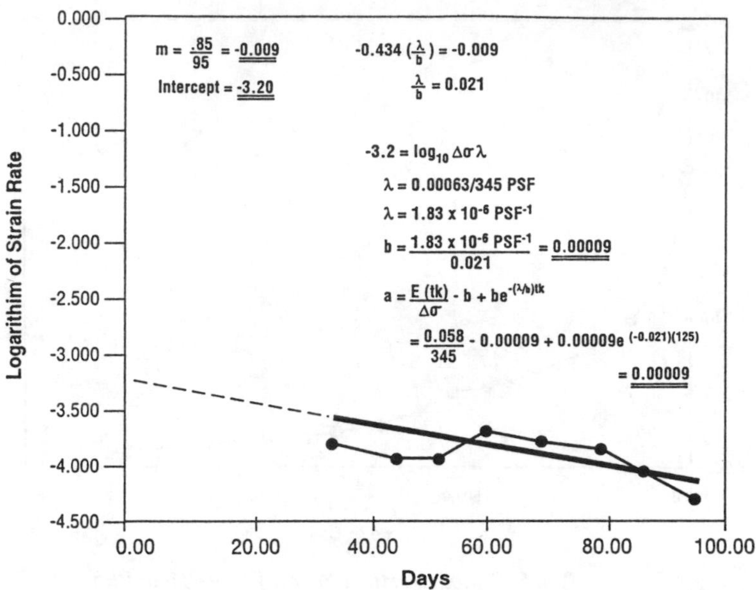

FIG. 7-*Estimated Rheological Parameter "b", "λ" and "a"*

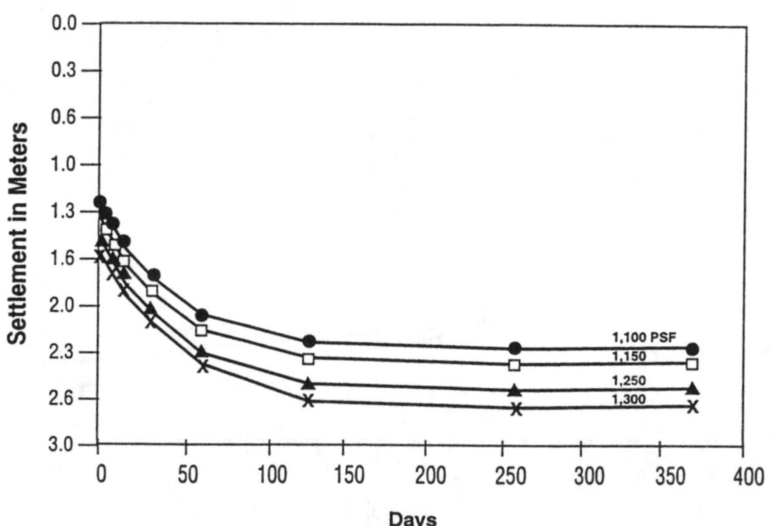

FIG. 8-*Typical Settlement Prediction Curve*

Results

Typical site-specific rheological parameters for Mash A and B are presented in Table 5. Using site-specific parameters and Equation 1, Settlement Prediction Curves were developed and it was determined when preloading was completed.

TABLE 5-*Typical Site-Specific Rheological Parameters*

Marsh	λ	***b***	***a***
A	1.8×10^{-6}	0.00009	0.00009
B	9×10^{-7}	0.00006	0.00006

When the preloading was determined to be completed, the preloads were stripped down to the pavement subgrade elevation and the pavement sections installed. The paving work was completed in the Spring of 1997, and to date there has been no measurable settlement of the pavement surface in the preloaded areas.

During pore pressure monitoring, the coefficient of consolidation, C_v, was back calculated, based on the amount of pore pressure dissipation, in accordance with Equation 2 and 3. The estimated apparent field coefficients of consolidation are presented in Table 6.

$$U_z = 1 - \frac{U}{Ui} \tag{2}$$

Where

U_z = Percent Consolidation and Mid-Point of Consolidating Layers
U = Pore Pressure Reading at time t,
Ui = Initial Pore Pressure Reading at t = o.

$$C_v = \frac{TH_{dr}^{\;2}}{t} \tag{3}$$

Where

C_v = Coefficient of Consolidation, m^2/s,
T = Time Factor,
H_{dr} = Length of the Drainage Path, m,
t = Time, sec.

TABLE 6-*Estimated Apparent Field Coefficients of Consolidation*

Stratum	**Type of Material**	**Apparent Field Coefficient of Consolidation, C_v (m^2/s)(1)**	**Apparent Field Coefficient of Consolidation, C_v (m^2/s)(2)**
1	Peat	1.4×10^{-6}	4.5×10^{-6}
2	Organic Silt	...(3)	...(3)

Footnotes:
(1) At Marsh A without vertical drains.
(2) At Marsh B with vertical drains at 1.8 meters on center.
(3) Vibrating wire piezometers were not installed in the organic silt layer.

Based on the results presented in Table 6, the estimated apparent field coefficient of consolidation for the peat is approximately the same as the laboratory coefficient of consolidation (1.4×10^{-6} m^2/s versus 1.9×10^{-6} m^2/s). It was not possible to compare the coefficient of consolidation for the organic silt because vibrating piezometers installed in the organic silt deposit malfunctioned. The apparent field coefficient of consolidation for the peat at Marsh B was about 3 times greater due to the installation of vertical drains.

Conclusions

Preloading was a cost-effective solution for constructing about 900 linear meters of new 2-lane highway for the reconstruction of STH 29 between the cities of Hatley and Wittenberg, Wisconsin. To date, there has been no measurable settlement of the pavement surface along the areas preloaded. Therefore, settlement prediction curves, developed using the Gibson and Lo rheological model, where reliable for predicting when preloading was complete and when the preload surcharge could be stripped.

Settlement plates spaced about ever 30 meters along the preloads were adequate for obtaining the settlement data necessary for generating the rheological Settlement Prediction Curves. Vibrating wire piezometers spaced about every 100 meters along the preload were useful for estimating the time rate of settlement, and for determining when the next preload could be placed. However, this information could also be determined by interpreting the settlement data strain versus time plots. Based on this, a wider vibrating wire piezometer spacing, say every 130 to 200 meters, is considered reasonable for providing back-up information for projects of similar size and type.

References

Canadian Foundation Engineering Manual, 3rd Edition, Canadian Geotechnical Society, pp. 261-263, 1992.

Gibson, R. E. and Lo, K.Y., "A Theory of Consolidation of Soils Exhibiting Secondary Compression", Acta Polytechnica Scandinavica, Ci10296, 1961, pp. 1-15.

Edil, T. B., "Use of Preloading For Construction on Peat", 13th Minnesota Conference on Soil Mechanics and Foundation Engineering, St. Paul, Minnesota, 1982.

Phil Flentje[1] and Robin Chowdhury[2]

Instrumentation for Slope Stability - Experience from an Urban Area

---

REFERENCE: Flentje, P. N., and Chowdhury, R. N., "**Instrumentation for Slope Stability—Experience from an Urban Area,**" *Field Instrumentation for Soil and Rock, ASTM STP 1358*, G. N. Durham and W. A. Marr, Eds., American Society for Testing and Materials, West Conshohocken, PA, 1999.

ABSTRACT: This paper describes the monitoring of several existing landslides in an urban area near Wollongong in the state of New South Wales, Australia. A brief overview of topography and geology is given and reference is made to the types of slope movement, processes and causal factors. Often the slope movements are extremely slow and imperceptible to the eye, and catastrophic failures are quite infrequent. However, cumulative movements at these slower rates do, over time, cause considerable distress to structures and disrupt residential areas and transport routes. Inclinometers and piezometers have been installed at a number of locations and monitoring of these has been very useful. The performance of instrumentation at different sites is discussed in relation to the monitoring of slope movements and pore pressures. Interval rates of inclinometer shear displacement have been compared with various periods of cumulative rainfall to assess the relationships.

KEYWORDS: landslides, subsurface shear movement, monitoring, inclinometer, piezometer, velocity of movement, pore water pressure

---

Introduction

Monitoring and observation of slopes is an essential part of the assessment of slope stability. Observational approaches facilitate the assessment of hazard and risk associated with slope movement and landsliding processes. The most important measurements associated with foundation investigations, natural slopes and earth dams

[1]Research Fellow, Department of Civil, Mining and Environmental Engineering, University of Wollongong, Wollongong, NSW, 2522 Phone +61 2 42213056 Fax +61 2 42213238 email: pflentje@uow.edu.au.

[2]Head of Department, Department of Civil, Mining and Environmental Engineering, University of Wollongong, Wollongong, NSW, 2522 Phone +61 2 42213037 Fax +61 2 42213238 email: robin_chowdhury@uow.edu.au.

concern pore water pressures, settlements, lateral movements and shear movements. Amongst these, the two most important for slope stability and landslide monitoring are subsurface shear movement and pore water pressures.

In geotechnical engineering applications, instrumentation, observation and monitoring may serve several purposes. For some projects it is necessary to check important assumptions concerning the basic geotechnical/geological models used for analysis and interpretation. In slope stability problems it may be necessary to verify the shape and location of the slip surfaces and the character of the movement, e.g., rotational or translational. Furthermore, it may be necessary to check the assumptions made in analysis and design about the magnitude of subsurface pore water pressures and their spatial and temporal variability.

Apart from checking design assumptions, continued monitoring can be a useful tool for updating of assessments concerning stability, hazard and risk. In addition, continued monitoring is essential for assessing the effectiveness of remedial measures after their installation and/or construction.

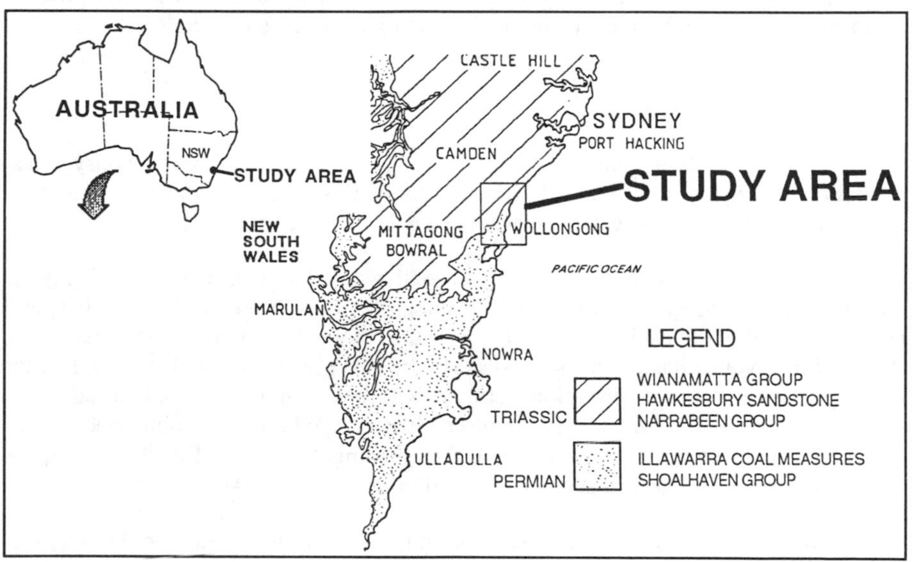

Figure 1 – *Study Area Location Plan showing the extent of Triassic and Permian sediments in the Southern half of the Sydney Basin*

This paper is concerned with monitoring of geotechnical instrumentation in sloping urban terrains. Subsurface shear movements were monitored at a number of existing inclinometer installations near the city of Wollongong, on the south coast of New South Wales, Australia. Subsurface shear movement was compared to antecedent rainfall patterns in a number of instances. Water levels in open standpipes were measured at several sites. In some instances subsurface pore pressures were also measured more accurately using vibrating wire piezometer installations. However, the emphasis during the research was on subsurface shear movements and their relationship to rainfall.

The Wollongong Area

The City of Wollongong, including the northern suburbs, is a major urban centre on the south coast of NSW with a population of approximately 200,000 people. A total of 328 past and present sites of land instability are known in an 87 km^2 area within the Wollongong City Council local government area. At least 60 houses have been damaged and a further 29 houses have been destroyed since the turn of the century by land instability. As well as affecting suburban areas, landslides affect the major road and rail transport routes which link Wollongong to Sydney and other regional centres (Shellshear 1890, Hanlon 1958, Bowman 1972 and Pitsis 1992).

Wollongong is situated on the coast approximately 80 km south of Sydney (Fig. 1). The local government area includes a highland plateau area to the west of the escarpment topped by a near vertical sandstone cliff line which grades down to the east into terraced lower slopes. The eastern and southern margins of the district comprise a narrow coastal plain which is flanked on the east by the Pacific Ocean.

Geological Setting

The Sydney Basin comprises a basin sequence of dominantly sedimentary rocks extending in age from approximately Middle Permian to at least Middle Triassic. The study area lies on the south eastern margin of the Sydney Basin (Fig. 1).

The geological bedrock sequence of the Illawarra district is essentially flat lying with a low angle dip, generally less than five degrees, towards the northwest. This gentle northwesterly dip is a result of the relative position of the district on the southeastern flanks of the Sydney Basin. The northwesterly dip is superimposed with relatively minor syn-depositional and post-depositional structuring (folding and faulting). Normal faulting within the Illawarra area is common, although the fault throws infrequently exceed 5 metres. The structural geology of Wollongong, Kiama and Robertson, mapped on 1:50000 sheets, has been discussed in detail in Bowman (1974).

The geological units encountered within the district, in ascending order, include the Illawarra Coal Measures (locally including intrusive/extrusive bodies collectively known as the Gerringong Volcanic facies), the Narrabeen Group and the Hawkesbury Sandstone. The geology of these units has been discussed at length in several publications such as Bowman (1974) and Herbert and Helby (1980).

Extending down from the base of the upper cliff line, the ground surface is often covered by debris of colluvial origin. This material comprises variably weathered bedrock fragments supported in a matrix of finer material dominantly weathered to sand, sandy clay and clay and brought downslope under the influence of gravity.

Instrumented Sites

Inclinometer monitoring records of forty two boreholes from 19 landslide sites have been examined and analysed. Records of movements prior to 1993 have been acquired from government authorities and from consulting engineers who were responsible for the initial drilling of boreholes and installation of the inclinometers. Monitoring at several sites was continued during the period 1993-1997 as part of a Landslide Hazard Assessment project. The inclinometer monitoring records comprise the recorded profiles of the borehole inclinometers and the date of each monitoring visit.

A sample of 10 boreholes each with a brief summary of monitoring coverage and movement data are listed in Table 1. This table also shows the borehole identification number and the site reference code of each landslide.

Table 1 – *A Sample of 10 Boreholes out of a total of 42 in the borehole inclinometer dataset with site reference code (SRC), monitoring coverage and rates of movement.*

Borehole Name	SRC	1st reading	last reading	number readings	depth to shear (m)	peak rate mm/day	average rate mm/day
2	026	28/03/93	9/12/96	12	15-16	0.03	0.02
4	026	28/03/93	9/12/96	12	8.5-9.5	0.03	0.02
6	026	28/03/93	9/12/96	12	11.5-12.5	0.04	0.02
3	064	9/03/89	8/08/96	30*	7.0-8.0	1.93	0.07
7	064	19/10/89	8/08/96	27*	1.5-2.5	2.09	0.06
WH2	134	14/12/90	16/05/96	16*	11.0-12.0	1.1	0.01
WH10	134	12/11/91	15/05/96	11*	13-13.5	0.03	0.004
WH13	134	12/11/91	16/05/96	11*	16.5-17.5	0.11	0.01
WH14	134	12/11/91	16/05/96	10*	3.4-3.9	0.08	0.01
WH15	134	18/02/92	16/05/96	9*	creep	0.03	0.002

* denotes that some readings were collected as part of this research project

It was considered necessary to adopt a standard numbering system for the 328 sites of instability incorporated into a landslide inventory of this area (Chowdhury and Flentje, 1997, 1998a). This landslide inventory was compiled during the research project mentioned above. Each site is identified with a 3 character Site Reference Code (SRC). This site reference code is unique to each location, is plotted for each site on a map record and constitutes the "primary" or "key" field in a land instability database.

In general, landslides in the Wollongong area can be summarised as follows:

- predominantly rainfall triggered but often a consequence of urbanisation
- 80 % of the total number of landslides have volumes in the range 500 m^3 to 50,000 m^3, but the maximum landslide volume is 600,000 m^3
- velocities are typically very slow to extremely slow according to the international scale (WP/WLI, 1993)
- depths to the slip surfaces range from 0.5 m to 20 m
- movements are rotational and/or translational

- landslide material generally comprises of clay and sandy-clay colluvium
- more than 50 % of the landslides can be classified as complex debris slump - debris flows (Varnes 1978)

In addition to inclinometer monitoring, open standpipe piezometers have been used to record standing water levels at some sites. Two vibrating wire piezometers (VWP) are installed in the head area of one landslide. These VWP have been monitored as part of this research project.

Inclinometer Profiles

Before giving details of three significant landslide sites in subsequent sections, some general comments on the installation of inclinometers and generated inclinometer profiles are provided here.

The inclinometers have been installed by various government organisations and engineering consulting firms as part of geotechnical investigations of landslide sites. These installations generally comprise 70 mm outside diameter plastic probe inclinometer casing installed within the full length of the borehole. The space between the casing and the wall of the borehole is backfilled with sand.

Inclinometer records used during this research project include the displacement at 0.5 m intervals along the length of the borehole, relative to the position of the bottom of the borehole. Cumulative displacements are summed from the base of the borehole to the top. Recording displacement at 0.5 m intervals provides optimum accuracy, such that probe inclinometers can measure changes in inclination of the order of 1.3 mm to 2.5 mm over 33 m lengths of inclinometer casing (Mikkelsen 1996), although one manufacturer claims a system accuracy of ± 6 mm per 25 m.

Analysis of each record enables the identification of time intervals between each subsequent monitoring visit and the magnitude of the movement during each such time interval. This enables the cumulative displacement in subsurface shear movement to be compared over time and, therefore, allows the interval rates of shear to be determined.

There are numerous different methods of reporting and discussing inclinometer profiles. For example, the New South Wales Government Railway Services Authority Geotechnical Services (RSA) employs an A^+/A^- and B^+/B^- axis nomenclature system with the A^+ axis orientated perpendicular to the railway line and pointing downslope, with the B^+ axis clockwise 90° from the A^+ axis. The RSA typically reports the A axis data only. The New South Wales Government Roads and Traffic Authority (RTA) employs an A/B and C/D axis nomenclature system with the A axis orientated along the direction of maximum slope and the C axis orientated 90° in a clockwise direction from the A axis. The RTA typically reports the profiles of both axis. Coffey Partners International (CPI) uses the same axis labelling system as that employed by the RSA, except the A^+ axis is usually orientated along the steepest inclination downslope. CPI reports both axes profiles, and then calculates and reports the maximum resultant displacement vector (magnitude and azimuth).

Composite 20 Year Rainfall Histogram and Antecedent Rainfall Curves

Rainfall is recognised as an important causal and triggering factor for slope instability. To assess the relationship between slope instability and rainfall within the study area daily rainfall totals have been used to determine daily rolling antecedent rainfall curves, for which Antecedent Rainfall Percentage Exceedance Time (ARPET) values have been determined (Chowdhury and Flentje 1998b). These antecedent rainfall curves and percentage exceedance values have been compared with monitored landslide shear displacement curves to examine upper and lower bound antecedent rainfall thresholds associated with landslide movement.

Daily rainfall totals have been collected from several rainfall stations to compile an unbroken 20 year composite record, which is currently being extended to cover a 100 year period. The existing 20 year record extends from 1 January 1977 to 31 December 1996. This period is significant not only for its duration, but also for some exceptional rains that have fallen during this period and several periods of significant land instability which it encompasses. The data were entered into a computer spreadsheet whereby antecedent rainfall for daily rolling periods of 7 days, 30 days, 60 days, 90 days and 120 days have been computed.

In the following sections three significant landslide sites are discussed in some detail.

The Coalcliff Terrace Landslide, Site 026

This landslide site covers an area of approximately 66,000 m^2, and has a volume of approximately 600,000 m^3 making it the largest volume landslide in the Wollongong study area. According to Cruden and Varnes (1996) this landslide can be described as an active, advancing, composite, extremely slow, moist to wet debris slump in the head area and a debris slide-debris flow lower down. The landslide is rotational in the head area of the site, but predominantly translational for most of the site below the head area. The depth of the colluvium material at this site is up to 16 m. Shear displacement is typically occurring along the bedrock/ colluvium interface.

This site is traversed by the South Coast Railway Line (a dual electric freight and passenger railway line) and a two lane main road. Ground movements have been reported at least every decade at this site since 1942. The site has been the subject of several detailed geotechnical investigations. During the most recent of these, 13 boreholes were drilled, 6 of which had inclinometers installed, and another 6 had a total of eleven pneumatic piezometers installed.

In late 1997 and early 1998 this site has had an elaborate series of remedial measures installed, and as a result may have been stabilised. These remedial engineering construction works include a 200 m long row of vertical drainage wells with an interconnecting series of subhorizontal, gravity-fed, drainage relief drives.

Inclinometer profiles (A axis only) recorded by the Railway Services Authority, from three borehole inclinometers at this site have been examined for rates of movement. One of these inclinometer profiles is shown in Figure 2. This profile displays a block style of movement with a depth to the slip surface of 9.25 m. Progressive movement at a depth of 4.5 m in borehole 4 is shown in Table 2.

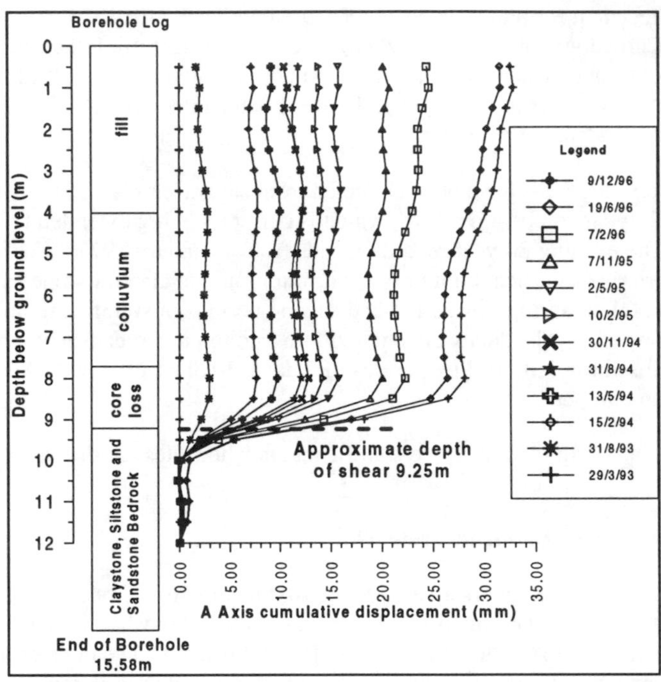

Figure 2 – *Summary Borehole Log and Inclinometer Profiles for Borehole 4, Site 26 at Coalcliff, near Wollongong in New South Wales, Australia*

Table 2 – *Borehole 4, Site 026, monitoring period and displacement (A axis data only) recorded at 4.5m depth*

Date of reading	Days (since start)	Interval (days)	Displacement (mm)		Rate of shear (mm/day)	
			Resultant	Interval	Interval	mean
28/03/93	0	0	0	0.00	0.000	0.000
31/08/93	156	156	2.81	2.81	0.018	0.018
15/02/94	324	168	7.61	4.80	0.029	0.023
13/05/94	411	87	9.67	2.06	0.024	0.024
30/08/94	520	109	11.86	2.19	0.020	0.023
10/11/94	592	72	12.04	0.18	0.003	0.020
10/02/95	684	92	13.8	1.76	0.019	0.020
2/05/95	765	81	15.49	1.69	0.021	0.020
7/11/95	954	189	19.61	4.12	0.022	0.021
7/02/96	1046	92	22.37	2.76	0.030	0.021
19/09/96	1271	225	27.74	5.37	0.024	0.022
9/12/96	1352	81	29.42	1.68	0.021	0.022

Total displacements indicated for boreholes 2, 4 and 6 over the period 28th March 1993 to 9th December 1996, are 27.6 mm, 29.4 mm and 29.0 mm respectively, all at average rates of approximately 0.02 mm per day, or, if extrapolated, 8 mm per year. On the WP/WLI (1993) velocity scale, such a rate of displacement is classified as extremely slow. Yet, RSA engineers maintaining the dual electric railway line that traverses this site have reported an annual maintenance cost exceeding tens of thousands of dollars.

Cumulative displacement and rate of shear for three boreholes at this site, including borehole 4, are shown in Figure 3 together with the 90 day antecedent rainfall curve. The cumulative displacement curves for the three boreholes show the continual movement at the site over the period of monitoring, 28th March 1993 to 9th December 1996. Periods where movement is lower than 0.015 mm per day (the average rate of movement is approximately 0.02 mm per day) correspond to periods of low antecedent rainfall, except Borehole 2 for the period November 95 to February 1996. Movement has continued at this site for the duration of the whole monitoring period.

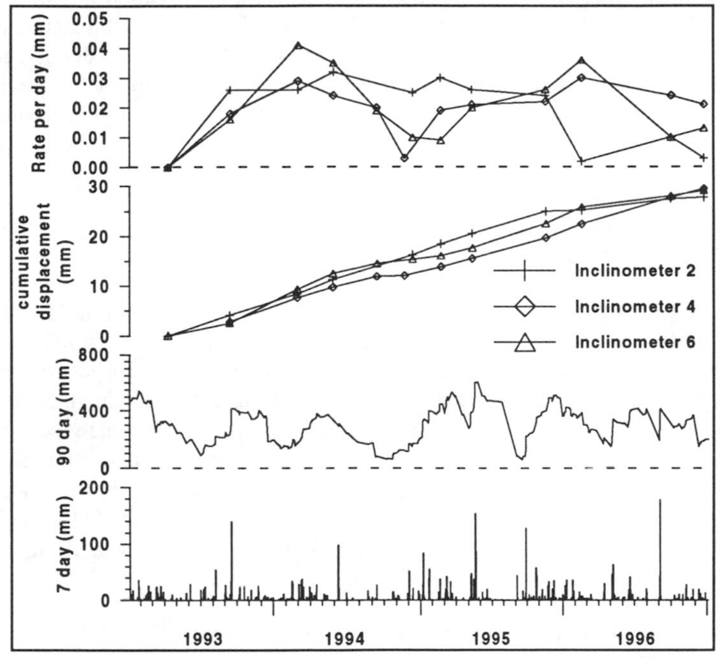

Figure 3 – *Site 026, daily and antecedent rainfall and shear displacement for 3 boreholes.*

Periods of accelerated movement, leading either to peaks or steep positive slopes on the rate per day curves, generally correspond to high values in the antecedent rainfall curves over the monitoring period.

The effects of the recently completed remedial stabilisation works on shear displacement at depth and ground movement at this site is yet to be assessed. This assessment will be largely based on data obtained from the instrumentation installed at this site, and on other ground surface survey data. The results of this assessment will be reported elsewhere.

A Landslide in Goodrich Street, Scarborough, Site 064

Site 064 covers an area of approximately 5000 m^2 and a volume of almost 18,000 m^3, making it the 77th largest landslide in the study area. The head area of this landslide is also traversed by the South Coast Railway Line. The area below and downslope of the railway side-fill embankment is residential land.

The most recent geotechnical investigation of this site was carried out by the then State Rail Authority of New South Wales, now a subsidiary of that organisation known as the Railway Services Authority Geotechnical Services, and by their consultants in the period from 1989 to 1991. These investigations followed various land instability "events" during the preceding 40 years. These events include scour wash outs, general track "subsidence", development of tension cracks between the two tracks, a train derailment and inundation of the site with debris resulting from ground movements upslope.

These investigations included the excavation of test pits, the drilling of numerous boreholes with installation of two inclinometers and two vibrating wire piezometers. Three additional open standpipes were also installed.

The monitoring results from all the instrumentation installed at this site, combined with daily rainfall and a daily rolling 90 day antecedent rainfall curve (rainfall station 7km away from the site) are shown in Figure 4. The inclinometer profiles (not included here) clearly define the depth to and style of movement at this site, which, combined with surface surveys of ground movement, allowed the approximate plan area and volume of this landslide to be calculated. The cumulative shear displacement and rates of shear displacement curves in Figure 4 clearly show the response to rainfall, and, in particular, the 90 day antecedent rainfall in the periods March - June 1989 and February - April 1990. A series of longitudinal subsurface drainage trenches were designed and installed at this site, at a cost of almost $400,000.

Monitoring of the site following the installation of these remedial works has been continued. This "post construction" monitoring has revealed a significant decrease in slope movement but the site has not been stabilised fully.

Interval rates of displacement following the first monitoring visit since completion of the remedial works have ranged from a maximum of 0.03 mm/day down to less than 0.01 mm/day. However, in the head area of the landslide, as indicated by inclinometer 3, 21 mm of displacement has occurred since the remedial works, whereas only 2.5 mm of displacement has been indicated in the toe area of the landslide, as indicated by the monitoring data from inclinometer 7. Monitoring data from the shallow vibrating wire piezometer, installed at a depth of 4.2 m in the head area of the landslide, in the vicinity of borehole inclinometer 3, has shown a significant increase in pore water pressure during 1994 and 1995. This build up of pore pressure may be part of the reason for continued movement in this area of the landslide.

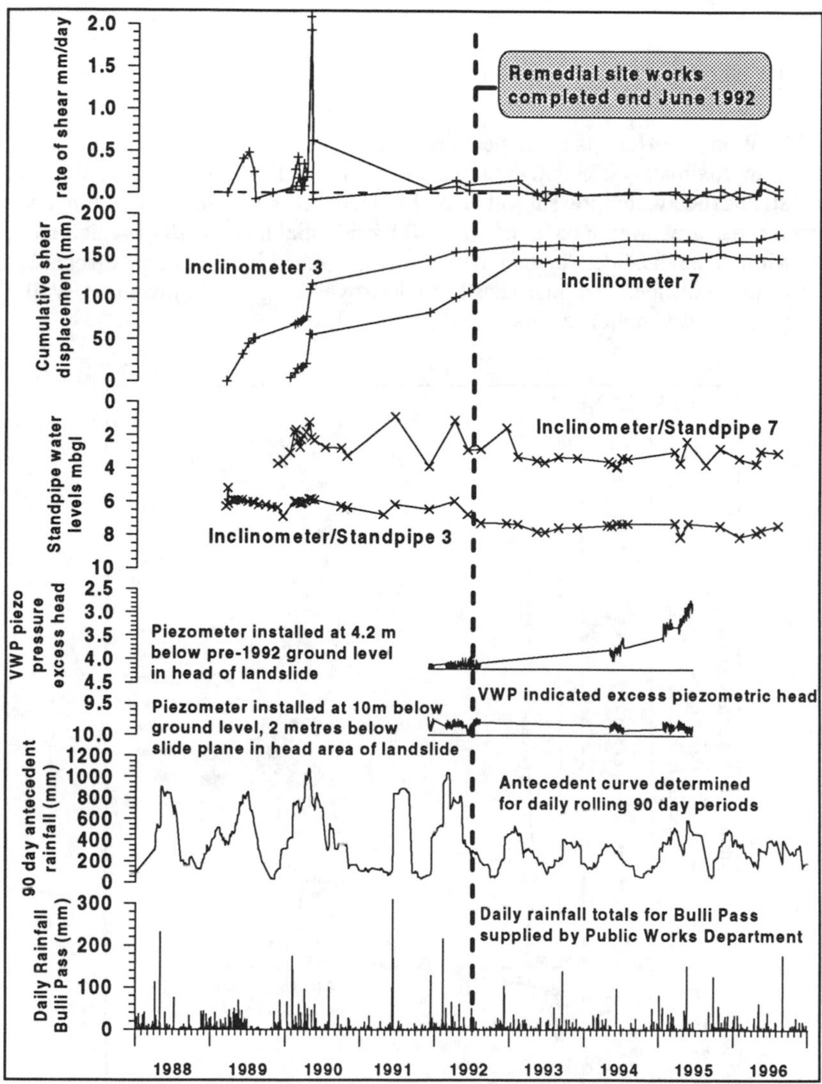

Figure 4 – *Site 064, MonitoringData for all of the Instrumentation Installed at this site plotted with Daily Rainfall Totals and 90 day Antecedent Rainfall Curve.*

It must also be noted that in other research work carried out by the writers and reported elsewhere (Flentje 1998, and Chowdhury and Flentje 1998b), the 90 day antecedent rainfall curve has been considered to show a close relationship to landslide movement at this site. Since the remedial works were completed in June 1992 the landslide site has not experienced 90 day antecedent rainfall magnitudes of similar scale to those which have been correlated with "failures" in 1989 and 1990. Hence, there is concern about the effectiveness of the remedial measures and monitoring of this site is continuing.

The Woonona Heights Landslide, Site 134

The Wonoona Heights Landslide has a plan area of approximately 20,500 m², a volume of approximately 225,000 m³ and is ranked in the land instability database as the 9th largest landslide within the subject area. The landslide site is located within a densely developed urban area, which contains up to 100 residential houses, 29 of which are situated within the landslide. Another 10 houses are located immediately adjacent to the margins of the landslide. One house has been destroyed by ground movement, whilst at least 19 have required major repairs.

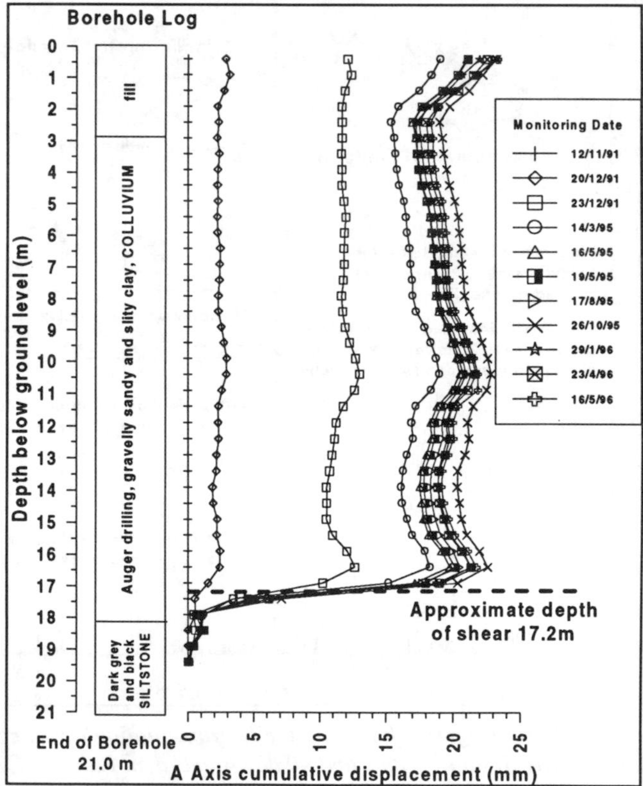

Figure 5 – *Summary borehole log and inclinometer profiles for borehole 13, Site 134 at Woonona Heights, near Wollongong in New South Wales, Australia.*

This landslide has a long history but is a subtle feature, very little damage being visible from the streets. According to a previous geotechnical report, disturbed ground can be observed in a 1948 black and white aerial photograph. Accelerated movement phases have been documented at this site in 1977, April 1990, March - June 1991 and November 1991 - March 1992, all of which have resulted in road and residential damage.

The site has recently been investigated on behalf of the local council by an international geotechnical consulting firm. The comprehensive investigation included the drilling of 22 boreholes, and included the installation and periodic monitoring of 7 inclinometers and numerous open standpipe piezometers. Following this investigation remedial works have been proposed. These works have not yet been implemented. Consequently, the site remains episodically active as an extremely to very slow moving landslide (WP/WLI, 1993).

This landslide is located within the area of subcrop of the Illawarra Coal Measures. The slide material comprises of a sequence of some fill and a gravelly and sandy-clay colluvium sliding over residual bedrock, which includes several coal seams. Monitoring of three inclinometers and piezometers has been carried out at this site. All the inclinometer profiles within the landslide show similar block styles of movement, with an approximate average depth of 11 m to the slip surface, the maximum depth being almost 18 m, as shown in Figure 5.

This monitoring has confirmed that movement at this site, under residual strength conditions, continues to occur when the piezometric levels associated with seepage reach about 1m below the ground surface (Fig. 6). The results of laboratory shear strength testing and stability analyses are outside the scope of this paper.

Further instrumentation and monitoring is required to investigate the lower half of the area of this landslide. In this area, the basal slip surface of this landslide is essentially horizontal and immediately overlies a highly fractured coal seam with numerous low strength claystone bands of tuffaceous origin. Additional inclinometers and piezometers would provide extremely useful information for the geotechnical model of this landslide and may facilitate the design of more economical remedial measures than those which have already been proposed.

Discussion

With the types of landslides common in this study area, an acceptable minimum inclinometer monitoring interval is 30 days. Additional monitoring visits are required in response to rainfall "events", with additional follow up visits one day to several weeks later, dependent upon the rainfall and the associated movement response. During periods of dry weather, the period between monitoring visits can be extended. The preferred 30 day period allows accurate interval rates of displacement to be calculated. More frequent monitoring does incur significant costs, and it is accepted that monitoring schedules will often be controlled by available financial resources.

Use of instrumentation for subsurface monitoring of slopes along the Illawarra escarpment has been very useful for establishing the character, range and velocity of slope movements. The episodic nature of landslide movements has been established and more research is required to determine the conditions under which catastrophic failures may take place. This is likely to be a very difficult area of research because catastrophic failures are rare and because instrumentation may not be installed at the particular site which undergoes catastrophic failure. Furthermore, if instrumentation existed at the site, it may not survive the large movements which occur during a catastrophic landslide.

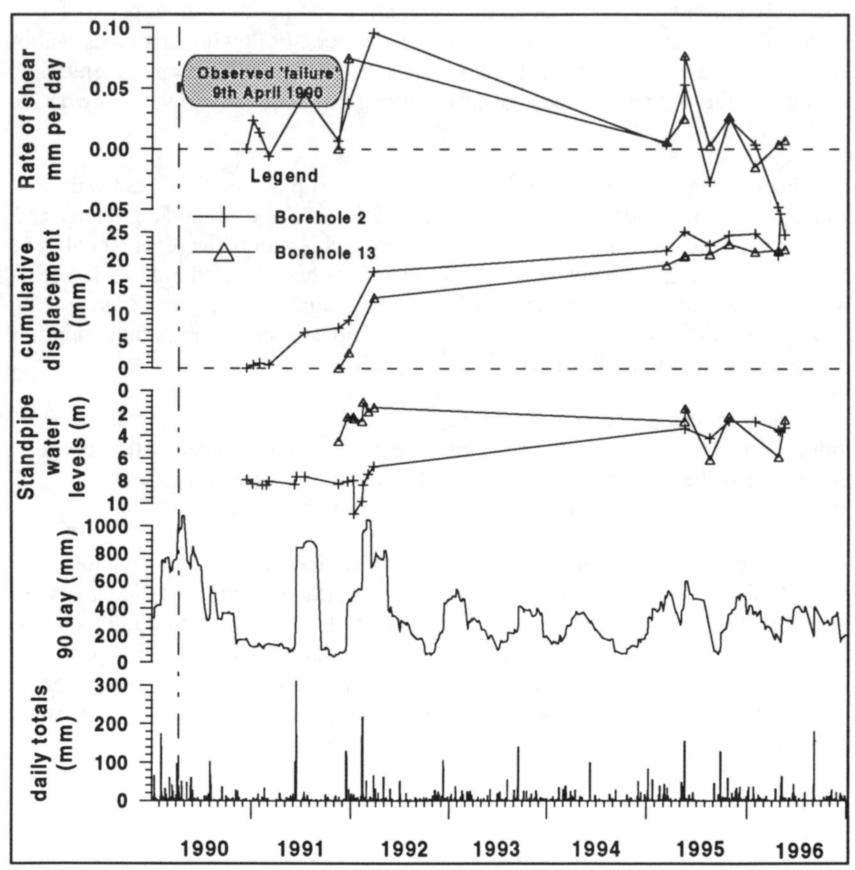

Figure 6 – *Site 134, daily and 90 day Antecedent Rainfall, Cumulative and Rate of Shear and Standpipe Water Levels.*

Some success has been achieved in correlating landslide movements to cumulative rainfall magnitudes and different levels of antecedent rainfall have been considered in this research. These relationships have been quantified on a site by site basis, as well as for the landslide "population" as a whole. It appears from this research that 90 day and 120 day antecedent periods correlate fairly well to peak or accelerating movement phases at many of the sites.

Open standpipes (slotted pipes, with the annulus back filled with sand over the entire borehole depth) have been widely used in the Illawarra region because of their low cost. Yet, it is recognised that the associated pore pressure data may not be reliable or accurate. Vibrating wire or pneumatic piezometers, whilst more expensive on a per unit basis, provide more reliable and accurate data.

Concluding Remarks

The use of inclinometers has proved to be immensely valuable for monitoring the performance of slow moving landslides in the Northern Illawarra Region of New South Wales, Australia. In fact this monitoring is considered to be the single most important factor in detailed site assessments of all such landslides regardless of their size and previous history. Experience has shown that simple and robust instrumentation is the best in these situations.

The inclinometers used at all sites have performed well and have proved to be reliable. Based on the experience gained during this research, movement rates of up to 17.6 mm per day and cumulative movements of up to 200 mm at slower rates of displacement can be measured with confidence. Rates of displacement as slow as 0.002 mm per day have been recorded over 90 day monitoring intervals. This level of accuracy is only relevant when considering the longer monitoring periods, e.g, more than a month. Measurements concerning extremely slow movements are not expected to be accurate or reproducible over very short monitoring periods, e.g, a few days. Of course, the use of inclinometers in faster moving landslides will require shorter intervals bewtween monitoring visits. In such cases, weekly or daily monitoring may be necessary.

References

Bowman, H. N., 1972. "Natural Slope Stability in the City of Greater Wollongong," *Records of the Geological Survey of New South Wales,* Vol. 14, Part 2, pp. 159 - 242.

Bowman, H. N., 1974. *Geology of the Wollongong, Kiama and Robertson 1:50,000 Sheets 9029-II 9028-I&IV Sheets*, Department of Mines, Geological Survey of New South Wales. Government Printer, New South Wales.

Chowdhury, R. N. and Flentje, P. N., 1997. "Relevance of Mapping for Slope Stability in the Greater Wollongong area, New South Wales, Australia," *Proceedings of the International Symposium on Engineering Geology of the Environment*, Marinos, P. G., Koukis, G. C., Tsiambaos, G. C. and Stournaras, G. C., (editors), Athens, Greece. June 23 - 27, pp. 569 - 574, Balkema, Rotterdam.

Chowdhury, R. N. & Flentje, P.N. 1998a. "A Landslide Database for Landslide Hazard Assessment." *Proceedings of the Second International Conference on Environmental Management.* Wollongong Australia. Sivakumar, M. &

Chowdhury, R. N., (editors), February 10 -13, Vol. 2, pp 1229 - 1237. Elsevier: Amsterdam.

Chowdhury, R. N. & Flentje, P. N. 1998b. "Effective Urban Landslide Hazard Assessment, " *Proceedings of the 8th Congress of the International Association of Engineering Geology and the Environment* (in press), 7p.

Cruden, D. M, and Varnes, D. J., 1996. "Landslide Types and Processes," Chapter 3, Turner and Schuster, 1996, pp 36 - 75.

Flentje, P. N., 1998. *Computer Based Landslide Hazard and Risk Assessment (Northern Illawarra Region of New South Wales, Australia),* unpublished, University of Wollongong, Department of Civil, Mining and Environmental Engineering Doctor of Philosophy thesis.

Hanlon, F. N., 1958. "Geology and Transport, with Special Reference to Landslides on the near South Coast of NSW," Presidential Address, Part II, *Journal of the Proceedings Royal Society of NSW*, Vol. 92, pp. 2-15.

Herbert, C., and Helby, R., (editors) 1980. *A Guide to the Sydney Basin*, Bulletin No. 26. Department of Mineral Resources, Geological Survey of New South Wales.

Mikkelsen, P., E., 1996. "Field Instrumentation," Chapter 2, in Turner and Schuster.

Pitsis, S. E., 1992. *Slope Instability along the Illawarra Escarpment*, University of New South Wales, Master of Engineering Science thesis, unpublished.

Shellshear, W., 1890. "On the Treatment of Slips on the Illawarra Railway at Stanwell Park," *Journal of the Proceedings of the Royal Society NSW*, Vol. 24 (1), pp. 58 - 62.

Turner, A. K. and Schuster, R. L., (editors), 1996. *Landslides, Investigation and Mitigation*, Special Report 247, Transportation Research Board, National Research Council. National Academy Press, Washington DC.

Varnes, D. J., 1978. "Slope Movement Types and Processes", Chapter 2, *Landslides, analysis and control.* Schuster, R. L. and Krizek, R. J., (editors).

WP/WLI (Working Party on World Landslide Inventory), 1993. "A Suggested Method for Describing the activity of a Landslide," *Bulletin of the International Association of Engineering Geology,* No. 47, pp 53 - 57.

Data Acquisition and Data Management

Craig H. Benson[1] and Peter J. Bosscher[2]

Remote Field Methods to Measure Frost Depth

REFERENCE: Benson, C. H. and Bosscher, P. J., "**Remote Field Methods to Measure Frost Depth,**" *Field Instrumentation for Soil and Rock, ASTM STP 1358*, G. N. Durham and W. A. Marr, Eds., American Society for Testing and Materials, West Conshohocken, PA, 1999.

ABSTRACT: This paper reviews the three common methods used to remotely measure frost depth: (1) soil temperature measurements, (2) soil electrical resistivity measurements, and (3) soil dielectric constant measurements. Soil temperature measurements are often made using thermocouples or thermistors. Specially designed probes are used to measure frost depth based on electrical resistivity. Inferences based on dielectric constant are usually made using time domain reflectometry. Frost depths based on soil temperatures are made equally well using thermistors or thermocouples. However, frost depths inferred from soil temperature measurements can be erroneous if the freezing point is depressed (due to unsaturated conditions or the presence of salts) or if the temperature profile is essentially isothermal and near 0°C. More reliable measurements of frost depth are made using electrical resistivity or dielectric constant measurements because the electrical properties of soil are affected more by phase change of the pore water than temperature.

KEYWORDS: frost, frost depth, freezing and thawing, thermocouples, thermistors, electrical resistivity, dielectric constant, time domain reflectometry, geophysics

---

In cold regions, seasonal freezing and thawing often have a significant impact on engineering behavior of earthen materials. For example, the strength and stiffness of pavements increase during freezing and decrease during thawing (Mahoney et al. 1985, Nordal and Hanson 1987, Janoo and Berg 1996, Jong et al. 1998). Also, the hydraulic conductivity of hydraulic barriers changes as pore water freezes and thaws (Chamberlain et al. 1990, 1995; Benson and Othman 1993). In other cases, soil is intentionally frozen for earth retention or to prevent contaminant migration (Tumeo and Davidson 1993, Andersland et al. 1996). In each of these situations, the depth or thickness of the frozen

[1]Associate Professor, Dept. of Civil & Environmental Engineering, University of Wisconsin-Madison, Madison, WI 53706, chbenson@facstaff.wisc.edu.

[2]Professor, Dept. of Civil & Environmental Engineering, University of Wisconsin-Madison, Madison, WI 53706, bosscher@engr.wisc.edu.

and/or thawed zones is important to characterize engineering behavior of the soil and the response of engineered systems.

Since the onset and rate of freezing are usually controlled by nature, frost depths are often measured remotely to reduce the number of site visits while maintaining a nearly continuous data record that includes critical events such as rapid thaws (e.g., Chamberlain et al. 1995, Jong et al. 1998). Three techniques are commonly used to remotely measure frost depths: (1) soil temperature measurements, (2) soil electrical resistivity measurements, and (3) soil dielectric constant measurements. Sensors and hardware used for each of these techniques can readily be deployed in the field, and connected to a datalogger and telecommunications system for automated data storage and subsequent downloading for analysis at the office.

A schematic of a typical setup is shown in Fig. 1. The setup consists of a datalogger equipped with a multiplexer, a cellular phone with a modem, and a probe (or a set of probes) equipped with a series of sensors used to measure soil properties related to frost depth. Additional sensors can be added to monitor meteorological conditions or other parameters relevant to the project. This simple system can be constructed with off-the-shelf hardware and software and can be installed and operated remotely. Solar panels and deep cycle batteries can provide the power needed for operation, and data observation and retrieval can be conducted from the office using a personal computer with a modem. Details regarding how to construct such a setup are described in Benson et al. (1994).

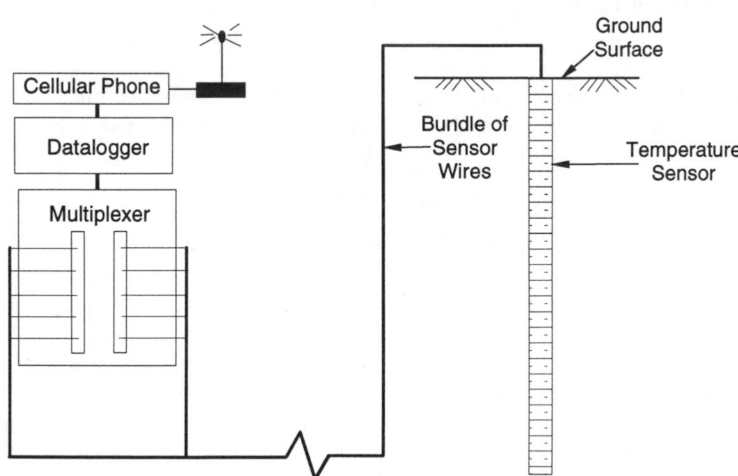

FIG. 1 - Schematic of typical field setup showing datalogger, cellular phone, multiplexer, and string of sensors buried in soil.

This paper reviews the three primary measurement techniques and provides recommendations for their application. The paper is based primarily on the authors' experience in remotely measuring frost depths, and its intent is to share this experience with others practicing cold regions engineering.

Soil Temperature Measurements

Frost depths are often inferred from soil temperature measurements because the free pore water in moist soils without significant salt concentration freezes at about 0°C. Soil temperatures are usually measured using thermocouples or thermistors.

Thermocouples

A thermocouple consists of two wires of dissimilar metal that are joined at the point where the temperature is to be measured. This point is called the "measuring junction." The junction of two dissimilar metals results in an electromotive force (e.g., voltage), which is known as the Seebeck effect. The voltage across the open end of the thermocouple (i.e., the "reférence point") is measured, and is proportional to the difference between the soil temperature at the "measuring junction" and the air temperature at the "reference point." Thus, when using thermocouples, the temperature at the reference point must be known or measured using another technique. The relationship between temperature and voltage is non-linear, and special polynomials are required to convert voltage to temperature (Burns et al. 1993). These polynomials are standardized by the National Institute for Standards and Technology (NIST) and are normally pre-programmed in modern dataloggers. Thus, the user only sees the measured temperature as data.

Type-T (copper-constantan) thermocouples are commonly used in practice because of their suitability for temperatures encountered in most natural environments and their affordability. Typically 20 AWG wire insulated with a PVC jacket is robust enough for field applications, although the authors have successfully used wire as thin as 24 AWG. Stranded wire is preferred because it is less likely to crack during installation or use. However, stranded thermocouple wire is more costly than solid wire.

Thermocouples are normally attached to a multiplexer so that measurements can be made at various depths using one channel on a datalogger. In this case, the reference temperature is the air temperature at the multiplexer connection. This temperature is usually made using a single precision thermistor or a resistance-temperature detector located adjacent to the multiplexer connections. To ensure that the reference temperature is representative of all the multiplexer connections, a metallic heat sink is placed over the connections and the enclosure in which the multiplexer is stored is heavily insulated and sealed against drafts.

A key advantage of using thermocouples is that they can be easily constructed and installed in the field. Thermocouple wire can be purchased in long spools and wire can be laid out from the measuring point to the multiplexer. The measuring junction is then formed by twisting the two wires together at the measurement point and soldering the junction using good quality electronics solder with flux. This seemingly minor advantage becomes significant during installation, because the wire length required between the measuring point and the multiplexer is rarely known a priori. Relatively long lengths are acceptable. The authors have used thermocouples with leads 80 m long with no loss in accuracy (Benson et al. 1994). Another advantage is that the measuring junction can be moist (or submerged) without affecting the measurement.

Thermistors

A thermistor is a temperature-sensitive resistor. Temperature measurements are made by inducing a small current across the resistor, and measuring the corresponding voltage drop. A calibration curve is then used to relate the voltage drop to temperature. The resistance-temperature characteristics of thermistors are highly dependent on the manufacturing process, yielding a unique calibration for each type of thermistor. Thermistor calibrations are normally well-behaved, which makes the conversion from voltage to temperature simple.

As with thermocouples, thermistors are normally connected to a multiplexer so that multiple measurements can be made from a single datalogger channel. However, a temperature reference is not necessary at the multiplexer connection and no special effort must be made to ensure all multiplexer connections have the same temperature.

The key advantage of using thermistors is that no temperature reference is required, which makes multiplexing easier. Thermistors are also more accurate than thermocouples, although the differences in the accuracy become important only when temperature variations in the milli-degree range become essential (Andersland and Ladanyi 1994). The primary disadvantage of thermistors is that the calibration is sensitive to the length of the lead wires, and thus thermistors are normally assembled and calibrated in the laboratory or factory with a fixed length of wire. In addition, the temperature sensitive resistor must be carefully sealed to prevent water entry with a sealant that permits easy heat transfer between the soil and resistor. Leaky thermistors almost always fail in geotechnical applications. In some cases, high freezing pressures can alter the calibration of thermistors, resulting in substantial error.

Comparison

In principle, thermocouples and thermistors should yield equally reliable temperature data for most geotechnical applications. However, engineers usually specify thermistors based on anecdotal reports suggesting that thermistors are more accurate than thermocouples. In a recent study the authors installed thermocouples and thermistors at various depths beneath three secondary highways in Wisconsin to assess whether the two devices yield similar temperatures (Benson et al. 1997, Jong et al. 1998). Temperatures were measured using both devices for 18 mos. and then compared.

Data are shown in Fig. 2 that were collected from a thermistor and a thermocouple installed 45 cm below the base of a pavement near Westby, Wisconsin. The subgrade was silty clay. Nearly identical temperatures were measured with both devices for soil temperatures ranging from -6°C to 25°C. In this case, type T thermocouples were used with a solid state multiplexer equipped with a platinum resistance-temperature detector. Nearly identical results were obtained at other depths and at the other sites (Jong 1998).

The data shown in Fig. 2 illustrate that thermocouples or thermistors can be used reliably to measure soil temperatures. Thus, selection of the appropriate device depends on cost, availability, and ease of installation. In general, thermocouples are less costly (even when thermocouple-specific multiplexers are used), are more easily installed, and are more likely to perform over extended periods even in very moist conditions. For example, Benson et al. (1994) made hourly measurements of soil temperatures at 30 locations with Type-T thermocouples for five years without a single failure.

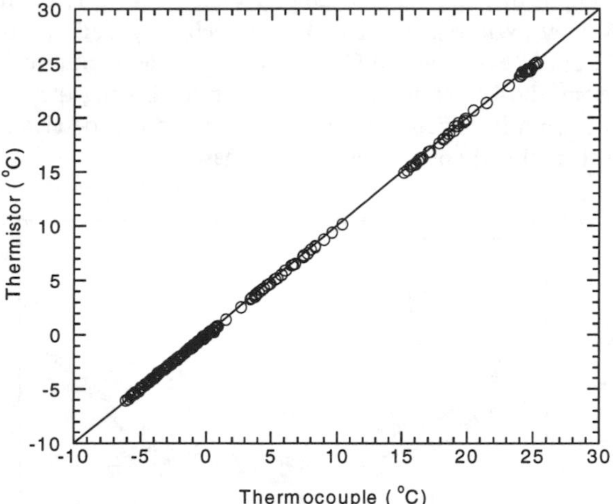

FIG. 2 - Comparison of temperatures measured with a thermistor and a thermocouple below the base of a pavement near Westby, Wisconsin.

Problems with Frost Depths Inferred from Soil Temperatures

Freezing Point Depression -- Freezing point depression is the reduction in the freezing temperature of pore water that occurs as soil becomes unsaturated or the pore water becomes contaminated with salts. In most cases where the soil is moist and the salt concentration is not unusual, the freezing point is depressed very little and the freezing temperature is 0°C for practical purposes. For example, Abichou (1993) found that the freezing point of a low plasticity glacial clay compacted near optimum water content was essentially 0°C. Kraus et al. (1997) report similar results for hydrated bentonite in a geosynthetic clay liner. In cases where the freezing point is depressed significantly, laboratory tests can be conducted to determine the freezing temperature and this temperature can then be used to infer frost depths from soil temperature data.

Freezing point depression is problematic when the salt concentration changes during the monitoring period, resulting in a change in the freezing temperature. For example, Leonards and Andersland (1960) show that contamination with a 1M solution of LiCl depressed the freezing point of a clay to –8°C. This situation can occur in roadway applications where deicing salts are employed. As deicing salts melt overlying ice, the salt-laden melt water infiltrates the subsurface. The frozen pore water then melts with essentially no change in the soil temperature. Consequently, the soil is believed to be frozen when it is actually thawed. Atkins (1979) describes an illustrative case history where this type of event occurred.

Isothermal Conditions -- Another significant problem associated with inferring frost depths from soil temperature data is that nearly isothermal conditions can make locating the frost depth difficult. For example, Fig. 3 shows temperatures beneath a pavement near Westby, Wisconsin. Throughout February and until mid-March, the temperature at 152 cm is essentially 0°C. Similarly, the temperature is about 0°C at a depth of 107 cm from mid-March until the end of April. During these periods the exact location of the frost depth is difficult to determine because it is determined indirectly via temperature rather than directly by monitoring for phase change.

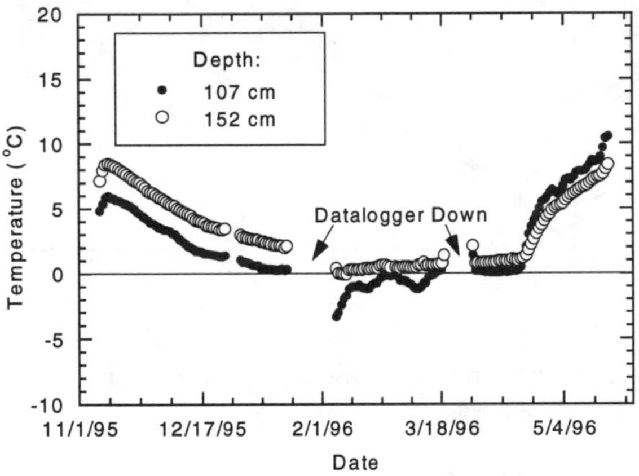

FIG. 3 - Temperatures at depths of 107 cm and 152 cm below the base of a pavement near Westby, Wisconsin (adapted from Jong 1998).

A more complicated condition can exist during spring thaws, when nearly isothermal conditions can exist throughout the near-surface soil profile at a temperature near 0°C (Atkins 1979). When this condition occurs, the frost depth is undefined because freezing conditions cannot be identified at any depth.

Electrical Resistivity Measurements

Another technique to measure frost depths is by measuring the vertical profile of electrical resistivity (ρ_e), or analogously electrical conductivity ($\sigma_e = 1/\rho_e$). This method is superior to temperature measurements because it provides a direct index of phase change. In particular, when water changes to ice, its electrical resistivity increases dramatically even though the temperature of the water may change only slightly. A similar effect occurs in most soils when they freeze or thaw.

Electrical resistivities of three compacted clays varying in plasticity are shown in Fig. 4 (Abu-Hassanein et al. 1995). The electrical resistivity (60 Hz) of these clays varies approximately one order of magnitude as the freezing point is crossed. Similar data are reported by Hoekstra and McNeill (1973). They report electrical resistivities for biotite granite, saturated sandy gravel, silt, and clay and in each case the electrical resistivity jumps at least one order of magnitude as the phase change occurs.

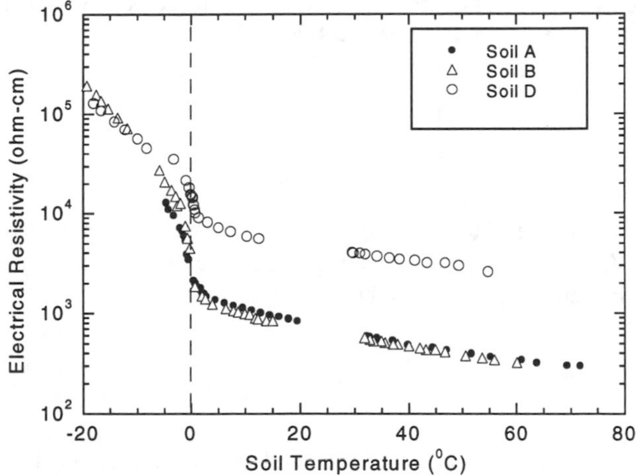

FIG. 4 - Electrical resistivity as a function of temperature for three compacted clays (adapted from Abu-Hassanein et al. 1995).

Probes

Probes for electrical resistivity measurements can be easily made from readily available construction materials or purchased from geotechnical equipment vendors. A detailed description of probe construction can be found in Atkins (1990). A brief summary is described here.

In its simplest form, an electrical resistivity probe consists of an electrical supply source, two electrodes buried in the soil, a measurement resistor, and a voltmeter (Fig. 5).

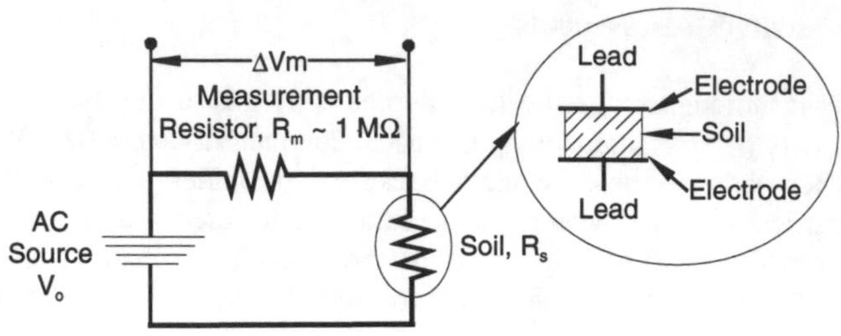

FIG. 5 - Circuit for Electrical Resistivity Probe.

Soil resistivity, R_s, is determined using Ohm's law:

$$V = IR \tag{1}$$

where V is the voltage across a resistor having resistance R with a current I. When Ohm's law is applied to the circuit shown in Fig. 5, R_s can be computed from:

$$R_s = R_m\left(\frac{V_o}{\Delta V_m} - 1\right) \tag{2}$$

where R_m is the resistance of the measurement resistor, V_o is the applied voltage from the source, and ΔV_m is the measured voltage drop across the measurement resistor. Resistivity (ρ_e) is then computed from the resistance, R_s, via:

$$\rho_e = R_s\left(\frac{A}{L}\right) \tag{3}$$

where A and L are the cross-sectional area and length of the soil through which the current flows. Often times A and L are ill-defined because current travels through the soil in a complex three-dimensional field. This problem is circumvented by ensuring that each set of electrodes has the same A and L, which results in the same constant of proportionality between R_s and ρ_e for each adjacent pair of electrodes. Also, inspection of Eq. 2 shows that the resistivity can be inferred simply by the voltage drop across the measurement resistor. Thus, voltage data can be used as a surrogate for resistivity without loss of generality when determining frost depths.

A suitable arrangement for the electrodes is described by Atkins (1990) and is illustrated in Fig. 6. A solid PVC rod is used to retain the electrodes in position. Any practical length rod can be used. The rod is solid so that it does not fill with water during thawing, which could short out the electrodes. PVC pipe or conduit should not be used unless it is filled with material that will not leak. A narrow slot is milled along the length of the rod to carry the wire (24 AWG stranded) for the electrodes. The electrodes are constructed by wrapping 14 AWG copper wire (i.e., ground wire for household electrical lines) around the PVC rod three times, and then twisting and soldering the two ends. One lead wire is also soldered to this junction. The spacing between each adjacent pair of electrodes is held constant so that L and A in Eq. 3 remain the same.

The probe is inserted in a pre-drilled hole having slightly smaller diameter than the rod. A rubber mallet can be used to tap the rod in place. A small hole must be used to ensure good contact exists between the electrodes and the soil and to ensure that an annulus is not present that can collect melt water. The top of the rod should be below ground surface to prevent water from intruding along the soil-rod interface. A small layer of bentonite can be placed above the rod to ensure water does not funnel down along the leads. The leads can daylight directly above the rod, or be routed through a shallow narrow trench to a multiplexer.

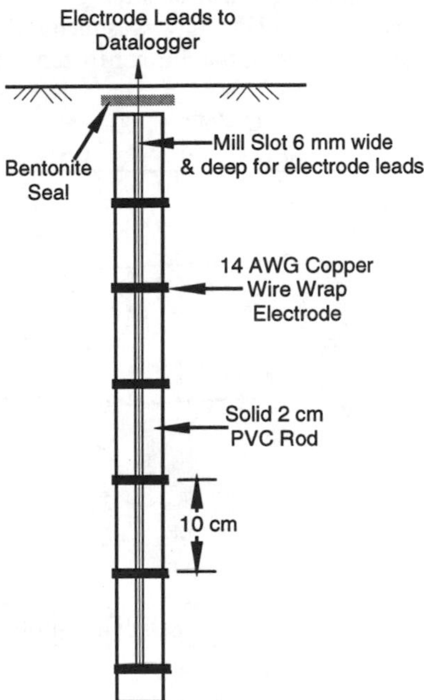

FIG. 6 - Schematic of frost depth probe with electrical resistivity electrodes (adapted from Atkins 1990).

Data Acquisition

Most modern dataloggers can be used to generate the source and to measure the voltage drop at the measurement resistor. The source needs to have a capacity of several milliamps, and must supply an AC signal to prevent polarization of the electrodes. Atkins (1979) suggests an AC frequency < 60 Hz and an operating peak to peak voltage of 3 V. Modern dataloggers generally can output this type of signal at frequencies less than 30 Hz.

A multiplexer is required to switch the electrode leads in and out of the circuit so that electrical resistivity can be measured at various depths along the rod. One single-ended channel is required for each electrode. Voltage across the measurement resistor is measured using a differential input channel on the datalogger each time the multiplexer connects to a pair of electrodes.

Sample Data

A profile of voltage measurements (i.e., ΔV_m) is shown in Fig. 7 that was obtained from a resistivity probe placed beneath a pavement near Rhinelander, Wisconsin. The data were collected on February 8, 1998. Two frozen zones exist. The deeper zone is from a deep freeze in early January where the frost depth had reached a depth of 160 cm. Top-down thawing followed during a warm period in late January, which thawed the upper 70 cm of soil. This thaw was then followed by a short freeze and a subsequent light thaw, which resulted in the thinner frozen zone near the surface.

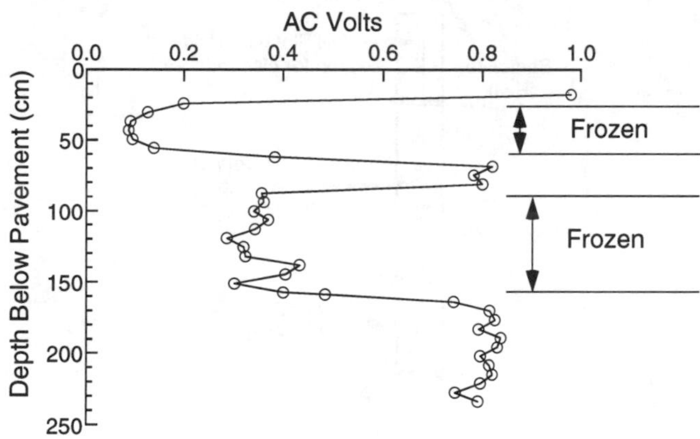

FIG. 7 - Voltage profile from frost depth resistivity probe beneath pavement near Rhinelander, Wisconsin.

A case history is described by Moo-Young et al. (1999) where resistivity probes were used to determine if frost penetrated the barrier layer in a landfill cover. The barrier layer was constructed with paper mill sludge. In this application, the resistivity probe was used to show that frost did not penetrate the paper mill sludge, whereas temperature measurements showed that a portion of the sludge layer dropped below 0°C during the winter months.

Advantages and Disadvantages

The primary advantage of the electrical resistivity technique is that the response being monitored is directly affected by the phase change and thus the frost depth can be accurately determined. Additionally, the probes are relatively simple to construct and can be connected to conventional multiplexers used with dataloggers. No special instrumentation is required and the cost of construction and installation is only slightly greater than the cost associated with thermocouples.

A disadvantage of the technique is that different soils have different electrical resistivity at the same degree of saturation due to differences in the electrolyte in the pore water and surface conductance (Andersland and Ladanyi 1994, Abu-Hassanein and Benson 1994). Thus, the voltage drop and electrical resistivity are not uniquely related to phase change. A clear understanding of the subsurface layering is necessary to employ the method. Another disadvantage is that gaps between the probe and soil invalidate the method.

Dielectric Methods

Like electrical resistivity, the dielectric constant of moist soil is significantly affected by freezing and thawing of the pore water. Thus, frost depth can be determined by monitoring the soil dielectric constant, K_a. The soil dielectric constant is the square-root volume-weighted average of the dielectric constants of the soil phases (Whalley 1993): air (K_o), solid (K_s), water (K_w), and ice (K_i):

$$\sqrt{K_a} = f_o\sqrt{K_o} + f_w\sqrt{K_w} + f_s\sqrt{K_s} + f_i\sqrt{K_i} \tag{4}$$

where f_o, f_w, f_s, and f_i are the volume fractions of air, water, solids and ice (i.e., $f_o + f_w + f_s + f_i = 1$). Because soil solids, pore water, air, and ice have different dielectric constants (3-10, 80, 1, and 3, respectively) (Topp et al. 1980, Patterson and Smith 1981), the apparent dielectric constant of the soil changes significantly as the unfrozen volume fraction of water changes.

The most common technique to measure K_a is time domain reflectometry (TDR), which is an electromagnetic technique similar to RADAR. A pulse generator sends electrical pulses down a coaxial cable. When the pulse encounters a change in the electrical properties of the cable, a portion of the pulse is reflected back towards the generator. The elapsed time between emission of the pulse and reception of the reflection is used to determine the speed of the pulse (V_p), which is a function of K_a:

$$V_p \approx \frac{c}{\sqrt{K_a}} \tag{5}$$

For soil measurement, V_p is determined by measuring the travel time of the electromagnetic pulse in a set of un-insulated conductors buried horizontally in the soil. The conductors are generally placed at the end of a coaxial cable and are referred to as a "waveguide." A three-rod waveguide is most common (Fig. 8), although two-rod waveguides are used. The rods are typically made from stainless steel and are 2 mm in diameter, 30 cm long, and separated by 2 cm (Benson and Bosscher 1998).

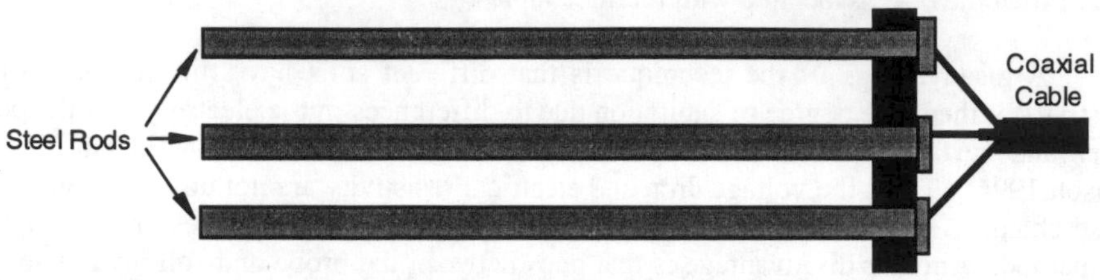

FIG. 8 - Three-rod TDR waveguide.

A typical TDR waveform from a waveguide is shown in Fig. 9 as a graph of voltage versus apparent distance. The waveform is the response recorded by the device that generates and receives the pulses. Such devices are commonly called "cable testers," because they are employed for testing utility cables. The apparent distance is determined based on the travel time of pulses measured by the cable tester and assuming V_p equals the speed of light (i.e., K_a = 1 in Eq. 5). The apparent length of the probe, L_a, is determined from the difference in the apparent distances corresponding to the inflection points in the waveform created by the beginning and ending points of the probe (Fig. 9).

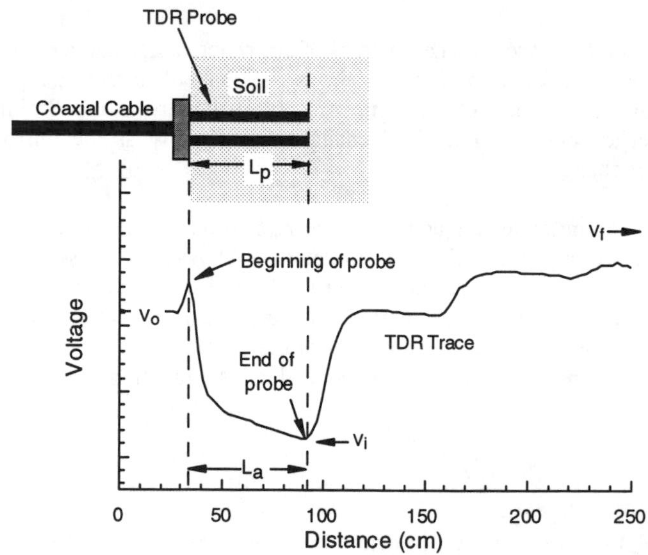

FIG. 9 - TDR Waveguide and Waveform (adapted from Suwansawat and Benson 1998).

The true length of the probe, L_p, is known. Thus, from Eq. 5,

$$K_a = \left(\frac{L_a}{L_p}\right)^2 \tag{6}$$

because L_a is determined assuming K_a =1 or effectively V_p = c.

K_a Response During Freezing

The change in K_a that occurs during freezing is illustrated in Fig. 10. The data were collected beneath a pavement near Unity, Wisconsin in a layer of clayey sand subgrade. A distinct drop in K_a occurs when the soil freezes because the dielectric constant of water is 80 and that of ice is 3. Similarly, a distinct jump in K_a occurs when the soil thaws.

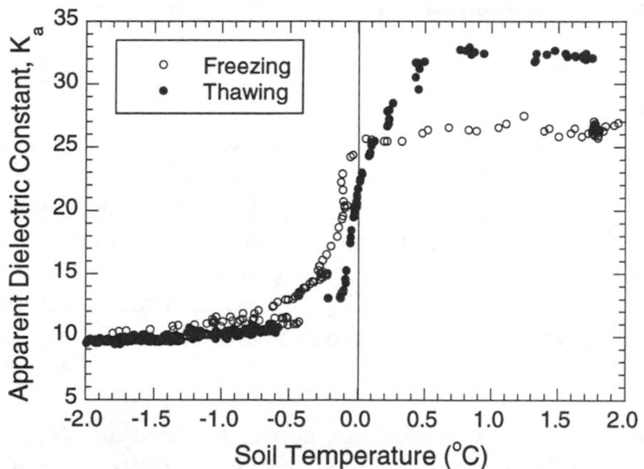

FIG. 10 - Apparent dielectric constant versus soil temperature beneath a pavement near Unity, Wisconsin (adapted from Jong et al. 1998).

Figure 11 shows how dielectric constant measurements are superior to temperature measurements for assessing phase change. The data were obtained 60 cm beneath a pavement in a silty clay subgrade at a site near Spirit, Wisconsin. The temperature slowly fell below 0°C over a three-week period in early December 1995 as the soil began to freeze. Based on the temperature data, frost could have arrived at a depth of 60 cm any day during this period. In contrast, the dielectric constant dropped dramatically, as the frost depth reached and passed 60 cm within two days in early December. Similar statements can be made regarding the thaw period, which began in early April 1996.

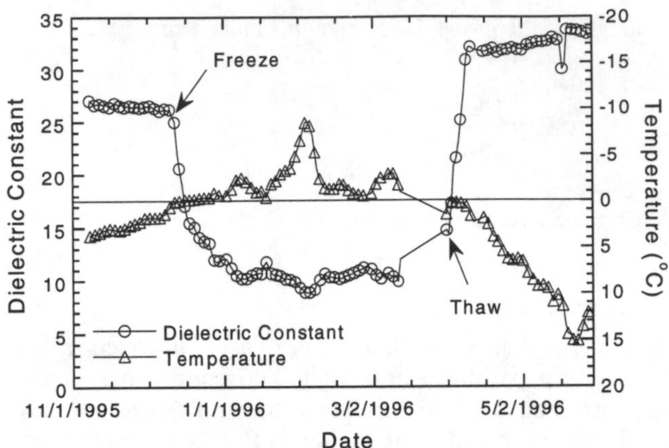

Fig. 11 - Dielectric constants and temperatures measured beneath a pavement near Spirit, Wisconsin (adapted from Benson et al. 1997).

Advantages and Disadvantages

The primary advantage of measuring frost depth by monitoring dielectric constant is that the phase change is clearly indicated by an abrupt change in K_a. And, unlike electrical resistivity measurements, similar K_a are obtained for most soils having similar water content. That is, for moist soils a nearly unique relationship exists between similar K_a and the phase of the water (K_a ~ 5-10 for frozen soil and ~ 15-25 for moist unfrozen soil). Another advantage of measuring K_a is that unfrozen water contents can be determined simultaneously with frost depth (Stein and Kane 1983, Smith and Tice 1988, Spaans and Baker 1995).

The primary disadvantage of dielectric constant measurements is the high cost of the equipment. A cable tester and its electronic interface typically costs about US$10,000. In addition, special coaxial multiplexers must be used that have a limited number of channels (8 - 10). These coaxial multiplexers are typically twice as expensive as traditional multiplexers used with dataloggers. The probes are also more difficult to construct, and require a large diameter hole (> 30 cm) for installation.

Summary and Conclusions

The three primary methods to measure frost depth have been reviewed in this paper: (1) temperature measurements made with thermocouples or thermistors, (2) electrical resistivity measurements and (3) dielectric constant measurements. The advantages and disadvantages of each measurement technique have been described, and their relative accuracy in predicting frost depths has been discussed using field data.

Techniques to incorporate the three methods in conjunction with field data acquisition equipment have also been described.

Comparison of the advantages and disadvantages of the three methods indicates that the best method to measure frost depth is to monitor the dielectric constant using time domain reflectometry (TDR). This method is extremely reliable, because the phase change from water to ice (and vice versa) is associated with an abrupt change in the dielectric constant. Moreover, essentially the same dielectric constants are obtained, regardless of soil type. The TDR method is also easy to implement, but the instrumentation is expensive. Electrical resistivity methods are the next most desirable because a large change in electrical resistivity of soil occurs when the pore water changes phase. The electrical resistivity method is also far cheaper to implement than the TDR method, but electrical resistivity is sensitive to soil type and installation quality. Measuring temperatures with thermocouples or thermistors is the least desirable method, but is also the least expensive and the simplest to implement. Either thermocouples or thermistors can be used, although installations using thermocouples typically are less expensive. The main drawback associated with the temperature method is that freezing and thawing temperatures in soil do not equal those of free water, and can change as the solute concentration of the pore water changes. Consequently, soil temperature can be an unreliable index of frost depth.

Acknowledgement

A variety of sponsors have supported the authors' research on field instrumentation including the National Science Foundation, the United States Environmental Protection Agency, Envotech Limited Partnership, NCASI, the Wisconsin Dept. of Transportation (WisDOT), and WMX, Inc. This support is gratefully acknowledged. The findings and opinions reported in this paper are solely those of the authors, and endorsement by the sponsors should not be implied. Brian Gaber of WisDOT provided the data from the resistivity probe in Rhinelander, Wisconsin.

References

Abichou, T. (1993), Field evaluation of geosynthetic insulation for protection of clay liners, MS Thesis, University of Wisconsin-Madison.

Abu-Hassanein, Z. and Benson, C., and Blotz, L. (1996), Electrical resistivity of compacted clays, *J. of Geotechnical Engineering,* 122(5), 397-407.

Abu-Hassanein, Z. and Benson, C. (1994), Using electrical resistivity for compaction control of compacted soil liners, *Proc., Tailings and Mine Waste '94,* Jan. 19-21, Fort Collins, CO, 177-189.

Andersland, O. and Ladanyi, B. (1994), *An Introduction to Frozen Ground Engineering*, Chapman and Hall, New York.

Andersland, O., Wiggert, D., and Davies, S. (1996), Hydraulic conductivity of frozen granular soils, *J. of Environmental Engineering*, ASCE, 122(3), 212-216.

Atkins, R. (1979), Determination of frost penetration by soil resistivity measurements, *Special Report 79-22*, US Army Cold Regions Research and Engineering Laboratory, Hanover, NH.

Atkins, R. (1990), Frost resistivity gages: assembly and installation instructions, *Technical Note*, US Army Cold Regions Research and Engineering Laboratory, Hanover, NH.

Benson, C. and Bosscher, P. (1998), Metallic time domain reflectometry (TDR) in geotechnics: a review, *Nondestructive and Automated Testing for Soil and Rock Properties, ASTM STP 1350*, W. A. Marr and C. E. Fairhurst, Eds., American Society for Testing and Materials, West Conshohocken, PA, 1998, in press.

Benson, C., Bosscher, P., and Jong, D. (1997), Predicting seasonal changes in pavement stiffness and capacity caused by freezing and thawing, Geotechnical Engineering Report 97-9, University of Wisconsin-Madison.

Benson, C., Bosscher, P., Lane, D., and Pliska, R. (1994), Monitoring system for hydrologic evaluation of landfill final covers, *Geotech. Testing J.*, ASTM, 17(2), 138-149.

Benson, C. and Othman, M. (1993), Hydraulic conductivity of compacted clay frozen and thawed in situ, *J. of Geotech. Eng.,* ASCE, 119(2), 276-294.

Burns, G., Scroger, M., and Strouse, G. (1993), Temperature-electromotive force reference functions and tables for the letter-designated thermocouple types based on the IPTS-90, *NIST Monograph 175*, Washington, DC.

Chamberlain, E., Erickson, A. and Benson, C. (1995), Effects of frost action on compacted clay barriers, *Geoenvironment 2000*, ASCE, GSP No. 46, 702-717.

Chamberlain, E., Iskander, I., and S. Hunsicker (1990), Effect of freeze-thaw cycles on the permeability and macrostructure of soils, *Proc. Frozen Soil Impacts on Agricultural, Range, and Forest Lands*, K. Cooley, ed., US Army Cold Regions Research and Engineering Laboratory, Hanover, NH.

Hoekstra, P. and McNeill, D. (1973), Electromagnetic probing in permafrost, *Proc. 2nd Intl. Conf. on Permafrost*, Yakutsk, USSR, National Academy of Sciences, Washington, DC, 517-526.

Janoo, V.C., and Berg, R.L. (1996), PCC airfield pavement response during thaw-weakening periods: a field study, *CRREL Report 96-12*, U.S. Army Corps of Engineers, Cold Regions Research and Engineering Laboratory, Hanover, NH.

Jong, D., Bosscher, P., and Benson, C. (1998), Field assessment of changes in pavement moduli caused by freezing and thawing, *Transportation Research Record*, in press.

Jong, D. (1998), Freeze-thaw effects on the stiffness of flexible pavement systems, MS Thesis, Dept. of Civil and Environmental Engineering, University of Wisconsin-Madison.

Kraus, J., Benson, C., Erickson, A., and Chamberlain, E. (1997), Freeze-thaw and hydraulic conductivity of bentonitic barriers, *J. of Geotech. and Geoenvironmental Eng.*, ASCE, 123(3) 229-238.

Leonards, G. and Andersland, O. (1960), The clay-water system and the shearing resistance of clays, *Proc. Research Conference of Shear Strength of Cohesive Soils*, ASCE, 793-818.

Mahoney, J., Lary, J., Sharma, J., and Jackson, N. (1985), Investigation of seasonal load restrictions in Washington State," *Transp. Research Record 1043*, Transportation Research Board, National Research Council, Washington, D.C., 58-67.

Moo-Young, H., LaPlante, C., Zimmie, T., and Quiroz, J. (1999), Field measurements of frost penetration into a landfill cover that uses a paper sludge barrier, *Field Instrumentation for Soil and Rock, ASTM STP 1358*, G. N. Durham and W. A. Marr, Eds., American Society for Testing and Materials, West Conshohocken, PA, 1999.

Nordal, R. and Hansen, E. (1987), *The Vormsund Test Road, Part 4: Summary Report,* Norwegian Road Research Laboratory.

Patterson, D. and Smith, M. (1981), The measurement of unfrozen water content by time domain reflectometry: results from laboratory tests, *Canadian Geotech. J.*, 18, 131-144.

Smith, M. and Tice, A. (1988), Measurement of the unfrozen water content of soils; comparison of NMR and TDR methods, Report 88-18, U.S. Army Cold Regions Research and Engineering Laboratory, Hanover, NH.

Spaans, E. and Baker, J. (1995), Examining the use of time domain reflectometry for measuring liquid water content in frozen soil, *Water Resources Research*, 31(12), 2917-2925.

Stein, J. and Kane, D. (1983), Monitoring the unfrozen water content of soil and snow using time domain reflectometry, *Water Resources Research*, 19(6), 1573-1584.

Suwansawat, S. and Benson, C. (1998), Cell for water content calibration using time domain reflectometry, *Geotech. Testing J.*, ASTM, in press.

Topp, G., Davis, J., and Annan, A. (1980), Electromagnetic determination of soil water content, *Water Resources Research*, 16(3), 574-582.

Tumeo, M. and Davidson, B. (1993), Hydrocarbon exclusion from groundwater during freezing, *J. of Environmental Engineering*, ASCE, 119(4), 715-724.

Whalley, W. (1993), Considerations on the use of time-domain reflectometry (TDR) for measuring soil water content, *J. of Soil Science*, 44(1), 1-9.

L. Randall Welch and Paul E. Fields

Integrated Automation of the New Waddell Dam Performance Data Acquisition System

Reference: Welch, L. R., and Fields, P. E., **"Integrated Automation of the New Waddell Dam Performance Data Acquisition System,"** *Field Instrumentation for Soil and Rock, ASTM STP 1358,* G.N. Durham and W.A. Marr, Eds., American Society for Testing and Materials, West Conshohocken, PA, 1999.

Abstract: New Waddell Dam, a key feature of the U.S. Bureau of Reclamation's Central Arizona Project, had elements of its dam safety data acquisition system incorporated into the design and construction. The instrumentation array is a reflection of the dam's large size and foundation complexity. Much of the instrumentation is automated. This automation was accomplished while maintaining independent communication connections to major divisions of the instrument array. Fiber optic cables are used to provide high quality data, free from voltage surges that could originate in a nearby powerplant switchyard or from lightning. The system has been working well but there are concerns with a lack of continued equipment manufacturer support.

Keywords: dam performance, automated dam monitoring, New Waddell Dam, fiber optic cable

Introduction

New Waddell Dam is 104 m high with a length of 1 433 m. It is located on the south end of the 1 360 x 10^6 m^3 Lake Pleasant reservoir about 40 km northwest of Phoenix, Arizona (Designers' Operating Criteria 1993). It is a zoned embankment structure having a thin central impervious core, internal chimney and blanket drainage zones and gravel cobble fill shells. Slopes are 2:1 downstream and 2 1/4:1 upstream. Construction was completed in 1992.

The dam is founded on bedrock except in the narrow valley center where the foundation excavation extends only to the top of a dense, permeable, older alluvium. Crest elevation, 527 m, provides 1.5 m of freeboard above the maximum water surface

Technical Specialist and Automation Specialist, respectively,
Structural Behavior and Instrumentation Group,
U. S. Bureau of Reclamation, P.O. Box 25007, Denver, CO 80225.

elevation. Crest width is 12 m and paved to enhance the ability to detect deformation cracks and to increase erosion resistance. A 183 m-long grouting gallery is beneath the embankment core on the right abutment just downstream of embankment centerline and upstream of a pump/generating (P/G) plant.

Hydraulic structures include two 4.3 m-diameter Central Arizona Project (CAP) outlet works tunnels (connect to the P/G Plant), a 4.9 m-diameter river outlet works tunnel, a 198 m-wide ogee crest overflow service spillway, and a 107 m-wide fuseplug dike auxiliary spillway. Outlet capacities are 122 m^3/s and 158 m^3/s, respectively. The service spillway, capacity 5 235 m^3/s, is on the reservoir rim and approximately 2 286 m west of the right end (looking downstream) of the dam. The auxiliary spillway, capacity 3 540 m^3/s, is on the right reservoir rim about 1 220 m west of the service spillway.

New Waddell Dam provides the primary storage reservoir for the CAP. It also stores and passes the flows of the Agua Fria River, part of which is allocated to the Maricopa County Municipal Water Conservation District.

FIG. 1-*New Waddell Dam*

CAP water is delivered to the dam, from the Colorado River at Lake Havasu, through nearly 260 km of the Hayden-Rhodes Aqueduct system and about 10 km of the reversible Waddell Canal. Power is generated from the releases of reservoir water through the Waddell Pumping-Generating Plant at the base of the dam on the right abutment. Figure 1 is a photo of the dam.

Foundation

The site location for New Waddell was not a first choice. It was selected after environmental considerations ruled out more favorable locations. Geology of the foundation has a complex history and structure. This includes a fault zone, three buried stream channels, layers of weak material and layers of high permeability. Foundation rock consists largely of volcanic material such as tuff and andesite.

Instrumentation

New Waddell Dam was provided with instruments to allow the measurement of embankment and foundation deformation, appurtenant structures deformation, collected seepage flow, and piezometric pressures. Specific portions of the embankment and foundation were selected for internal deformation measurement and provisions for monitoring general surface deformation of the embankment were also made. Structural measurement points were installed in both CAP outlet works tunnels, intake towers, and service spillway walls. In addition, tiltmeters were installed on each intake tower and each of the three associated access-bridge piers. Weirs were installed in the toe drain system to allow monitoring of the seepage trends of individual segments of the foundation. Arrays of piezometers were installed at five cross sections and along the downstream embankment toe. Data from those piezometers show the effect of cutoffs and drainage features upon the piezometric pressure distribution.

Automation of a portion of the instrumentation was desirable for two reasons. 1) Due to the difficult nature of the foundation, many instruments were installed to check design assumptions. (Monitoring during first filling was necessarily frequent and the number of instruments involved required a large manpower commitment even with automation.) 2) The dam is upstream of a large and rapidly growing suburban area. Consequently, future dam safety considerations are likely to continue requiring the highest level of instrument monitoring and dam surveillance.

Generally, nine areas of the embankment are instrumented (Prudhom 1995). Other than the embankment surface, downstream toe, and appurtenant structures, these areas include the following six instrument sections:

- Section A, a single zone of highly fractured rock (an inactive fault zone) beneath the far right embankment and downstream of a secant pile cutoff wall.

- Section B, a shallow but steep-sided ancient stream channel backfilled beneath the embankment in the vicinity of the outlet works tunnel.

- Section C, downstream of a shallow cutoff wall beneath the central right portion of the embankment.

- Section D, downstream of a deep cutoff wall beneath the central left portion of the embankment.

- Section E, a zone of possible weak tuff material beneath the downstream embankment shell at the foot of the left abutment.

- Section F, a large mass of core material overlying a steep left abutment slope.

Below is a table of the phenomena measured and instruments used to obtain the measurements:

TABLE--*New Waddell Dam instrumentation*

Measurement		Instrument
Deformations	Embankment Surface	Measurement Points Reference Piers
	Embankment Internal	Inclinometers Extensometers*
	Foundation Internal	Inclinometers Shear Indicators*
	Appurtenant Structures	Measurement Points Tiltmeters*
Piezometric Level	Cutoff Wall and Drain Efficiency	Vibrating-wire Piezometers*
	Foundation Water Pressure	Vibrating-wire Piezometers*
	Seepage Exit Gradient	Porous-Tube Piezometers*
	Local Groundwater Level	Observation Wells
Seepage Quantity	Toe Drain Flow	V-notch Weirs*
	Seeps and Springs	Various Devices

* Automated Readings

Design provided ready accommodation for an automated data acquisition system. The layout and components are such that most of the piezometric level monitoring, much of the seepage quantity monitoring, and a significant portion of the internal and structural deformation monitoring were easily automated. Five instrument terminals service the six instrument sections listed above. A fiber optic cable system connects these instrument terminals to a central monitoring station in the switchyard control house. Fiber optic cable was chosen for its indifference to lightning and radio interference and superior data transmission. Automated installations are indicated by an asterisk in the table above and are the only instruments discussed below.

Figure 2 shows general locations of the embankment performance instruments with an emphasis on the automated installations. Measurement points, observation wells and other non automated instruments are not shown.

Internal Embankment Deformations

Measurements of internal embankment deformations are obtained from inclinometers and embedded extensometer installations. Only the extensometers are automated. They allow measurements of lateral compression and extension along lines parallel to the embankment centerline, at 6 m intervals, at the elevations of the installations. Locations of the extensometers are in instrument section F.

The possibility of embankment cracking, caused by an abrupt change in steepness of the left abutment profile, is monitored with an embedded extensometer installation. The installation is a series of extensometers placed end-to-end to span the entire mass of the core material that may be subjected to tensile stress. It is high in the core and along the embankment centerline.

Each individual interval of the extensometer installation (assembly) is 6 m long and can measure up to +/- 75 mm of horizontal displacement parallel to the embankment centerline. The left end of the installation is anchored into a stable rock portion of the foundation creating a reference point. Horizontal displacements of the flanges on both ends of each extensometer interval are measured relative to the stationary reference point to provide absolute movements.

Readings of all the extensometer intervals are obtained at a central location, instrument terminal F. A switch box within the terminal allows selecting individual intervals. Each interval is identified by the number of the flange on the right end. Readings are displayed as a percent of the 150 mm full range. Measurement resolution is 0.15 mm.

Internal Foundation Deformation

Measurements of internal foundation deformations are obtained from three shear indicators and four inclinometer casing installations. The shear-indicators show the elevation of horizontal displacement beneath the downstream shell. Use of them reduces the needed frequency of inclinometer readings. The shear-indicators consist of a parallel electrical circuit of resistors. Readings are simply a count of the number of resistors in the circuit from each end. A shear (break) in the circuit results in lower reading values.

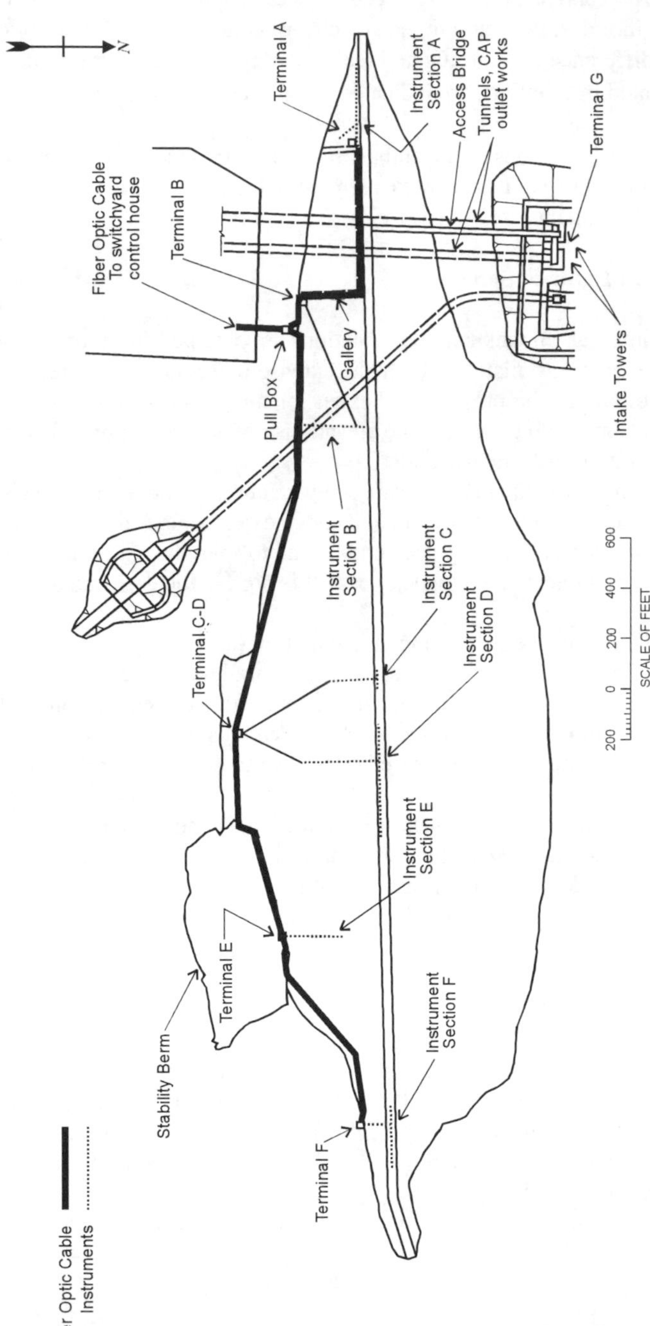

FIG. 2 - *Instrumentation Cable & Terminal Location*

Shear Indicators

All shear indicators extend 30 m into possibly weak foundation material, at instrumentation section E. These shear indicators, along with nearby piezometers to show foundation pore pressure, make up instrument section E. Like the inclinometers, these installations provide data on horizontal deflections, however, no measurement of the amount of shear is given. They are capable of indicating that shear of 3 mm or more has occurred and approximately at what depth it occurred.

Shear indicators are read quickly and thus more frequently than inclinometer casings. If shear is shown, the manual reading frequency of nearby inclinometers would be increased. Polyvinyl chloride (PVC) casing was installed within the shear-indicator borehole to protect the installation from minor displacements that could result from caving or other readjustment of stress that occurs only near the borehole. Because of the specified spacing of electrical resistance elements (0.3 m) the shear zone elevation bounds can be determined within 0.3 m.

Appurtenant Structures Deformation

Tiltmeters are installed atop each of three piers supporting the CAP intake tower access bridge and atop the two intake towers. The bridge pier installations are placed horizontally on the upper surface of the bridge pier cap, between the bridge girders. Tower tiltmeters are on the floors inside each respective control house.

Each tiltmeter contains two orthogonally fixed electrolytic transducers to allow measurements of tilt in two planes. The resolution is 0.2 arc second with a repeatability of 0.4 arc second. The range is ±4.6°. A power supply unit and switch box that services all the tiltmeters are in the right CAP tower. This unit operates on 105-205 VAC with internal backup batteries. A portable digital readout unit can be connected to the switch box to obtain readings. These readings, even when not automated, are obtained easily and much more frequently than surveys of the associated structural measurement points. Use of the tiltmeters reduces the number of costly surveys that would otherwise be required to monitor the stability of the tall structures associated with the CAP tunnels.

Seepage Quantity

Seepage is collected and measured by a drain system that runs within and parallel to the downstream toe of the embankment. The surface of the embankment foundation downstream of the core is divided into sections by concrete baffle walls contained within the blanket drain. Each such section drains through a single V-notch weir placed within a toe drain inspection well located at the lowest corner of the area defined by the baffle walls. Seven weirs are on the right abutment and two on the left.

Accuracy of the measurement depends upon flow, however, the measurements are expected to be accurate to the nearest 10 L/s for flows between 75 and 1 500 L/s. Besides a staff gauge, an electrical (vibrating-wire) pressure transducer is installed upstream of each weir. Cables from the weir transducers are routed to the nearest instrument terminal. This allows the weirs to be read remotely at the instrument terminals and be automated.

The weir transducers are placed deep into the weir pool so that a zero reading on the weir crest would still result in a significant amount of pressure on the transducer. That was done to avoid the problems of obtaining accurate readings for near zero flows. The depth range of these transducers is zero to 3.4 m, however, their calibration is based only upon the transducer behavior in the zero to 0.6 m range.

Cutoff Wall and Drain Efficiency

The primary performance monitoring at instrumented sections, A through D, is for the efficiency of foundation-seepage cutoff features.

Efficiency of the cutoff walls is determined by the hydrostatic pressure distribution within the foundation and lower portions of the embankment core. At instrument section A, it is the efficiency of a secant pile cutoff wall that is of interest. At section B, it is the efficiency of a grout curtain and backfill of an excavated buried channel that is of interest. An absolute determination of efficiencies will not be available because, to preserve core integrity by avoiding the use of cable trenches upstream of centerline, no instruments are installed upstream of the embankment centerline. Four piezometers distributed between instrument sections B, C and D and embedded in trenches in the lower core are the only piezometers embedded in the core. They show the piezometric effect resulting from seepage passing over the top of cutoff features.

Only vibrating-wire piezometers (VWs) are used. Closed piezometers are necessary to provide a practical means to monitor piezometric levels beneath such a large embankment. Cables run downstream from the transducers, within the sandy portion of the drainage blanket, to the nearest instrument terminal.

All vibrating-wire piezometers and transducers at the site contain thermistors. This allows the application of temperature corrections. For piezometers, temperature corrections are insignificant due to the low in situ temperature variation and instrument insensitivity to such variations. However, using seasonal variations of temperature data does provide a sense of speed and/or quantity of seepage flow. Only larger flows moving relatively fast can significantly affect the surrounding ground temperatures.

At sections A through D, the piezometers closer than 15 m to the embankment centerline and deep, below elevation 488 m, have a range of zero to 168 m of pressure head (17 of these high range transducers are in use). All others beneath the embankment have a lesser range of zero to 67 m. Specified accuracy of all vibrating-wire piezometers was 0.5 percent of full range.

Instrument section A has three piezometers in each of five drill holes beneath the downstream shell and downstream of a secant pile cutoff wall. In a plan view the array forms a general "T"-shape. The cap of the "T" is a plane of piezometers located near the wall to detect wall leakage. The stem is placed to map the energy dissipation of seepage as it moves downstream of the wall. (This pattern of piezometers is used downstream of each of the three foundation cutoff walls). One other piezometer is placed within the base of the chimney drain above each "T" section. At depth, these piezometers show the upward gradient. The shallow piezometers, including the chimney drain installation, show the capacity of the drainage system to convey collected seepage.

Instrument section B has two shallow foundation piezometers and four embankment piezometers. The embankment piezometers provide data from a horizontal slice of the piezometric pressure distribution and with the foundation installations provide an indication of flow direction. Effectiveness of the drainage features is shown by the three downstream piezometers in an upper row. Foundation installations are high range transducers beneath the core.

At instrument section C two columns of two piezometers each are installed in drill holes just downstream of the cutoff wall beneath the downstream portion of the core and beneath the drainage blanket. Also, two columns of two piezometers are installed at two separate offsets downstream of the cutoff wall. This foundation array allows a mapping of the piezometric pressure distribution in the region of the foundation and core just downstream of the cutoff wall and reveals any foundation piezometric levels greater than the tailwater elevation. One piezometer is embedded in the lower core downstream of the cutoff wall. Six piezometers of this group of 10 are 15 m or more downstream and have a range of zero to 67 m of pressure head. The other four have a zero to 167 m range.

Instrument section D is similar to sections A and C. However, the array is larger and contains 18 piezometers. The downstream side of the cutoff wall is monitored with three elevations of piezometers. Each piezometer is installed in a drill hole and has a high range. There is also a line of two embedded in the lower core (monitoring wall overflows) and one column of three under the downstream core edge and one column of three under the drainage blanket. As in section C, the array shows where pressure is greater than tailwater, the direction of flow, and the ability of the blanket drain to convey collected seepage.

Foundation Pore Pressure

Instrument section E contains three columns of three piezometers each beneath the downstream shell. These vibrating-wire piezometers, installed in drill holes, are for monitoring pore water pressure within possibly weak tuffaceous foundation materials. Each column of 0- to 60 m range piezometers is near a shear indicator. The drill holes extend to a depth of 30 m.

Seepage Exit Gradients

Seventeen porous-tube piezometers (PTs) placed approximately at a 30 to 60 m spacing are along the downstream embankment toe. These locations are on the abutments above tailwater. All are used to detect the presence of water in zones of higher permeability and to allow an estimate of the exit gradients if seepage does occur. In some cases these toe installations are complimentary to the instrumented sections. All behave as standpipe installations, however, each is outfitted with a vented vibrating-wire transducer having a range of zero to 34 m of head. Cables from these transducers are routed to the nearest instrument terminal. The transducers allow the remote standpipes to be read more quickly and at a central location. If a transducer malfunctions, manual readings can be taken and the transducers easily replaced.

Data Acquisition System

As shown on Figure 2, there are six instrument terminals. Each contains a conventional 120-volt power supply with a battery backup. These terminals allow centralizing the reading of 177 instruments. Switch boxes that allow manual as well as automated readings for all of the following instrument installations are found in the terminals:

- 83 vibrating-wire piezometers
- 9 weir vibrating-wire transducers
- 69 thermistors
- 3 shear indicators
- 8 extensometers intervals
- 5 tiltmeters

These terminals are designated as A through G to correspond to the instrument sections. Two instrumentation sections, C and D, terminate at a single terminal designated as C-D. Terminals A and B are bolted on the walls within the right abutment grouting gallery; A is in the west end near the junction with a west access adit; B is near the east entrance. Terminals C-D, E and F are within concrete manhole structures as shown on Figure 3. Terminal G is the tiltmeter wiring panel in the right CAP intake tower control house. All terminals are supplied with electrical power, grounding, and protection from transient voltage and electromagnetic interference (EMI). EMI protection is circuitry within the switch boxes that contains gas discharge tubes. This is only limited protection but helps insure that an electrical potential that arises at the terminal will not be transmitted to the transducers.

Each terminal other than G is connected by a separate heavy-duty fiber optic(FO)cable to a point under a raised floor in the P/G plant switchyard control house. On each FO cable end is a bidirectional signal conversion interface device called an optical-transceiver. They are AC powered and convert electrical signals to optical signals and vice versa. Each is provided with a water-resistant enclosure.

The optical fiber is multi-mode having a 62.5 micron core specified to have no signal point loss greater than 0.2 decibels. Signal attenuation is no more than 3.5 decibels per kilometer with a bandwidth of at least 160 megahertz. No splices were allowed. Each FO cable was field tested twice, once on the reel after delivery and again immediately after installation. Optical time domain reflectometry was used on both cable ends in the testing to verify cable length, signal attenuation and that there were no discontinuities or breaks. The optical fiber is embedded in a loose tube baffle and has a minimum bend radius of at most 150 mm. "ST" connectors were used in the coupling with the transceivers.

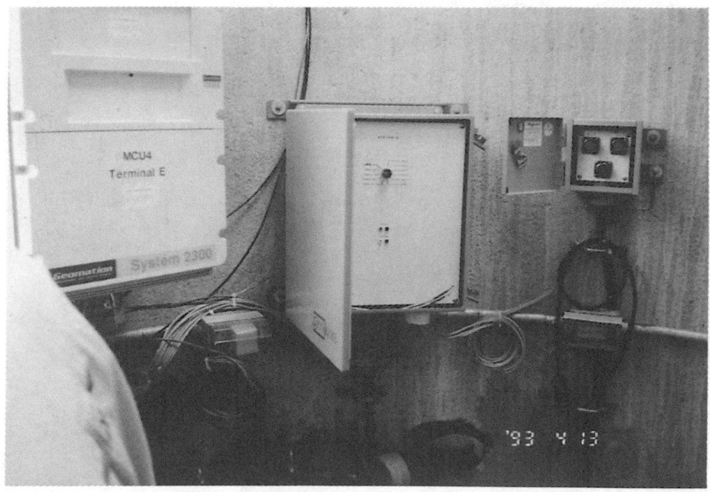

FIG. 3-*Typical terminal well equipment layout.*

FO cable routing is through a 75 mm diameter buried flexible polyethylene electrical duct. No elbows or other joints were needed to make duct bends. The duct lays flat when uncoiled, was lubricated internally and contained a 6 mm rope for pulling the FO cable.

Terminal G is connected to the central network monitor by UHF radio through terminal F.

During construction, automation equipment was installed at each terminal location as instrumentation installations were completed. This provided an opportunity to evaluate and test the automation equipment prior to first filling. Reclamation personnel performed all system installation, programming and testing.

A measurement and control unit (MCU) capable of reading up to 70 instruments is located at each terminal. The MCUs are capable of reading a variety of instruments depending upon which multiplexer modules are installed. Each is programmable and independently performs all physical measurements and computations based on user specified parameters according to specified scheduling and sequencing information. During construction the MCUs were linked to the network monitoring station (NMS) located in the switchyard control house by means of UHF radios. This was done to reduce the potential for fiber optic cable damage while heavy construction was underway. As previously noted, all but one of the MCUs are networked to the NMS with the fiber optic cable. Each wired MCU is connected separately to the NMS with a duplex cable (two optical fibers). The duplex feature assured independent two-way operation of each MCU while allowing the use of fewer and simpler fiber optic transceivers.

The NMS consist of a microcomputer with telephone modem, WWV time-code receiver, and Geostationary Orbiting Environmental Satellite (GOES) transmitter.

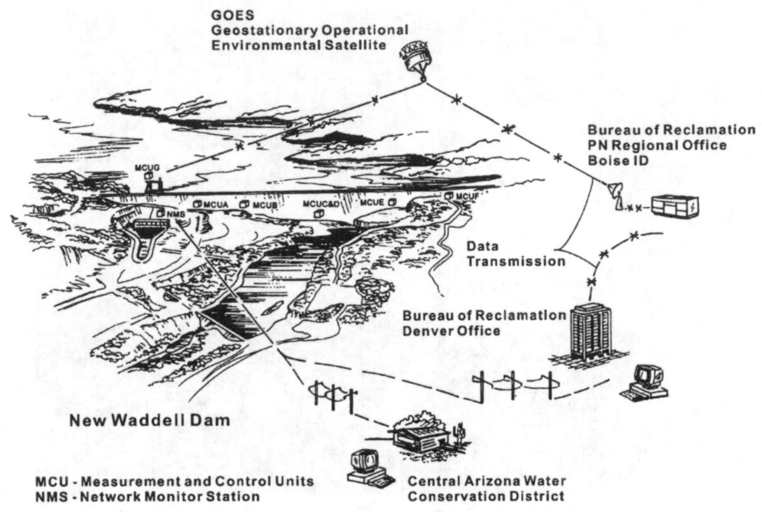

FIG. 4-*Communication links to automated instruments.*

Communications software is installed on the NMS to allow downloading by the Project Office in Phoenix, Arizona and Reclamation's Technical Service Center in Denver via telephone modems. The NMS is the logging destination for all measurement data from the MCUs. Data can be logged to the database, screen printer, and the GOES transmitter. Remote users dialing into the system may request real time measurements, reprogram, transfer files and trouble shoot. Figure 4 illustrates the communication links.

It is noteworthy that the MCU and NMS equipment is no longer supported by the manufacturer. Although the equipment continues to perform well, there will come a time when failures will begin to be a problem. Exact replacements are no longer available.

Conclusion

The performance instrumentation at New Waddell Dam represents a state-of-the-art system for monitoring the behavior of a large embankment dam on a complex foundation. It represents the first case for Reclamation where full advantage was taken of the opportunity to design an infrastructure specifically for the automation equipment. Unique aspects, in Reclamation experience, included the embedment of embankment extensometers, the use of shear indicators beneath the embankment and the use of fiber optic signal cable for performance data acquisition. All of those components are integrated into the automation system. Automation has worked well and has been a necessary feature in this age of "downsizing" and growing populations exposed to dam hazards. A specialized technician was needed on site two for two days to check fiber optic cable quality. However, the cable proved to be sturdy and there were no other complications. Even though this automation has been working well, designers of long term monitoring systems for dams are cautioned to consider the ability of automation equipment manufactures to provide continued service.

References

"Designers' Operating Criteria, New Waddell Dam," U.S. Bureau of Reclamation, Engineering and Research Center, Denver, Colorado, March 1993

Prudhom, B.D., July 1995, "Final Instrumentation Report (L-16) - New Waddell Dam - Stage 2," U.S. Bureau of Reclamation, Phoenix Area Office, Phoenix, AZ.

William S. Brokaw[1]

Using Geomatics in the Acquisition and Management of Field Data

---

REFERENCE: Brokaw, W. S., **"Using Geomatics in the Acquisition and Management of Field Data,"** *Field Instrumentation for Soil and Rock, ASTM STP 1358*, G.N. Durham and W. A. Marr, Eds., American Society for Testing and Materials, West Conshohocken, PA, 1999.

ABSTRACT: The term Geomatics refers to the integration of geographic information systems (GIS) with global positioning satellite (GPS) technology, remote sensing and communication technology, (i.e., the Internet). A project involving the collection of environmental data at a site follows a series of consecutive events. The initial step is to identify project parameters and determine data needs. Data capture methods (i.e. collection and analyses) are selected and then implemented in the field. Once the data are generated, organization and communication are essential in making the data useful. The objective of using Geomatics is to facilitate the acquisition and management of data and to provide a mechanism for effectively communicating this information.

Before a field work is initiated, valuable information about the site and surrounding area can be obtained via the Internet. Homepages maintained by the US Geographic Survey, US Census Bureau and environmental organizations provide maps and other regional data in digital format, enabling a project manager to assess a wide range of regional conditions. Aerial and satellite photographs can be obtained to further understand the site and surrounding area. As site-specific data are generated, the geographic coordinates of each sampling point can be accurately and rapidly determined with GPS receivers. Using GIS software, a project manager can efficiently integrate the GPS geographic coordinate data with analytical data and the information obtained from the Internet-aerial photographs. GIS serves as the primary data management engine, organizing and providing access to all information generated as part of the project. The Internet can be utilized to disseminate the information to concerned parties.

KEYWORDS: geomatics, GIS, GPS, internet

Geomatics, a term more commonly used in Canada than in the US, refers to a set of technologies used to manage spatially related data. The homepage for the Ministry of

[1]Scientific Associate, LFR, Raritan, New Jersey 08869.

Natural Resources of Canada defines Geomatics as "...the science and technology of gathering, analyzing, interpreting, distributing and using geographic information.." (www.geocan.NRCCan.gc.ca). The University of Florida homepage adds "...it is a new profession that includes activities previously known as surveying, mapping, photogrammetry and geographic information systems (GIS)." (www./surv.ufl.edu:443).

Geomatics technologies are efficient and powerful tools that can be used to acquire, manage and communicate environmental information. Over the last few decades, advances in hardware and software took these computer tools from the mainframe to the desktop computer. The natural progression is to take these tools from the desktop into the field, where the data originate.

By nature, environmental projects revolve around data and progress in three general phases.

1. Desk-based collection and processing of existing data to plan the field work,
2. Field-based data collection, verification, updating, and
3. Use of data in decision making. (Carver et al. 1996).

An environmental investigation is often an iterative process. The above phases may be conducted repeatedly for a single site, especially when monitoring is involved. Geomatics integrates well with iterative activities due to its many components and flexibility.

This paper follows the flow path of a typical field project involving the assessment of a known or potentially contaminated site. Opportunities are identified within each phase where Geomatics can be incorporated into the process.

Planning

Sampling projects are undertaken when there is a need to make a decision about a site. It may be a purchase decision (i.e. a due diligence investigation), to comply with a regulatory directive or to determine if a remedial action is needed. The data acquisition objectives are identified and then a scope of work is developed and agreed upon by the parties involved.

Implementation of a project usually begins in the office, on the desktop. Data needs are identified and procedures formulated to acquire the data. It is also important to consider the ultimate use of the data. Will they be used to develop fate and transport models or to assess risk? Is there likely to be a long-term investigation, or is the project simply a brief, short-term investigation?

During the initial planning phase, existing data are acquired and evaluated such as:

1. Surface and subsurface hydrologic data,
2. Topography and planemetrics,
3. Historic aerials and orthophotos,
4. Historic maps and surveys,

5. Soil logs and geotechnical data, and
6. Previous laboratory result reports (Douglas 1995).

To take advantage of Geomatics, pertinent, existing information may need to be converted into digital format. Hardcopy data, from preliminary assessments or prior investigation should be organized into a database management system (DBMS), 1995. The DBMS allows a user to sort, retrieve, analyze, delete and update the data as needed. Paper maps, drawings and photographs can be digitized or scanned. When possible, original computer files such as computer-aided design (CAD) drawings or orthophotos should be obtained (if they exist). The scale, orientation and projection of the maps and drawings all need to be known so that the maps and images can be properly integrated.

The source and date of information should be carefully recorded. In addition, users should be aware of the accuracy and precision of the maps and images that are utilized. Accuracy is how correct, or free from error, the data are. Precision is a measure of how finely one is able to measure differences in either position or value (Madry 1996). These attributes of existing data (source, age, accuracy and precision) allow a user to assess the quality and reliability of the data.

When the data and other information are in digital format they can be incorporated into a geographic information system (GIS) which couples the information in the DBMS with the maps, images and other graphics. In addition to providing the organizational and analytical power of a DBMS, GIS allows a user to *visualize* the information. GIS provides an easy method to render numeric data, allowing a user to better understand and interpret the information. Over half the neurons in a human brain are devoted to processing optical information. People have a natural ability for recognizing and interpreting visual patterns and visualization comes naturally to users (Buttenfield 1996).

Another advantage of GIS is that users can incorporate existing regional data with site-specific data. GIS projects are layered in a manner similar to CAD drawings. The layers, often referred to as coverages, contain the graphic images (points, lines, polygons, pixels) that are mathematically linked to attributes. Many state and federal environmental agencies maintain Internet sites containing public-domain coverages, satellite images, aerial photos and other data that can be incorporated into a GIS project. These data may cover large regions but often contain details that are useful on a site-by-site basis. Information such as soil type, geology, elevation, topography, surrounding land use and surface hydrology is useful when planning a sampling event.

As opposed to hardcopy, site-specific maps and drawings, GIS coverages are already in digital format and usually projected in a geographic coordinate system, either latitude/longitude or a Cartesian system (e.g. state plane feet or Universe Transverse Mercador). Digitized site-specific maps and data can be projected into the same coordinate system and become an integral part of these coverages.

In addition to allowing individuals to obtain valuable data from a multitude of sources, including governments, universities and private vendors, the Internet also serves as a conduit for information between users. E-mail allows direct transfer of digital data between private parties. Files containing images and drawings are often too large to be copied on a disk and Internet transfers have become the most efficient way to transfer the files.

Once sufficient information has been collected and organized, a *virtual* concept of a site can be generated and then used for planning purposes. This may be as simple as selecting the location for a few soil samples or as complex as generating extensive regional data for a comprehensive risk assessment.

Acquisition

What ultimately happens in the field is the least predictable element of a data acquisition project. This lack of predictability is the strongest argument for advanced planning. During the planning phase, when the data requirements are identified, the data acquisition methodologies are selected for use in the field.

Field instruments associated with Geomatics include dataloggers, portable computers, GPS receivers and cellular phones. Measurement and sample collection tools vary widely but all share the attribute of generating data. Geomatics provides the link between the measurement and management of data and thus can reduce the lag time between the generation and use of these data.

During the planning phase, simple considerations should be taken into account, such as creating ways to make written field notes compatible with the records and fields of the project's DBMS. Forms or checklists can be used to assure that all pertinent data are properly recorded.

Dataloggers and portable computers can be used in the field to add information directly into the DBMS as it is generated. Numerous database tables are often generated for various aspects of a project and can be integrated using the simple relational database techniques such as keeping at least one field in common between various database sets. For example, a table (database set) used to record sample locations and descriptions can share the sample name field that is used for the analytical results report database.

Global positioning systems (GPS) receive satellite-generated signals and record this information in a database. This database is usually formatted prior to collecting the satellite signals. Attributes (i.e. labels or descriptions) such as site name and the name of the GPS point can be entered into the database prior to using the GPS receivers in the field. The database fields for the GPS coordinate points can be kept consistent with the fields used in other database sets (e.g. well sample results). When the positional coordinates are then established, the GPS database can be linked to other field databases allowing use of data as they are generated.

GIS further integrates the positional data with the site data. GIS can create graphic images from the numeric values in the site-specific databases (i.e. model the data), merge this information with data generated during the planning phase and better illustrate site conditions for field crews. This can give field crews the ability to redesign sampling strategies in the field as new information is collected (Carver et al. 1996). Trends and patterns may emerge that were not anticipated or were not evident until the data could be visualized.

There are two basic forms of GIS, raster and vector. Both can be used by field crews for on-site analysis of data. In raster GIS, data and other attributes are assigned to cells, or pixels. Each point on the computer screen can have a host of information attached to it. Raster is especially useful for cell-based modeling, such as contouring contaminant isopleths and can be used to take advantage of photogrammetry information such as satellite or orthophoto images. In vector GIS the data and other attributes are assigned to points, lines and polygons.

Much of the GIS data available from government agencies are in vector GIS format. Vector-based GIS allows a user to integrate site CAD drawings and regional GIS coverages with the data that are generated in the field. GIS software programs are increasingly combining both vector and raster capabilities into the same program, giving users the advantages provided by both systems.

Although having database, GPS and GIS in the field can add new dimensions to the data gathering capabilities, there may simply not be enough time in a typical field day to both collect and analyze the data as they are generated. However, with the use of cellular phones and the Internet, field capabilities can be extended through an electronic link, a two-way link. Data and images can be transmitted via E-mail as they are generated and then analyzed and interpreted by others. The results of the analyses can be E-mailed back or even posted on a homepage and accessed for use by the field crews.

Geomatics can dramatically shorten the time interval between the generation and the use of data and create a new paradigm (Huxhold and Levensohn 1995), a conceptual framework around which a project is implemented. It adds a new element and requires new skills. It requires field crews and project managers to think differently about the way they implement their projects.

Even if field personnel do not use Geomatics tools themselves, they should be trained in the fundamentals of Geomatics for two reasons. One, they will understand why they are being required to implement new and perhaps unfamiliar procedures in the field and two, they can contribute their expertise to the process. What may seem logical and simple in the office may actually be ridiculously impractical to implement in the real world. One of the objectives of using Geomatics in the field is to collect and store the data (as generated) for the benefit of all users. The more field personnel understand the process, the better they will be in making the effort successful.

Example Project

A recent ground water assessment project conducted in New Jersey illustrates how even a simple project can be accelerated by incorporating GPS, GIS and the Internet into the data acquisition process.

Buyers were considering the purchase of a four-acre light industrial site. However a previous Phase I assessment indicated potential environmental impacts. A contract closing was scheduled for the next week, but there were no data on the ground water conditions at the site. Samples needed to be collected, but there was not sufficient time to install and sample monitoring wells. Furthermore, the site was in full operation and access was limited. The parties involved agreed to use temporary wells to obtain ground water samples for analysis.

While waiting for the utility markout, available GIS data were gathered for the site. Soil, geology, surface hydrology and land use coverages were obtained through the Internet and from NJDEP. Digital topographic maps and aerial photos were obtained through the USGS homepage and commercial vendors.

These layers of information, plus a CAD-generated outline of the site, were compiled in a GIS using commercially available software and a virtual model of the site and surrounding area was created. Seven temporary well point locations were selected and a simple DBMS created. Using GPS, the precise state plane coordinates and elevations were determined for each of the seven points on the site, establishing an x, y, z coordinate system for the site. These coordinates were added to the DBMS.

When the utility markout was complete, a direct-push soil boring rig was used to insert seven temporary well points during the morning. The depths to water and field parameters (pH, conductivity, and headspace) were measured in each well point and the data entered into a datalogger. Then, ground water samples were collected for analysis by an offsite laboratory. All of these attributes were added to the DBMS and a profile of the site's ground water was established. Ground water elevation contours and direction of flow were determined and correlated with headspace and conductivity data.

Based on these findings, arrangements were made with the laboratory to expedite certain samples. The following morning, while the lab worked on the samples, a second round of headspace analyses, field parameters and water level measurements was taken and logged. Because the data from each round of measurements were compiled into a DBMS and linked to GIS maps, the data from the two rounds of measurements could be readily compared, both in tabular form and visually.

By that afternoon, when the temporary wells were being abandoned, the preliminary laboratory data were available. Instead of faxing analytical sheets to the office, the lab simply downloaded the pertinent data from their laboratory information management system (LIMS) into a database file and E-mailed them to the project manager.

Using a cellular phone and a modem, the project manager (who was still in the field) accessed his mailbox, downloaded the file and added the data to the existing DBMS. During the road trip back to the office, various maps were generated using GIS that illustrated ground water conditions at the site.

The next day (one day before contract signing), the maps were plotted and tables were generated using pertinent data from the DBMS. This visualization and tabulation of the data allowed for a rapid peer review process. A senior hydro-geo engineer was able to examine all of the project data, including the GIS coverages and aerials obtained before the fieldwork, within a single software program. The maps were converted to bitmaps and transmitted to the clients and their attorneys for review and discussion.

Upon viewing the information, the clients were able to quickly see site conditions and gained a better understanding of the risks they were facing. The site was not pristine but it also was not severely contaminated. The contract was signed.

Advantages and Disadvantages

If improperly implemented, a Geomatics program can be a nightmare and a waste of energy. The wrong software, inadequate hardware or lack of training can result in time-consuming mistakes. Moving data from paper into digital format can potentially be a daunting task involving a considerable cost (DeFina et al. 1997).

Conversely, a properly implemented Geomatics program can provide a significant competitive advantage. It can accelerate the organization, review and statistical manipulation of spatially related information and enhance the decision making process (DeFina et al. 1997).

Training is a critical element of a successful program. Personnel need to become familiar with and understand the fundamentals of Geomatics in order to properly integrate it with existing methodologies.

A USEPA study (1994) concluded that typical Geomatics software programs become more cost effective with use. The magnitude and number of projects the software was applied to greatly influenced the cost and effectiveness.

When field-based Geomatics was used for a remote mapping of a proposed national park in Siberia, a team of researchers gave Geomatics a true acid test. They identified a series of advantages and disadvantages which are itemized below (Carver et al.. 1996).

Advantages:

1. Interactive development of sampling strategy through visualization and feedback.
2. On-the-spot environmental modeling and feedback through field-based verification.
3. Greater appreciation of the problem in the context of the processes operating.
4. Improved confidence in the data and greater user awareness of its limitations.
5. Reduction on the number of field visits and associated project costs.
6. Ability to integrate local ideas and knowledge at the start of the analyses.
7. Increased ability to convince local decision makers of the approach adopted.

Disadvantages:

1. Logistics problems associated with power sources and transportation/protection of equipment in the field.
2. Lack of technical backup facilities/services.
3. Problems of data availability, sensitivity, and security in some countries/regions.
4. Education and training of field teams in GIS concepts and techniques.
5. Long fieldwork preparation lead times.
6. It is very hard work.

While Siberia is more remote than most areas, the advantages and disadvantages itemized above are applicable to many field situations. Ideally, the disadvantages are offset by the advantages.

Geomatics has the potential of producing an enormous amount of additional data for a particular project and at the same time provides an organizational structure for this information. Information can be reeled in from a wide range of sources but these sources will have varying degrees of accuracy and relevance. As data are acquired, the metadata (the data about the data used in a Geomatics project) need to be kept current. Adequately recording and maintaining a metadata requires a higher level of discipline and additional effort above and beyond the scope of the project itself.

Geomatics allows the user to take further advantage of the abilities of computers for data acquisition, data management and analyses. Field and laboratory data can be rapidly queried and visually displayed to test hypotheses or to illustrate site conditions. What used to take hours or days for an individual to do with separate software programs can be done in seconds with an integrated Geomatics system, assuming proper protocols are established and followed.

Summary

Most environmental scientists and engineers generating data in the field already use many components of Geomatics such as databases, digital maps and E-mail. These components often are merged into GIS in the office where the data are ultimately stored and used. Implementing Geomatics in the field simply extends these capabilities and is an example of the evolution of information management.

Geomatics integrates a myriad of tools.

1. Internet transfer protocols such as http (the world wide web) and file transfer protocol (ftp) for acquiring and communicating data.
2. Raster and vector generated images such as orthophotos and CAD drawings.
3. Relational databases, spatial database engines and structured query language (SQL).
4. Satellite photogrammetry and global positioning systems.
5. GIS topology, which creates the mathematical relationships between graphic spatial elements and associated attributes.
6. Printers and plotters that produce visual representations that can be intuitively understood.

The most basic Geomatics tools are a desktop computer, a modem with Internet access, plus database and GIS software. CAD software, GPS equipment, cellular phones, dataloggers and portable computers help to complete the picture but are not essential, especially in the initial stages of implementing Geomatics.

However, computer skills are essential. One needs to possess a basic understanding of computers and software systems in order to apply these tools. The learning curve may be steep in certain areas but can result in a higher level of attainment.

As these technologies develop, they become more seamless and easier to use. Word processing software can now produce Internet compatible documents with a click of a mouse. CAD programs are increasingly adding GIS capabilities and object oriented programming, such as Visual Basics is greatly simplifying the use of database programs.

By adopting a global, inclusive approach with respect to personnel, these tools can be utilized by individuals at all levels of expertise. While GIS and database experts are useful in helping to set up systems, they may not at all understand how the data are generated and how it is ultimately used by decision makers. Whatever systems are set up, they should be designed with maximum flexibility to incorporate improvements. Those using the systems should be encouraged to provide input improve them and those who run and maintain these systems should constantly seek out this advise and offer solutions and new innovations.

The overall objective is to increase efficiency. Increasing efficiency means accomplishing more with less energy.

References

Buttenfield, B., "Scientific Visualization for Environmental Modeling: Interactive and Proactive Graphics". *GIS and Environmental Modeling: Progress and Research Issues*, Fort Collins, CO, GIS World Books, 1996, pp. 463-467.

Carver, S., Heywood, I., Cornelius, S., and Sear, D., "Evaluating Field-Based GIS for Environmental Characterization, Modeling and Decision Support". *GIS and Environmental Modeling: Progress and Research Issues*, Fort Collins, CO, GIS World Books , 1996, pp. 43-47.

Date, C.J., *An Introduction to Database Systems*. USA, Addison-Wesley Publishing Company, 1995, p. 2.

DeFina, J. , Maitin, I., and Gray, A., "Site-wide collection of remediation data in support of environmental quality objectives", *http://www.earthsoft.com/articles,* 1997.

Douglas, W. J., *Environmental GIS: Applications to Industrial Facilities*, Boca Raton FL, CRC Press Inc. 1995, p. 41.

Huxhold, W.E. and A. Levinsohn, *Managing Geographic Information System Projects.* New York. Oxford University Press, 1995, pp. 5, 31.

Madry, S., *The Fundamentals of Geomatics*, unpublished draft, 1996, p. 81.

Ministry of Natural Resources of Canada, Internet Homepage, *http://www.geo.NRCan.gc/geomatics/htmle/gen-g01.html,* 1997.

United States Environmental Protection Agency, *GIS/Key™ Environmental Data Management System: Innovative Technology Evaluation Report*, Cincinnati OH, USEPA, 1994, p. 107.

University of Florida Internet Homepage, *http://www.surv.ufl.edu:443,* 1997.

David T. Hansen[1]

Documentation of Data Collected from Field Instrumentation for Spatial Applications

---

REFERENCE: Hansen, D. T., "**Documentation of Data Collected from Field Instrumentation for Spatial Applications,**" *Field Instrumentation for Soil and Rock, ASTM STP 1358*, G.N. Durham and W.A. Marr, Eds., American Society for Testing and Materials, West Conshohocken, PA, 1999.

ABSTRACT: Data from field instruments are increasingly being incorporated into Geographic Information Systems (GIS). This adds a geospatial component to time series data and permits analysis and display with other geospatial data. From the perspective of GIS analysis, two issues are of importance. The first is the identification and description of the instrument location. The second is the description of the data stream from the sensor package and the access to that data for analysis and display. They are the basis for effective and proper use of the data from the instrumented site.

The spatial location of the sensor or sensor array in latitude and longitude or in a map projection system is a critical parameter for the use of the data in GIS. The accuracy of the spatial locations of sensors is important in identifying the spatial analysis that is appropriate for the data. Besides geographic coordinate information, local coordinate information available for the site is valuable information for locating the sites and instruments. Local coordinate information often is more accurately known than the geographic position. Local coordinate information can be used in computer modeling of

[1]Soil Scientist/GIS Specialist, U.S. Bureau of Reclamation, Mid Pacific Region, 2800 Cottage Way, Sacramento, CA 95825-1898.

the sensor data independently and can be used to refine the initial values of geographic coordinates.

To effectively use data captured from the instruments, information on the structure and contents of the data base is required. Sensor data may then be carried into the GIS data base management system (DBMS) or accessed from the DBMS set up for the time series data. For effective use and application of the data, all require information on the parameters being captured. For GIS data, this information is described as elements in ASTM D 5714-95 Content Specifications for Geospatial Metadata.

KEYWORDS: geospatial, metadata, site investigation, data base management system, datum, field instrumentation, time series data

---

Geographic information systems (GIS) are valuable tools for displaying and analyzing time series data. This includes data collected via field instrumentation. In GIS, the location of the field instruments is part of a data base that also has access to the parameters being capture by the instrument sensors. This location or geographic position permits display, manipulation and spatial analysis of the data. When site location is carried into the same coordinate system as other data sets, information from these other data sets can be incorporated into the analysis. This includes construction of cross sections, computer visualization, and surface analysis of the data values.

Spatial location includes horizontal position and vertical elevation (or depth) in a defined spatial reference system. The spatial reference system may either be a local system or a global system. In a local system, the instrument location is identified based on measurements relative to other features at the site. Usually, horizontal location is identified globally as the instrumented site's latitude and longitude. These spherical coordinates are converted into a planar map projection system for analysis with other data. The vertical position (elevation or depth) of the instrument is identified with respect to a local datum for the site or to an established vertical datum.

Instrument position is one set of parameters from what is often a complex data set captured from field instruments. Data from an instrumented site may represent several physical and chemical parameters which are collected continuously or at intervals. The data base management system (DBMS) must be able to manage this data stream and usually has tools for the manipulation, analysis and reporting these data. GIS is a data base system for the management, analysis, and display of geographic information. It operates on spatial entities representing location and attributes associated with those entities. Often in practice, the DBMS of the GIS system is linked to a separate DBMS storing data captured from the instrumented site. Recognizing the structure and definitions of fields in each DBMS improves the ability to access data for query, analysis, and display.

Information on the data captured and stored from the field instruments and coordinate systems identifying the instrument location is metadata. Several standards are in effect which identify metadata elements for describing these characteristics. Metadata describing GIS data sets have been identified in ASTM Specifications for Content of Digital Geospatial Metadata (D 5714). The elements identified in that guide provide detailed descriptions of the spatial representation of the instrument location and data captured from the instruments. Other guides identify elements that should be considered in site selection for instrumentation, description of the site, or description of the parameters being measured. These include the following:

- ASTM Guide for Site Characterization for Environmental Purposes with Emphasis on Soil, Rock, the Vadose Zone and Ground Water (D 5730),
- ASTM Practice for a Minimum Set of Data Elements to Describe a Soil Sampling Site (D 5911),
- ASTM Practice for the Minimum Set of Data Elements to Identify a Ground Water Site (D 5254),
- ASTM Guide for Selection of Data Elements for Ground-Water Investigations (D 5474),
- ASTM Guide for the Set of Data Elements to Describe a Ground-Water Site, Part 1 - Additional Identification Descriptors (D 5408),
- ASTM Guide for the Set of Data Elements to Describe a Ground-Water Site, Part 2 - Physical Descriptors (D 5409),
- ASTM Guide for the Set of Data Elements to Describe a Ground-Water Site, Part 3 - Usage Descriptors (D 5410).

Many elements in these guides are complimentary with D 5714. In addition, ASTM Guide for Continual On-Line Monitoring Systems for Water Quality Analysis (D 3864) identifies items that should be considered in the installation of field instruments. While the focus of this paper is on information needed for the management and application of time series data within a GIS environment, these standards need to be reviewed with respect to field instrumentation for soil and rock.

The Spatial Dimension

In GIS, the instrumented site is represented as a spatial feature. The feature may be represented either in vector or raster format. The format for representing the site or instrument is dependent on the types of spatial analysis that are to be performed. Spatial location including horizontal and vertical position are attributes stored and maintained by GIS. Figure 1 shows sampling locations represented as points in vector format for a portion of the San Francisco Bay and the Sacramento - San Joaquin Delta Estuary. Monitoring at these locations are coordinated through the Interagency Ecological Program (IEP). They represent a variety of sampling conditions from trawling surveys to instrumented sites for a variety of chemical, physical, and biological parameters. IEP provides summaries of this data at Web site, wwwiep.ca.gov. Representation of the sites as points is the simplest implementation of time series data in GIS where site location alone is shown. These locations are displayed with other spatial data showing San Francisco Bay and the Sacramento - San Joaquin Delta along with the global coordinate system of latitude and longitude.

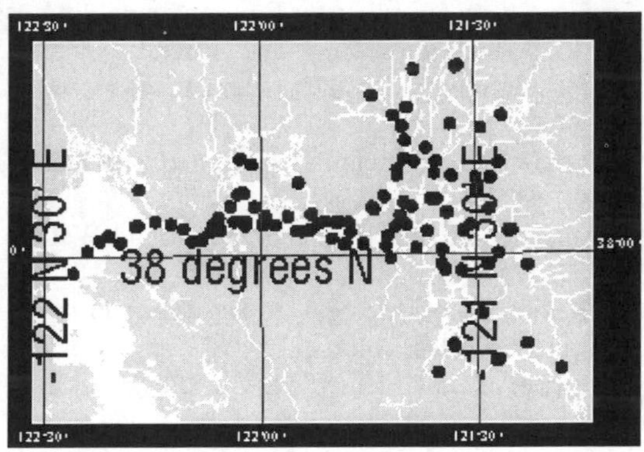

FIG 1--Site Locations in San Francisco Bay and the Sacramento - San Joaquin Delta Estuary

Coordinate Systems

Location in GIS is carried in defined coordinate systems with horizontal and vertical components. A common coordinate system for the spatial representation of features in GIS permits display and spatial analysis of different data sets as shown in Figure 1. The parameters of this coordinate system and the accuracy in the placement of the sample sites in the coordinate system are of major importance for successful operation of GIS analysis. These parameters include the definition of the datums or reference points for the coordinate system and other characteristics of the coordinate system. Control for these coordinate systems is based on networks of control points at the local, regional, and national level. Control systems at the national level are coordinated and reported by the National Geodetic Survey (NGS).

Horizontal Position-- For horizontal position, latitude and longitude are the recommended coordinates to report the position of the instrumented site location on the surface of the earth. It is a universally recognized system for determining positions on the globe. This is the recommended practice in D 5911 for identifying a soil sampling site (D and D 5254 for identifying a ground-water site. This global coordinate system is dependent on the model of the ellipsoid that is used. For the United States, there are two different ellipsoid models in active use. These are either the Clarke 1866 model or the WGS 1984 model. These ellipsoid models are the basis for the North American Datum of 1927 and the North American Datum of 1984. When coordinates of latitude and longitude are reported, the datum on which those values are based is a key parameter to be identified.

Many maps in active use in the United States such as the topographic series produced by the United States Geological Survey (USGS) until very recently were produced with map projections based on the 1927 datum. Positions captured with Global Positioning Systems (GPS) are frequently based on the 1984 datum. The datums are based on different spheroids of the earth and affect the values reported for latitude and longitude across the United States. In Delta area of California, identification of the incorrect datum for values of latitude and longitude will displace the site by several hundred meters. GIS data for an area frequently includes horizontal coordinates based on both systems. For proper display and analysis of data in GIS, the underlying datums for each of the data sets must be identified and converted to a common datum. Peter Dana of the University of Texas has a Web page describing geodetic datums.

In practice, the locations of instrumented sites are usually carried in a planar map coordinate system rather than latitude and longitude. This permits the direct measurement of distance, area, or direction. However, when representing a spherical surface in a planar system not all properties of distance, direction, area, or shape of features can be preserved throughout the geographic extent of a project. For a GIS project, a map projection system selected for a project area is determined by which properties are of most importance for that project area. Data to be used in that project is then transformed into the common map projection. Peter Dana has Web pages describing map projection systems. Snyder and Snyder and Voxland provide comprehensive descriptions of the parameters and effects of a variety of map projections.

Vertical Position-- The vertical position (depth, height, altitude, or elevation) of the instrument location is generally a major characteristic for the application and use of the data. As identified in D5911, D5254, and D5714, the reported vertical position is dependent on the vertical datum on which the measurement is based. The reported value is often assumed to be reported as elevation with respect to mean sea level. Mean sea level is based on measurements at regional tide gage stations and in this sense it is a local or regional datum. Besides local and regional vertical datums, two vertical datums are in common use in North America and the Delta area of California. In 1929, the National American Vertical Datum (NAVD29) set a fixed datum for the United States. This is the datum used for most USGS 7.5 minute topographic sheets. Since mean sea level actually varies over the Earth, the North American Vertical Datum (NAVD88) was developed based on adjustments back to mean sea level at Father Point, Canada in 1988. For the Delta area of central California, there is about a one meter difference in elevations based on these two datums. In addition to differences in elevations based on these two datums, the Delta area is subject to active subsidence. Subsidence in the Delta region is primarily due to compaction, erosion, and decomposition of organic soils from agricultural development over the past 100 years. Other areas of California as well as the United States are subject to active vertical shifts from seismic events or ground water withdrawal.

In the fall of 1997, several agencies cooperated in improving the network of stations set for NAVD88 by the National Geodetic Survey (NGS). This project set stations identifying elevation based on NAVD88. Global positioning systems (GPS) were used to adjust and refine the network for vertical elevations with less than a 2 cm difference in elevation across the network. At the same time, it improved the network of horizontal control within the region. Unpublished presentations by Don D'Onofrio of NGS, Marti Ikehara of USGS and Monte Lorenz of the United States Bureau of Reclamation (USBR), indicate a difference of less than 0.5 cm to more than 30 cm in height form elevations shown on the USGS 7.5 minute map series after they had been converted to NAVD88.

These differences in elevation (or depth) due to local or national datums and changes in surface elevations from subsidence have major implications for surface modeling and water modeling in the Delta area. Existing elevations and measurements from stream gages and tidal gages, as well as ground elevations need to corrected to a common datum. Development of this control network by NGS, USGS, and USBR provide the control for adjusting existing GIS data based on other datums to the NAD83 and NAVD88 datums.

GIS Coordinate Accuracy--Where site positions can be clearly located on the USGS 7.5 minute map series, locations have been expected to be within about 10 to 20 meters in the horizontal position and 1 to 10 meters in the vertical position. This map series is constructed to meet National Map Accuracy Standards and these are the errors expected for well defined locations at that map scale. Increasingly, global positioning systems (GPS) or similar devices are being used to directly record the position of sites. Currently with GPS units, the horizontal position of the site can be captured more accurately in a shorter period of time than the vertical position. Often software in the GPS unit permit the reporting of the location either in latitude and longitude or a map projection. Presently, GPS can rapidly report horizontal positions within 100 meters and with survey grade equipment and proper procedures within the centimeter level. At present, the vertical position reported by a GPS unit will not be as accurate as the horizontal position unless it can be tied into a NAVD88 station. The National Geodetic Survey (NGS) provides access to geodetic control information for both horizontal and vertical control at the Web site www.ngs.noaa.gov and provides data sheets on known NGS control at http://sinbad.ngs.noaa.gov/FORMS/ds_area.html.

The horizontal and vertical coordinate systems of a GIS data set are reported as part of the metadata in the spatial reference information section of D5714. This information includes parameters defining these coordinate systems such as map projections and datums. The resolution of the coordinate values and the accuracy of the coordinate values are also to be reported.

Local Measurement Systems-- At the time of instrument installation the relative positions of instrument have often been located by survey or measurement relative to local features. This relative position is often more accurate than the coordinates that are carried in GIS and may be very important for the analysis of the data collected by the instruments. D 5730 provides guidance on locating site locations. D5474 identifies elements for further definition of the site location by a map and description on the position of the ground-water monitoring. These relative positions for the local site configuration should be recognized for the site. This is particularly true for horizontal position where instruments are placed for sampling particular depths in the water column or particular stratums. These positions can be carried into GIS for spatial modeling and visualization independently of other data that is in a map projection. These local measurements can also be used to adjust GIS stored coordinates where the local coordinate system can be tied to a survey control station with known geographic coordinates. Elements for identifying and describing local coordinate systems for the position of GIS spatial entities are identified in D 5714.

The Data Base Dimension

Attributes associated with spatial features in GIS provide access to data that will be used in any spatial analysis. Depending on the structure defined for the GIS data base, the spatial feature may identify individual instruments or clusters of instruments that are reporting a variety of physical and chemical parameters. Basic attributes will identify the instrumented location and instruments or clusters of instruments being represented. Attributes will include a unique identifier for linking to the data base structure storing the parameters or summary values captured from the instruments. Unique identifiers provide the link or key fields for accessing the data for individual parameters captured from the instruments. Data captured from the instruments are usually stored in separate tables in a data base structure designed specifically for time series data. Typically, this is in date stamp format.

To use GIS as a tool in the display and development of models utilizing time series data, an understanding of the data structures and DBMS is required. This is particularly true where GIS is being used as a tool to visualize or model values from the data base. Of particular importance are:

- Table Names - Identification of tables carrying time series data or data summaries from the instrumented sites.
- Key Fields - Field that carry the unique identifiers providing access to data for the site, instrument, and parameter values. These fields serve as the basis for relates or joins between tables.
- Date and Time Representation - Time stamp field and other information on:
    - Time period represented by sensor data.
    - Frequency of measurement.
- Attributes - Additional fields containing date and time, measured values or summaries. This includes attributes that will be used for queries, analysis and display. The attribute description includes:
    - Data type
    - Field width
    - Domain of valid values
    - Units of measurement
    - Measurement resolution or precision of reported values
    - Special values or flags for:
        - No data conditions,
        - Censored data values,
        - Calculated values,
        - Other special conditions

This information provides the basis for query, analysis, and display of the data captured from the instrumented site.

With access to the time series data, GIS can be used to query and display the results of analysis. Figure 2 shows a portion of IEP data for the Chipps Island site in the Delta. Chipps Island has been a key point for monitoring water quality conditions in the Delta. The graphs have both average monthly and average daily salinity values for the months of February and June between 1958 and 1994. Winter months are typically periods of high river flows as the result of seasonal storms from the Pacific Ocean and salinity levels generally remain low. The winters of 1976, 1977, 1989, 1990, and 1991 represent drought periods with low volumes of run off in the Sacramento and San Joaquin River systems. Low river flows and corresponding higher average salinity levels are typical during the summer months. The variation in average daily values are due to tidal flows from San Francisco Bay.

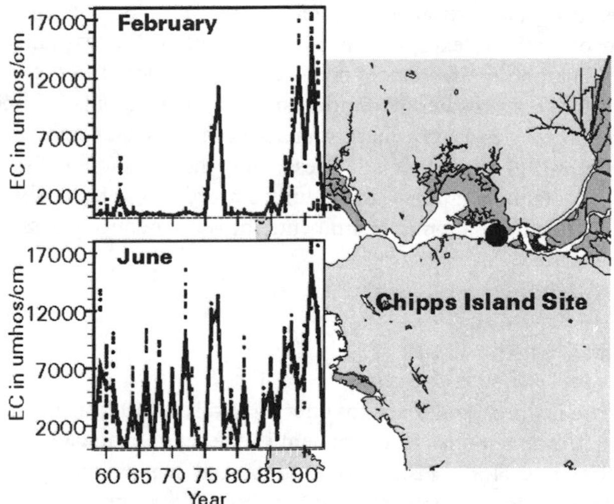

FIG 2 -- Salinity Values at the Chipps Island Site in the Delta.

Development of this display is based on information from IEP describing the data files, fields containing the recorded data, and special values. For GIS data, D5714 identifies this information as elements of the entity and attribute section of the metadata reported for a GIS data set. Date and time information is expressed in the format YYYYMMDD and HHMMSSSS as identified in ANSI standards for time dependent information. The four character requirement for year avoids potential confusion between the year 2000 and 1900 for computer processing. The time element includes specifications to identify local time and its relationship to Universal Time (Greenwich Mean Time). In addition, D5714 identifies a data quality assessment for the attributes that are carried into a GIS system. For ground-water monitoring sites, D5714 complement elements in D5474 for identifying the types of data, frequency of sampling, and data quality. Similar elements are also identified in D3864 for on-line monitoring of water treatment systems.

Elements in D5714 represent only information for documenting and managing data actually used and accessed in GIS. It can identify the purpose and limitations for the digital representation of data from the instrumented site. Other guides such as D5474, D5730, D3864, D5409, and D5410 identify considerations for the design and installation of instrument systems. ASTM Guide for Continual On-Line Monitoring Systems for Water Quality Analysis (D 3864) is focused on monitoring systems installed in a laboratory or for water system control rather than field installation. It identifies items that should be considered such as instrument calibration, validation of instrument reported values, and review of outlier or exceptional values reported by the instrument. It also includes guidance on reporting instrument precision. D5474 identifies the elements that should be reported for ground-water monitoring sites. It complements D5409 which identifies the elements for describing the measurement of physical parameters at a site and D5410 which identifies the elements for describing the measurement of chemical and other parameters at a site. In addition to these guides, Gordon and Katzenbach (1983) provide guidance on the installation of instruments for water quality assessment.

Summary

The data base is the central focus for query and display of time series data from instrumented sites. The location of the instrument is one set of characteristics which may be represented in GIS in vector or raster format for spatial display, query, and analysis. When utilizing GIS as a tool, it is important to recognize that a specialized DBMS may be required with special tools for maintaining and for analysis of time series data. The GIS component of any overall system is used to store and maintain the spatial location for the site. As part of the GIS data set representing the site, GIS attribute tables provide access to the records of the time series data for query, analysis, and display in GIS.

The universal system used to identify the location of the instrumented site are latitude and longitude. In actual GIS display and analysis, a Cartesian coordinate system is used in a defined map projection. This permits the display and analysis of data from the instrumented site with other GIS data layers. The altitude or elevation of the site is an additional attribute locating the position of the site. For effective use of these coordinates for the site, the datums used for latitude and longitude and for elevation need to be identified. Where the instrument location is carried into a map projection system, the parameters which define the map projection system need to be identified. These are required elements in D5714 which identifies metadata for describing GIS data sets. The local coordinate systems for site instrument position or local measurements of instrument location are important to maintain. As national and regional survey control networks improve, the local coordinates can be tied into map projection systems with other GIS data. The local coordinate values can be carried into GIS and used in modeling the data independently of other GIS data.

For effective use in GIS additional information is required on the data structure and format of the time series data. This metadata is identified in the entity and attribute section of D5714. This information identifies the tables and fields representing the time series data. It includes information on the period of record of the data, the frequency of data values, units of measure, and resolution of the reported data values. This metadata may have application in describing the DBMS independent of the GIS data set. It should be reviewed for application to time series data. D5714 complements other standard guides for site characterization, describing a soil sampling site, and for describing a ground-water site. These guides and the standard guide on continual on-line monitoring systems for water analysis contain additional information that should be considered in field instrumentation.

Gordon, A. Brice, and Max Katzenbach, 1983, *Guidelines for Use of Water Quality Monitors*, U.S. Geological Survey, Open File Report 83-081

Snyder, John P., 1983, *Map Projections used by the U.S. Geological Survey*, Bulletin 1532. Second Edition, Washington D.C.: U.S. Government Printing Office.

Snyder, John P., 1987, *Map Projections - A Working Manual*, U.S. Geological Survey Professional Paper 1453, Washington D.C.: U.S. Government Printing Office.

Snyder, John P.,and Philip M. Voxland, 1989, *An Album of Map Projections*, U.S. Geological Survey Professional Paper bulletin 1453, Washington D.C.: U.S. Government Printing Office.

On Line:

Dana, Peter H., "Geodetic Datum Overview", University of Texas Department of Geography, *http://www.utexas.edu/depts/grg/gcraft/notes/datum/datum.html* (October 14, 1997) - Description of datums.

Dana, Peter H., "Global Positioning Systems Overview", University of Texas, http://www.utexas.edu/depts/grg/gcraft/notes/gps/gps.html (February 15, 1998) - Description of GPS.

Dana, Peter H., “Map Projections Overview”, University of Texas, *http://www.utexas.edu/depts/grg/gcraft/notes/mapproj_ftoc.html* (October 14, 1997) - Description of map projections.

IEP “Interagency Ecological Program Home Page”, *http://wwwiep.water.ca.gov/* (Accessed: February 2, 1998) - Interagency Ecological Program for the Delta Estuary in Calfornia. Provides access to data files on reported a variety of physical, chemical, and biological parameters collected by the agencies.

NGS, “Data Sheet by Area”, *http://sinbad.ngs.noaa.gov/FORMS/ds_area.html* (Accessed: February 2, 1998) - National Geodetic Survey data sheets of control point information for the United States.

M. Hawkes,[1] and W. Allen Marr[2]

Data Acquisition and Management for Geotechnical Instrumentation on the Central Artery/Tunnel Project

---

Reference: Hawkes, M. and Marr, W. Allen **"Data Acquisition and Management for Geotechnical Instrumentation on the Central Artery/Tunnel Project,"** *Field Instrumentation for Soil and Rock, ASTM STP 1358*, G. Durham and W.A. Marr, Eds., American Society for Testing and Materials, West Conshohocken, PA, 1999.

Abstract: Boston's Central Artery/Tunnel Project has implemented a unique method of subcontracting for monitoring geotechnical instrumentation. The M026J contract includes a provision for the collection of 1.3 million readings over a six-year period. The contract uses a fixed unit cost per reading. This paper describes some aspects of the data acquisition and management processes that handle the flow of data for the project.

Keywords: instrumentation, data management, databases

Introduction

The Central Artery/Tunnel Project in Boston is one of the largest, most complex and technologically challenging highway projects in American history. The project design reduces traffic congestion and improves mobility in one of America's oldest and most congested major cities, improve the environment and support economic growth for New Englanders. The project has two major components:

> An eight-to-ten-lane underground expressway replaces an existing six-lane elevated highway. The new expressway is directly beneath the existing road, culminating at its northern limit in a 14-lane, two-bridge crossing of the Charles River. When the underground highway is finished, the crumbling elevated road will be demolished.
>
> I-90 (the Massachusetts Turnpike) will be extended from its current terminus south of downtown Boston through a tunnel beneath South Boston and Boston Harbor to Logan Airport. The first link in this new connection – the four-lane Ted Williams Tunnel under the harbor – was finished in December 1995.

---

[1] Manager and Geotechnical Instrumentation Engineer, GEOCOMP-Brown, JV – CA/T Instrumentation Monitoring Contract M026J – 421 Dorchester Ave., South Boston, MA 02127.

[2] Chief Engineer, GEOCOMP Corporation, 1145 Massachusetts Ave., Boxborough, MA 01719.

To put this highway improvement in the ground in a city like Boston amounts to one of the largest, most technically difficult and environmentally challenging infrastructure projects ever undertaken in the United States. The Project spans 7.5 miles of highway, 160 lane miles in all, about half in tunnels. The Project will place 3.8 million cubic yards of concrete – the equivalent of 2,350 acres, one foot thick – and excavate 13 million cubic yards of soil. The larger of the two Charles River bridges, a ten-lane cable-stayed bridge, will be the widest ever built and the first to use an asymmetrical design.

The Project also includes four major highway interchanges to connect the new roadways with the existing regional highway system. At Logan Airport, a new interchange will carry traffic between I-90 and Route 1A as well as onto the airport road system. In South Boston, a mostly underground interchange will carry traffic between I-90 and the fast-developing waterfront and convention center area. At the northern limit of the Project, a new interchange will connect I-93 north of the Charles River to the Tobin Bridge, Storrow Drive, and the new underground highway.

At the southern end of the underground highway, the interchange between I-90 and I-93 will be completely rebuilt on six levels, two subterranean to connect with the underground Central Artery and the Turnpike extension through South Boston. The interchange will carry a total of 28 routes, including High Occupancy Vehicle lanes, and channel traffic to and from Logan Airport to the east. A fifth interchange, at Massachusetts Avenue on I-93, will function as a part of the larger I-90/I-93 Interchange when the Project is finished but today is already helping improve Southeast Expressway traffic flow following early phases of reconstruction.

The Project has been under construction since late 1991. As of November 1998, final design is 98 percent complete, construction 46 percent complete. The next construction milestone, a bridge across the Charles River connecting I-93 in Charlestown with Leverett Circle and Storrow Drive, will be reached in1999. The I-90 extension through South Boston to the Ted Williams Tunnel and Logan Airport will open in 2001. The northbound lanes of the underground highway replacing the elevated Central Artery open
in 2002, the southbound lanes shortly thereafter in 2003. The entire project will be finished in 2004, including demolition of the elevated highway and restoration of the surface. The final cost of the project is $10.8 billion.

The 7.5-mile alignment is generally constructed using cut-and -cover excavations. The excavations are supported by variations of cast-in-place concrete diaphragm walls that are tied back or braced. The alignment passes directly through the downtown area of Boston in the same location as an existing elevated highway. Construction must occur while maintaining existing traffic flow and without affecting a large number of adjoining structures.

A joint venture of Bechtel and Parsons Brinkerhoff (B/PB) is the Management Consultant (MC) to the Massachusetts Transportation Authority, the owner. B/PB is also the Construction Manger.

The subsurface conditions along the alignment consist of glacial stratigraphy, the most prominent strata being organic silt deposits underlain by Boston blue clay. A variety of jet grouting, soil mixing, ground freezing and dewatering methods are being used to improve the subsurface conditions of these strata prior to construction.

A substantial geotechnical instrumentation program was developed by the designers to monitor the performance of constructed components and adjacent facilities. Currently, the instrumentation system includes monitoring groundwater pressures, deformations, loads, strains and vibrations at approximately 3000 locations. Movements of a large number of reference points are also monitored by surveying; however these are done by a separate contractor and are not considered in this paper.

A computerized application manages the large amount of data collected from these instruments and supports the project's need for rapid turnover of reduced data. The following sections present an overview of the system developed to collect, validate, store and report the geotechnical instrumentation data for the CA/T project.

Project Requirements and Objectives

The project consists of over 40 separate design and construction contracts. To obtain consistency in instrumentation and data management procedures across the entire project, the Management Consultant defined the primary objectives of the instrumentation, developed a project wide specification, and defined a means to collect and process the data [1]. Of major concern to the project is to put in place a system that produces reliable data from the geotechnical instrumentation, as it is needed, in the form that it is needed.

The large size of the excavations for the project, the proximity to numerous existing structures that are sensitive to movements, and the variable subsurface conditions along the alignment were factors pushing the need for rapid turn around of data. To meet the goals of the MC, readings taken on one day need to be reduced and reported by 08:00AM the next day. This is to permit all of the involved parties (engineers, contractors, and abutters) to have access to the data within 24 hours. The MC decided to award a single contract to monitor all geotechnical instrumentation for a six-year period. This was to provide a consistent data management process over the entire contract for the principal earthwork portions of the project. That contract was structured as a unit price contract, i.e. the Contractor would be paid a set price for each successful reading on each type of instrument. The work would require an estimated 1.3 million readings. The work was put out for bid.

In April 1997, the joint venture of GEOCOMP Corporation (Boxborough, MA.) and TLB Associates (Millersville, Maryland), was awarded the geotechnical instrumentation contract [2]. The total contract amount was approximately $10,000,000 for the six-year period. This represented approximately 60% of the engineer's estimate of the contract value.

Table 1 summarizes the types and quantities of instruments that are currently monitored. Typical monitoring frequencies are weekly, except for daily seismograph.

Table 1 - *Geotechnical Instruments*

Instrument Type	**Quantity** [1]
Seismographs	45 units, 149 locations
Convergence gages	135
Crack gages	992
Dial gages	4
Inclinometers	290
Multi-position borehole extensometers	6
Multi-point heave gages	23
Observation Wells	278
Probe extensometers	165
Shear displacement gages	6
Single-point borehole extensometers	9
Tilt meters	43
Vibrating wire load cells	28
Vibrating wire piezometers	725
Vibrating wire strain gages	12
Total	**2910**

[1] as of 06/30/98

Project Concept

GCB's approach to the job is to automate all aspects of the work that make financial sense. The overwhelming cost for the work is labor. Consequently every aspect of the data collection and reporting process was reviewed to determine what labor saving measures could be applied. Since we are only paid for readings when we provide a correct reading, errors must be kept to a minimum. On time readings are required to avoid penalties. Therefore, we desired approaches that would help us perform our work quickly. The project deliverables are tabulations of reduced data for each portion of the project, tabulations and graphs of instrument data which have readings that exceed pre-established threshold values, and placement of the raw data into the MC's project database.

Since the data must be provided to the Client in electronic form, it is immediately logical that data should be put into electronic form as early in the process as possible. After reviewing the possible approaches, we decided to employ handheld industrial computers to collect the data from each instrument. These devices could be used to collect data from a wide variety of instruments and readout devices. They reduce the opportunities for human errors and decrease the time required to get the data into the Client's database.

Contractors
Designers
Abutters

Distribute Data

Project Management
Bechtel/Parsons Brinckerhoff
Oracle Database

Consolidate data into daily reports
and load into Project Oracle Database
using network link over a phone line

GECOMP-Brown J.V.
Field Database

Transfer data to Field Database
using serial communication (RS232)

GECOMP-Brown J.V.
Personnel Data Recorder

Collect and validate data
using custom software

Measured Parameters
strain, pressure
displacement , load

Figure 1 - *Data Flow*

DATA FLOW

Figure 1 shows a schematic of the structure of the data collection and reporting process. The ultimate destination for the data is on the MC's computer network, and mostly in an ORACLE database system. Seismograph and Inclinometer data files are stored in series of electronic folders (directories) and files corresponding to construction contracts, instrument identifications and dates.

All data from the previous 24 hours must be reported to the Client by 08:00AM. Most data collection occurs in the daytime because of limits on access times to some locations and difficulties getting technicians to work night shifts. The majority of the data are collected during the construction work hours, typically from 06:00 to 16:00. As the work got underway, it became clear that seismograph data needed to be available the same day it was collected to permit the project staff to address complaints faster. Seismograph data are collected every morning and reported in the early afternoon of the same day.

Collecting Readings

A field data acquisition device (Personnel Data Recorder or PDR) collects instrumentation readings in electronic format. The PSION Workabout [3] is the primary data acquisition device. Figure 2 gives the specification for the Workabout.

Processor:	16 bit NEC V30H running at 7.68 MHz (80C86 compatible)
Internal RAM	256K and 1MB options
Internal ROM	2MB masked ROM containing operating system
Solid State Disks	2 drives accept Flash or RAM SSDs, (up to 16MB)
Display	backlite 240 x 100 pixels, gray-scale, graphics LCD
Keyboard	57 key alpha numeric layout as standard
Sound	Piezo Buzzer
Power	Internal: two AA alkaline cells or NiCd pack Backup: lithium cell
Interface	RS232 (19200 baud) and tethered barcode wand

Figure 2 – *PSION Workabout Personnel Data Recorder*

We chose this device as our basic data logging unit because of its ruggedness, its small size, its ability to run for weeks on a single battery, its ability to be programmed in a language similar to VisualBasic, and its large data storage capacity. We wrote software for the unit to collect and validate data for all the instruments read on the project. The entire database of instrument readings is kept on each PDR and used to check new readings immediately when they are obtained.

Data are obtained from the instruments manually, electronically, or automatically. Regardless of the method of collecting, the data ultimately resides in the computer data management system. Electronic collection of data in a format that interfaces with the data management system in the office is the most efficient method. The PDR provides a common gateway for all data into the data management system. The PDR provides the following features:

- determines the anticipated data range for the current instrument, using several methods.
- validates the current reading against the anticipated range.
- flags the reading to indicate which data range method was used.
- tags the reading with the technicians id.
- tags the date and time for the reading.
- stores reading.
- transfers reading to the data management system

Manually Collected Data - Manually collected readings are entered into the PDR using an alphanumeric-numeric keypad. Manually collected readings include readings from tape measures, dial gages, observation wells, probe extensometers, convergence gages, and crack gages. Manually collected readings also include a category of data that describe the status of instruments that may not be readable. The PDR provides a selection of typical remarks the technician can choose from without having to type the remark.

Electronically Collected Data - For electronically collected data, the PDR is connected directly to the instrument readout unit. This method is used to read vibrating wire strain gages and piezometers, inclinometers, and tiltmeters. The PDR is connected to the readout unit using a standard 9-pin RS232 serial and programmed to communicate with the readout unit. Electronically collected data are not subject to data entry errors, thereby eliminating one possible source of error.

Automatically Collected Data - Automatically collected data are collected and stored electronically. The readout-data storage device is typically referred to as a datalogger. The datalogger is connected to several instruments. It collects data at specified intervals. The data can be transferred to another location by radio or telephone links or by connecting a computer directly to the datalogger. Dataloggers are used in the following special situations on this project:

- collecting multiple readings from individual piezometers during pumping tests.
- collecting multiple readings from multiple instruments in difficult to access locations, for example in the subway tunnels.
- collecting "real-time" readings.

Surprisingly with the exception of Seismograph data, there are very few occasions where dataloggers are used on the CA/T project. We examined the potential to use data loggers on a large scale for this project. However, since we on average collect data from a sensor once per week, the cost savings provided by using dataloggers do not cover the cost of the added electronics.

Evaluating Readings

An immediate requirement for the PDR is to help determine whether a reading is good or not and whether a reading exceeds established threshold values. If the reading for a critical instrument exceeds established threshold values, project personnel must be notified within 15 minutes. However, an erroneous reading could exceed the threshold value and cause an invalid alarm. It becomes important that ways be found to identify erroneous readings as quickly as possible to minimize the number of false alarms and to minimize the cost associated with the bad reading. Minimizing the cost is especially important since we are not paid for an erroneous reading.

There are four possible categories for readings collected in the field:

- Good readings that fit previous trends
- Good readings that do not fit previous trends
- Erroneous readings that fit previous trends
- Erroneous readings that do not fit previous trends

The PDR can be programmed to compare the current reading with previous trends. A reading may fit previous trends and still be in error; however this will not trigger a false alarm. Conversely, a reading that doesn't fit previous trends may be good. These are the important readings because they are the ones that indicate change in performance which may be the warning of an impending failure.

Questionable or erroneous readings are due to:

- Misread or inaccurate instrumentation reading.
- Incorrectly keying numbers into the data acquisition device, for example mistakenly keying a 6 as using and adjacent key such as 5, 9 or 3.
- Misreading written data, for example reading a 5 as a 3, or a 1 as a 7.
- Erratic or unanticipated instrumentation responses resulting from construction activity.

Questionable readings due to construction activity can often be identified from field observations (jet grouting, pile driving, pumping test) or discussion with the engineers. Usually the readings are confirmed by taking an additional verification reading. All erroneous readings, except for reproducible equipment malfunctions can be eliminated by additional verification of the reading.

The PDR is programmed to set an acceptable data range for the current reading. If the reading is within the established range the reading is considered as valid. Valid readings are processed with no further quantitative checking.

Anticipated reading ranges are determined using one of the following methods:

1) plus-minus two standard deviations from the mean using the most recent eight readings to compute mean and standard deviation.

2) maximum and minimum of the most recent eight readings.
3) predetermined anticipated value plus-minus an acceptable error.
4) some other value determined by the specific conditions for the instrument.

The data window for a reading is computed "on-the-fly" by the PDR using the database of previous readings. The technician can switch between the different methods for setting the range. However, the ranging method is stored with the reading so that an engineer can further evaluate the reading. If the reading is determined to fall outside the acceptable range, the technician is instructed to take a second reading. If the second reading is also verified to fall outside the anticipated range, the PDR instructs the technician what action to take. Possible actions include telephoning the resident engineer immediately, telephoning the instrumentation superintendent, or tagging the instrument for further evaluation.

Data range computations using the previous readings (mean ± 2sd, and max/min) have proven to be not useful for instruments with a greater potential for erroneous readings. For example probe extensometer instruments using an electric reed switch which trigger a beeper in the presence of the magnetic reference anchor, are prone to erroneous readings. If the data range is set using previous readings, there is a tendency for the range to grow to an unacceptable size. Windows based on mean ± 2sd have grown more than 1 foot for these instruments. This is clearly unacceptable and therefor a data range using a predetermined reading with a range of ± 2/100 foot is used exclusively for probe extensometers. An engineer further checks these data before loading them into the MCs database.

Pore pressure instruments, such as vibrating wire piezometers, located in the zone of influence of pumping wells, jet grouting, and pile driving can display large changes in readings that may or may not fit the previous history. These instruments may change readings by a 20 or 30-foot pressure head in less than twenty-four hours. Data entry ranges for these instruments are typically 1 or 2 feet. Clearly this is a problem from the perspective of the field technician who is usually not aware of the cause of such apparently erratic readings. These readings are usually entered using data range based on the specific conditions of the instrument. The reading is flagged, and the flag results in a recheck of the reading.

Processing Readings

The PDR program dumps all data for the current date and selected instrument type into a daily data file. At the end of the day each technician creates the daily dump files and places the PDR and form listing the quantity and types of data dumps in an inbox for further processing. The daily dump files are transferred and loaded into the M026J Database Management system (the field database) using standard serial communications.

During uploading from the PDR, the database automatically assigns a unique identification integer to the reading. The identifying integer is referred to as the ReadingId. The ReadingId's are manipulated as supersets in the database using another integer called RepNo (report number). All data loaded or entered into the database are assigned ReadingId's and RepNo's.

The reason that report number and reading identification number are used instead of reading dates, is that database queries using integers execute several times faster than the same query using dates. Only for the database administrator can see or use the ReadingId's. RepNo's are used to select and process sets of readings collected on a specified date corresponding to the RepNo.

Transmitting Data

Data is transmitted to the Project Oracle system using standard telephone lines and modems. The data are transmitted in bulk grouped by RepNo and instrument type. All the data for each report and instrument type are downloaded from the Field Database into a single data file for transmitting to the Project Management. The main data transmission is usually preformed in one session late in the afternoon. A second session is performed the next morning to transfer all data collected during the night.

Once the data files are transmitted, they are loaded into the Project Oracle database system, using utility programs. It is the responsibility of the Project Management to further distribute data to Resident Engineers, Construction Contractors, Designers and Abutters.

Documentation

Documentation is accomplished using a series of hardcopy reports generated at each data processing stage. The first report is generated directly from the PDR daily dump file when it is loaded into the Field Database. The second report is generated when the Field Database downloads the data into a single file (for each instrument type) for transmission to the Project Management's network. A third report is generated from the Project Oracle Database detailing any problems during transfer. The final report is generated from the Project Oracle Database containing the reduced readings. The final report is transmitted to the Project Management with a formal transmittal letter listing the instrument types and construction sites that are reported. The transmittal letter also references the RepNo from which the data are processed.

All reports are placed in a single hardcopy file containing all documentation and field operations, comments associated with the report, and the day's activities.

Scheduling and Tracking

The Field Database system contains an element to schedule and track readings. This element was built to better respond to the Client's requests for a constantly changing set of instruments to be read. We receive work orders each Thursday for readings to take in the upcoming week. These work orders are then modified on a daily basis depending on what may occur in the project. Readings are scheduled by flagging an instrument as unread in the work order section of the Field Database system. A report number (date) and a technician is assigned to each unread instrument. The work order system creates reports of unread instruments that are distributed as daily assignments to the field crew. In a typical week, 1200 instruments are assigned to 10 field technicians.

When the readings are uploaded from the PDRs, the work order status is updated. The Report number, technician initials, and "unread" status are updated. At any time reports of read and unread instruments can be generated. If a technician was unable to complete his day's assignment, the unread instruments are reassigned to a new Report number.

Reporting

Most data reports are generated from the readings after they have been loaded into the Project Oracle Database system. Reports are generated in this manner because they can be recreated by anyone who is granted access to the Project Oracle system. The Oracle system is the final database destination for most of the readings. Once the readings are loaded in Oracle, they are secure from further modification.

Problems

We continue to modify our methods of identifying data that are in error. Primary sources of errors result from taking inappropriate actions when the data fall out side of an acceptable range, or not accurately determining the acceptable range. The final reports are printed in draft. To determine if erroneous readings are loaded into the Project Oracle database, and engineer checks the draft reports. Any erroneous readings are deleted from the Project Oracle and Field Database. By using a more reliable validation of data when it is collected, we hope to eliminate this laborious task.

At any particular time up to 18% of the instruments cannot be read. Reasons for not reading instruments are a result of access problems or instrument damage. We track reading problems and provide reports summarizing problem instruments. Because we are paid only for instruments that are read, there is an incentive to work with the Project Management to reduce the percentage of unread instruments.

The fast turnaround from data collection to final reporting and the large quantity of data put severe demands on the data collections and processing system. We have experienced failures in most of the electronic hardware, including network server, printer, copier, instrument readout units and PDR's. We have developed backup and redundant systems for all critical components of the data processing system. We have identified and worked around problems with instrumentation readout units.

A large amount of communication is required to design, install, initialize, collect, process, report and interpret instrumentation data. Add to this process a large number of contractors, designers, abutters and other interested parties and potential for communications problems increase. Communication problems arise because the projects data collection requirements change rapidly (daily, sometimes hourly). We use a system of pagers and cell telephones to communicate with the field personnel and to allow the field personnel to communicate with the Clients resident engineers. Project Management distributes verified data to the different parties using the Project Oracle database.

Future Trend

We expect to continue to look for more cost-effective ways to complete our work. Since just about all aspects of processing the raw data and providing it to the client have been automated to the extent possible, future efforts have to focus on ways to obtain valid readings. We continue to focus on how to reduce the number of invalid or failed readings. Recently, we reassigned a technician to focus on this task. He is to review each instrument that we have had problems obtaining valid readings for and seek out ways to repair, restore or retire the instrument.

Better communications technology may help us improve data flow. Instead of waiting until the end of the day, the technician might transmit readings several times a day. This would allow more time to prepare the reports and potentially make it possible to provide the reports on a shorter schedule. Options now available include; direct cabling from the instrument to the office, telephone links, or radio links. These types of connections are only viable in special situations requiring multiple readings from multiple instruments.

Low cost instruments with built-in low power radio communications using cellular technology may provide some assistance. These instruments communicate through one another to transmit readings over distances that could encompass the entire CA/T project. Devices of this type would be especially useful in locations where access is difficult or where frequent readings are required.

To reduce the possibilities for error, we are considering adding unique electronic identifications to each instrument. A device such as the iButton [4] (Figure 3) gives a unique number to the instrument that can be read with the PDR. This eliminates the possibility for the technician to mix up instrument numbers and also verifies that the technician actually got to the instrument. With the PDR, it can also help track the time at which the instrument was reached, permitting us to more closely consider the time requirements for each activity

Dallas Semiconductor iButton DS1990A (DS1420) 64-Bit ROM iButton: Contains a unique, unalterable, 64-bit unique registration number engraved both on the silicon chip and on the steel lid of the button

Figure 3 – *Dallas iButton for Instrumentation Identification*

Although we are providing all data in electronic format, we also provide a hardcopy report each day. This report varies between 2 and 6" of paper per copy per day. As computers become more pervasive on the project, we anticipate being able to dispense with the hardcopy report at some point.

Summary and Conclusions

The Central Artery/Tunnel project involves one of the most extensive geotechnical instrumentation systems ever deployed. The size of the excavations, the proximity of movement sensitive structures, and the sensitize nature of the abutters place great demands for a comprehensive, accurate and timely data management system. The project's management team must have timely, accurate information at all times to help maintain the safety of the project, address issues with the contractors, and answer the questions of concerned abutters.

Computers have significantly eased the data tracking and reporting tasks associated with the instrumentation monitoring. Data entry in the office is eliminated. The PDRs are a valuable tool for us to obtain accurate data in a cost-effective manner. However, there is still a considerable amount of labor required to routinely collect the data and validate questionable readings.

Automatically collected data are desirable for reducing manpower, collecting large quantities of data, and supporting a rapid reporting system. Technically, it is feasible. However, on the Central Artery/Tunnel an automated data collection system is not financially feasible because of the large cost of installing and maintaining such a system in a large congested construction environment.

After one year of work, we are continuing to refine our data collection and management techniques. We currently track the number of unsuccessful reading attempts so we can isolate the cause of the problems and work to increase the successes. Our database is being expanded to track other items that affect our cost and performance. For example, we have added provisions to track the maintenance activities on each piece of readout equipment. This permits us to identify trends in equipment performance so that we can look for ways to decrease our maintenance costs.

We believe that our efforts on this project are showing that geotechnical instrumentation can be monitored in a timely and cost effective manner. We consistently place data into the project team's possession within 24 hours of data collection. This is done with a delicate balance of dedicated technicians and electronic equipment. That balance is constantly altered to adjust to the changing needs of the project. We think the approach being used on this project can serve as a model for other large geotechnical instrumentation projects.

Acknowledgements

The authors would like to acknowledge and thank the Massachusetts Transportation Authority for permission to publish this paper and Bechtel/Parsons Brinkerhoff for their cooperation. In particular we recognize the cooperation of Dr. Thom Neff, Mr. Charles Daugherty and Mr. Dave Druss, all of the B/PB management team. We also acknowledge the dedicated efforts of our instrumentation team lead by Mr. Terry Knox of TLB Associates. This paper was prepared with the support of GEOCOMP Corporation without public funds. It contains the views and opinions of the authors and does not represent the views or opinion by the CA/T project.

References

[1] Bechtel/Parsons Brinckerhoff, "Geotechnical Instrumentation", Concept Report No. 2AB24, prepared for the Massachusetts Department of Public Works, 1991.

[2] Massachusetts Department of Public Works, "Central Artery/Tunnel Project, Geotechnical Instrumentation Monitoring (M026J)", Section 7210.505, Addendum 18, 1996.

[3] Psion Inc., "Psion Workabout: user guide", 150 Baker Ave., Concord, MA 01742, 1998.

[4] Dallas Semiconductor, Inc. "Data Sheet : DS1990A Serial Number iButton ", www.ibutton.com, 1998.

Pamela Stinnette, Ph.D., P.E.[1]

Data Management System For Organic Soil

REFERENCE: Stinnette, Pamela, **"Data Management System for Organic Soil,"** *Field Instrumentation for Soil and Rock, ASTM STP 1358*, G. N. Durham and W. A. Marr, Eds., American Society for Testing and Materials, West Conshohocken, PA, 1999.

ABSTRACT: A Data Management System for Organic Soil (DMSOS) has been developed that enables the acquisition, management and analysis of organic soil data as well as the presentation of results to be conducted effectively through a common interface. This development was in response to the data management needs of research investigating the engineering properties of organic soil and its extension to the stabilization of organic soil through dynamic replacement (DR). It is shown how the above functions are implemented efficiently using *Windows*-based software to perform comprehensive data management and analysis of data gathered from both laboratory and field tests. When the engineering properties of a given organic soil deposit are needed, a built-in Computer Advisor for Organic Soil Projects (CAOSP) predicts the properties from DMSOS based correlations. A unique and useful feature of the CAOSP is its ability to estimate the anticipated ultimate settlement of an organic soil deposit given the loading conditions and the moisture or organic content. Also incorporated in the DMSOS is a quality control system that utilizes computerized data acquisition/data management techniques in order to evaluate the degree of improvement of an organic soil layer at a given stage of treatment using DR.

KEYWORDS: database management systems, data acquisition, organic soil, consolidation, spreadsheets, ultimate settlement

---

Generally, geotechnical projects require site investigation, laboratory testing, model studies and field monitoring. Each of these phases produces extensive data which must be acquired, stored, analyzed and interpreted. In order to achieve an accurate measurement of test data with high resolution, state-of-the-art data acquisition techniques must be utilized. Once collected, the data should be stored in a database using a standard format. Analysis of the acquired data can then be performed using appropriate software.

---

[1] Assistant Professor, Hillsborough Community College, 1206 N. Park Road, Plant City, Florida 33566.

Currently there are database management systems (DBMS) being used in many commercial and business enterprises. These systems have found application in the area of geotechnical engineering as well. Geotechnical DBMS have been reported by researchers such as, Adams et al. (1993), Benoit et al. (1993), Ishii et al. (1992), Lee et al. (1993) and Sykora and Koester (1991).

Organic soil is considered an undesirable foundation material due to its low strength and high compressibility. Therefore, the Florida Department of Transportation (FDOT) and researchers at the University of South Florida (USF) cooperatively investigated: (1) the geotechnical properties of Florida organic soil; and (2) the feasibility of stabilizing it through the use of DR with sand. Since the quantity of data collected for the above research was voluminous, the development of a Data Management System for Organic Soil (DMSOS) was essential for effective implementation of the current and subsequent research projects on organic soil. Although, the work presented in this paper addresses the setup of the basic DMSOS structure from laboratory and field tests including DR, conducted on Florida organic soils, its scope could easily be extended to include all soil types.

In order to develop the data management system, a common interface was created using *Microsoft Visual Basic* programming. The *Microsoft Visual Basic* user interface of the DMSOS operates in a *Windows 3.1* integrated environment which supports *Microsoft Office Applications*. These applications include: *Access*, a relational database, *Excel*, a spreadsheet program and *Query*, a structured query language (SQL) that enables access and retrieval of data stored in the *Access* database to other applications such as *Excel*. The exclusive use of *Microsoft* products provides consistency and automation among the applications which results in their seamless integration in the DMSOS.

The above facilities were effectively utilized to store all measured properties of organic soil samples collected from test sites and formulate any possible correlations. A unique feature of the system is the inclusion of a Computer Advisor for Organic Soil Projects (CAOSP). This program, also written in *Visual Basic*, assists the user by providing the ability to estimate index properties and the ultimate settlement of any organic soil site from DMSOS established correlations. Additionally, the optimum energy required for effective field implementation of the DR technique can be evaluated through the use of the developed DR quality control system which is also incorporated in the DMSOS.

In order to illustrate the functionality of the DMSOS, the paper will be divided into three parts. Part I gives an overview of the DMSOS structure and details the acquisition of test data, data reduction, data entry in the *Access* database, data analysis and presentation. Part II describes the capabilities of the CAOSP such as prediction of index properties and ultimate settlement of an organic soil deposit under given loading conditions, based on DMSOS data and correlations. Finally, Part III outlines the DR quality control system integrated into the DMSOS.

PART I DMSOS STRUCTURE

Visual Basic 3.0 was selected to develop the common user interface for the DMSOS. This software offers database communication protocol to many of the commercially available database management systems. Additionally, as it is a *Windows*-based product, the resulting user interface can be graphical and it will be able to take advantage of the *Windows 3.1* operating environment. "Buttons" are commonly used graphical objects found in this operating environment. The on-screen buttons can be clicked or double-clicked with a mouse causing any number of actions to be performed. They have been incorporated in the DMSOS to facilitate an intuitive graphical user interface (GUI).

Figure 1 shows the opening screen of the DMSOS created using *Visual Basic 3.0* programming. An overview of the structural blocks and their functionality in the DMSOS is shown in Figure 2.

Data Acquisition

Due to the advancement in microcomputer technology, the use of data acquisition has become commonplace in both geotechnical laboratory testing and field monitoring. Many of the laboratory and field investigations on organic soil have utilized automated data acquisition. The data storage and simultaneous screen presentation during testing can be completely automated by a microcomputer equipped with a multifunction input/output interface board and appropriate software. The core of the described laboratory data acquisition system includes a 486-33 MHZ microcomputer with 20 megabytes of random access memory (RAM). The computer interface with the transducers is accomplished with the AT-MIO-16F-5 multifunction interface board. It contains a 12-bit analog to digital converter (ADC) which samples up to 16 separate single-ended (SE) analog inputs or 8 channels of differential input (DI) at 200 kilo-samples per second.

Visual Basic 3.0 was also selected for interfacing the data acquisition board to the microcomputer. *Visual Basic* communicates with the selected data acquisition board through the use of a dynamic link library (DLL) that includes functions for controlling the data acquisition board for the AT microcomputer. Through the use of this DLL, executable programs have been developed using *Visual Basic* for both laboratory and field tests. The laboratory tests include (but are not limited to): 1) the consolidation test which utilizes a linear variable differential transformer (LVDT) for displacement measurement and a pore pressure transducer, and 2) the triaxial compression test which utilizes a LVDT, a load cell and a pore pressure transducer. Data acquisition during field DR records data from an accelerometer, laser photoelectric cells and dynamic pore pressure transducers. Figure 3 shows a list of the available test options for data acquisition once the DATA ACQUISITION button has been clicked on the MAIN MENU. As an example, a double-click on DYNAMIC REPLACEMENT will bring up the user interface shown in Figure 4. It is on this screen that the user provides pertinent test information such as an output filename, if and how long the residual pore pressure and stress are to be monitored, and the data recording time interval.

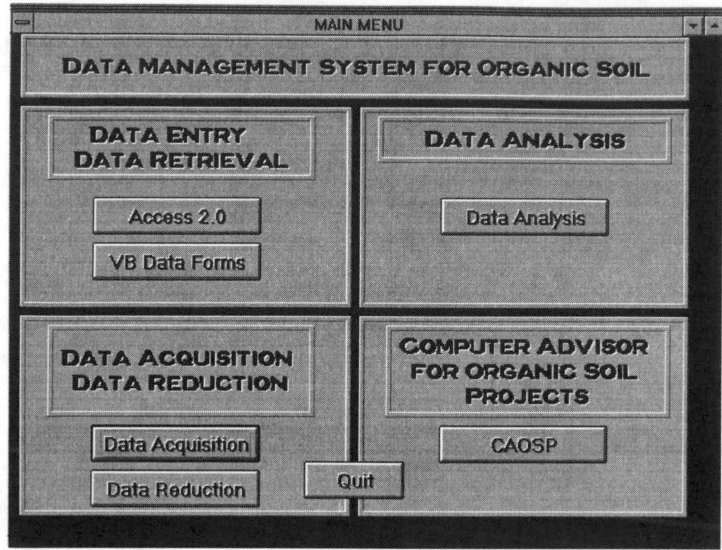

Figure 1 Opening screen of the DMSOS.

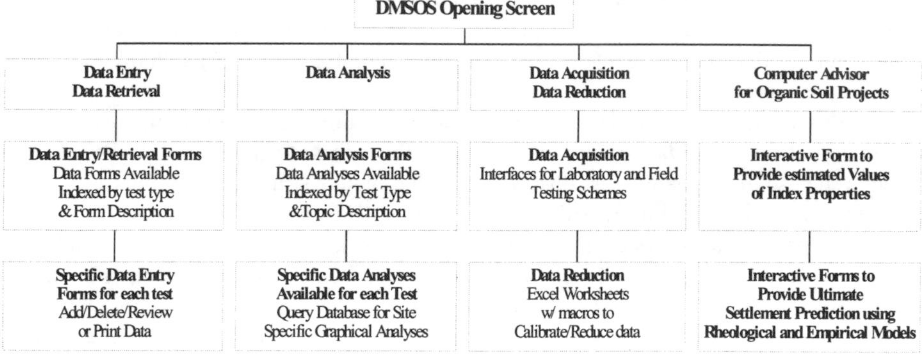

Figure 2 DMSOS block functionality flowchart.

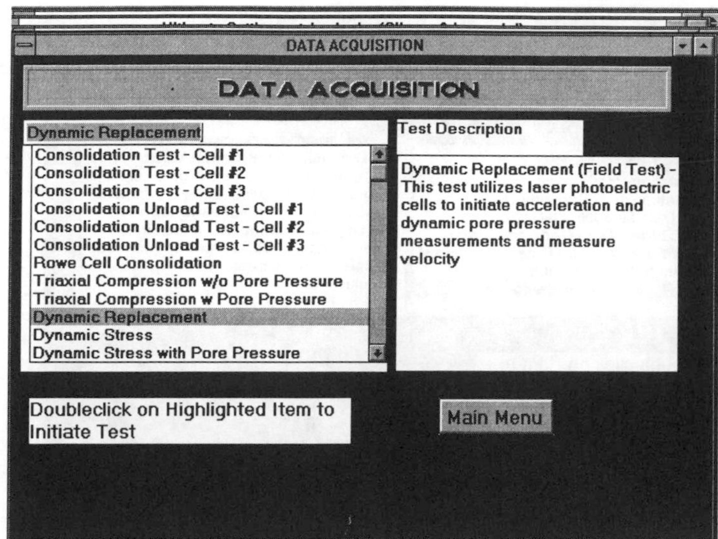

Figure 3 Available data acquisition tests screen.

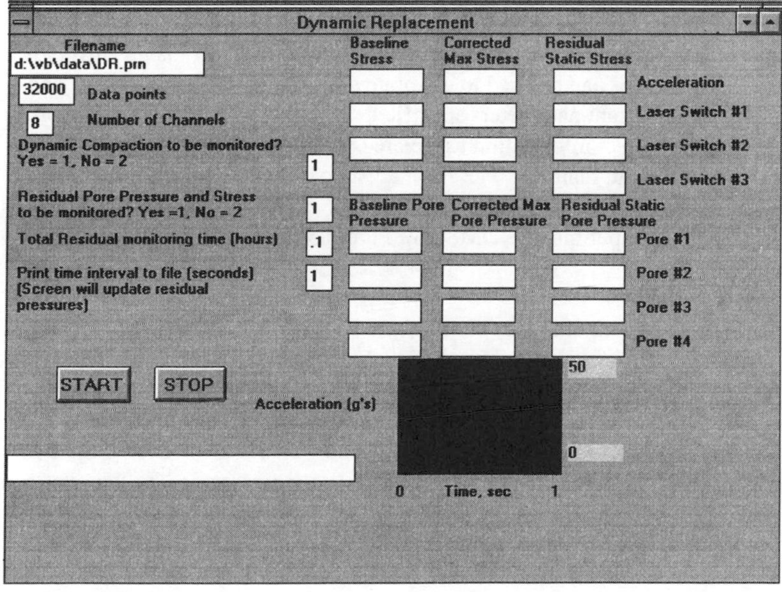

Figure 4 Dynamic replacement user interface.

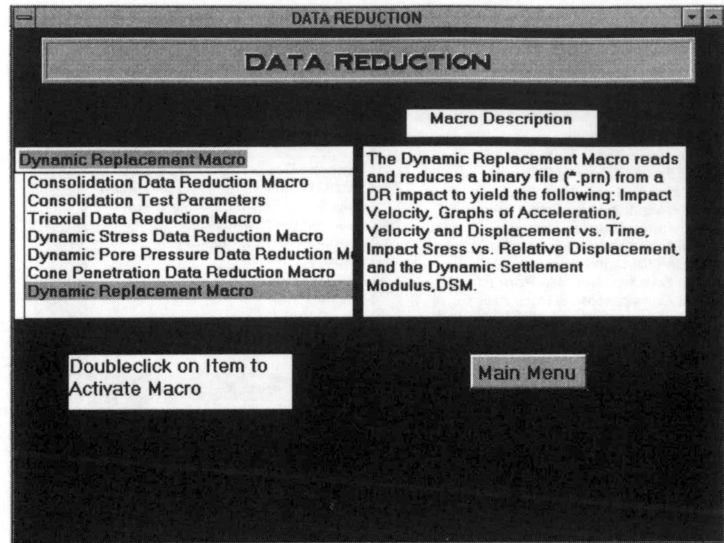

Figure 5 Data reduction screen of the DMSOS.

Data Reduction

After data (i.e., consolidation, triaxial, DR) have been acquired using data acquisition techniques and stored in the computer (*.prn file), the data may then need to be reduced depending on the specific test conducted. This is accomplished using *Excel* worksheets specific to each test type which include macros written in *Visual Basic for Applications*. These macros can be used to calibrate data that have been acquired and also plot and extract important parameters specific to each test. For instance, a macro may prompt the user to select information associated with the test such as the specific transducer(s), sample height, diameter, etc. from a list box. Once this information has been selected, calibration and analysis are then initiated. Specific results of the analysis can then be extracted for input into the DMSOS. Figure 5 shows the DATA REDUCTION SCREEN of the DMSOS.

Data Entry into the *Access* Database

The structure of the *Access* database is founded on a relational model (RM). Although other data models, such as object oriented and feature-based geographic information system (GIS) data models are available which have both spatial and non-spatial components, the relational model proved to be sufficient for the DMSOS, as the spatial component for the present study could be adequately represented by the relational model.

The RM is presently the most commonly used type of DBMS since it allows sufficient flexibility in data representation and provides access to its data for different

Microsoft Access

File Edit View Format Records Window Help

Table: Incremental Consolidation Results

ID	Site Identificatic	Pressure (kPa)	Void Ratio
1	SR951B	31.9	2.6
2	SR951B	63.79	2.347
3	SR951B	127.59	1.960184
4	SR951B	255.18	1.68
5	SR951B	510.35	1.380138
6	SR951B	1020	1.160651
7	SR951A	31.9	2.321247
8	SR951A	63.79	1.981775
9	SR951A	127.59	1.57216
10	SR951A	255.18	1.250127
11	SR951A	510.35	0.905617
12	SR951A	1020	0.666901
13	SR597A	31.9	1.482737
14	SR597A	63.79	1.416253
15	SR597A	127.59	1.31196
16	SR597A	255.18	1.216548
17	SR597A	510.35	1.085103
18	SR597A	1020	0.95497
19	S951B	31.9	3.088678
20	S951B	63.79	2.600955
21	S951B	127.59	2.103817
22	S951B	255.18	1.720136

Record: 1 of 60

Database: GDMS

Datasheet View

Figure 6 Incremental consolidation results table.

applications in a program-independent fashion (Rasdorf and Spainhour 1993). The relational DBMS is also well suited for representing tabular data. Although many relational DBMS are available, such as *dBase 5.0* and *Approach 3.0*, *Microsoft's Access 2.0* was selected as the Relational Database Management System (RDMS) for the DMSOS. This selection was based on the flexibility of the program and the ease of integration with the other *Microsoft Office* products. By using the *Visual Basic 3.0* programming language, access can be made directly to the *Microsoft Access* database. Additionally, coupled with the use of appropriate programming techniques, a high degree of control is afforded over how data is stored in the *Access* database.

The National Geotechnical Experimentation's Site (NGES) (Benoit et al. 1993) has developed a data dictionary (standard format for data entry) for its Central Data Repository (CDR) database. This data dictionary has been incorporated in the DMSOS to the extent that a smooth data exchange between the DMSOS and the CDR database will be facilitated. Data may be entered in the database in a number of formats. Inaddition to data being in the common numerical or text form, a data field may represent an embedded object such as a static picture (i.e, *.pcx) or a motion video (i.e., *.avi). Data entry into the DMSOS can be achieved by using one of two approaches: (1) direct data input or (2) data entry forms. Data may also be entered directly into the database through the use of *Access 2.0*. Figure 6 shows a portion of the *Incremental Consolidation Results Table*. Data are stored tabularly with each row representing a new

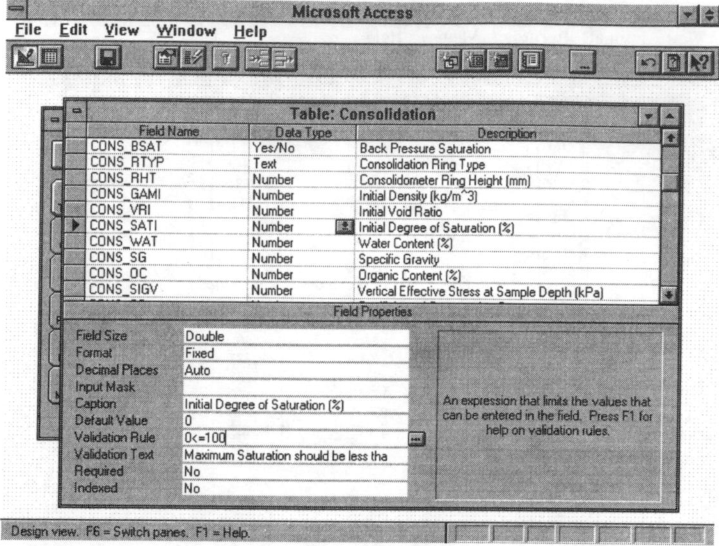

Figure 7 Design mode for the consolidation table.

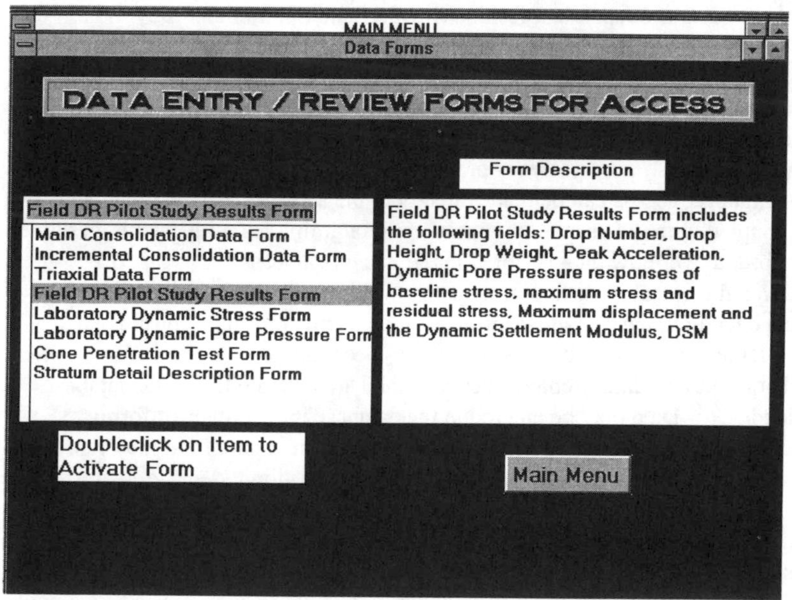

Figure 8 Available data entry / retrieval forms.

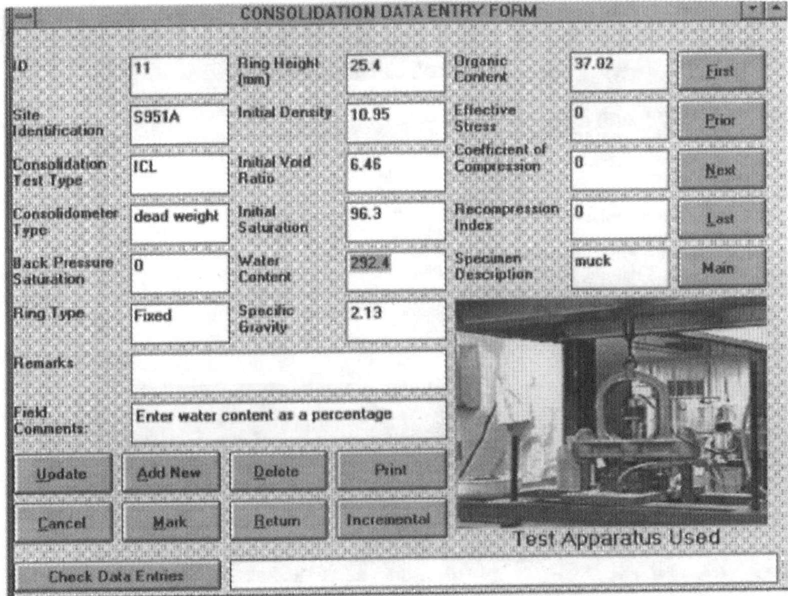

Figure 9 Consolidation data entry form.

record. For instance, records 1-6 correspond to oedometer specimen "SR 951B" which underwent six loading stages of testing. Restrictions can also be made on the data type, data format and on the input of a particular data field in the form of a "validation rule" incorporated in the *Access* database. In order to illustrate this concept the "Saturation" data field (CONS_SATI) is selected from the design mode for the *Consolidation Table* (Figure 7). The data type is seen to be "number" with a "fixed" format. The validation rule displayed reads "0 <= 100." This rule ensures that the data entered in this field stay within the prescribed limits. If the user enters data outside of these limits, the validation text "Maximum saturation should be less than 100% (0<=100)" is displayed. Although this mode of direct data entry is available, it is not envisioned as the primary method due to the development of data entry forms which provide a GUI for ease of data entry from the DMSOS. Data that already reside in the database may also be retrieved, reviewed and printed using these forms. As the different tests require separate tables for storage, there will also be separate forms for data entry and/or review. To access each of the various data entry forms, a list of all available data forms is provided to the user on a separate form with a description of the fields associated with that form. Figure 8 shows the data entry/review forms that are available in the DMSOS after the VB DATA FORMS button has been depressed on the MAIN MENU. The FIELD DR PILOT STUDY RESULTS FORM is highlighted and hence a list of included data fields are shown in the FORM DESCRIPTION box. As an example, a double-click on the MAIN CONSOLIDATION DATA FORM reveals the CONSOLIDATION DATA ENTRY FORM (Figure 9). This form may then be used for new data entry, or the review of pertinent consolidation results. As seen in

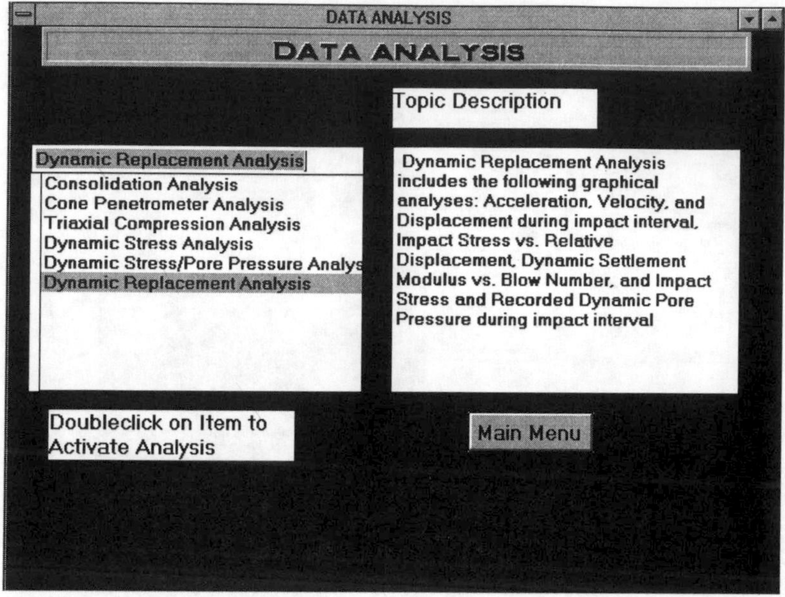

Figure 10 Data analysis available.

Figure 9, a static picture of the test apparatus used is displayed on the form. The reference to this picture and all of the pertinent data fields displayed are stored in the *Access* database. A separate data form is required for entry of data associated with each incremental loading stage and may be accessed by clicking on the INCREMENTAL button.

The data entry forms, developed using *Visual Basic 3.0*, provide additional guidance for data entered into the database. As an example, the "water content" data field is highlighted on the CONSOLIDATION DATA ENTRY FORM (Figure 9). The field comments read "Enter water content as a percentage," thus ensuring that the user is aware of the input data format. Additionally, when the CHECK DATA ENTRIES button is clicked, pertinent data fields are checked against previously established correlations to assure data integrity. The user is prompted of the results in the adjacent box. Coupled with *Access*, this will help maintain data integrity and ensure that data stored is in the appropriate units/form etc. and follows standard testing convention.

Data Analysis and Presentation

Data stored in the *Access* database is retrieved by using *Query*. *Query* enables dynamic communication between *Excel* and the *Access* database by directly querying the *Access* database, extracting the information sought, and then importing the data into the *Excel* spreadsheet. The data can then be manipulated and analyzed using macros programmed with *Excel's Visual Basic for Applications*. Hence, with little noticeable delay and

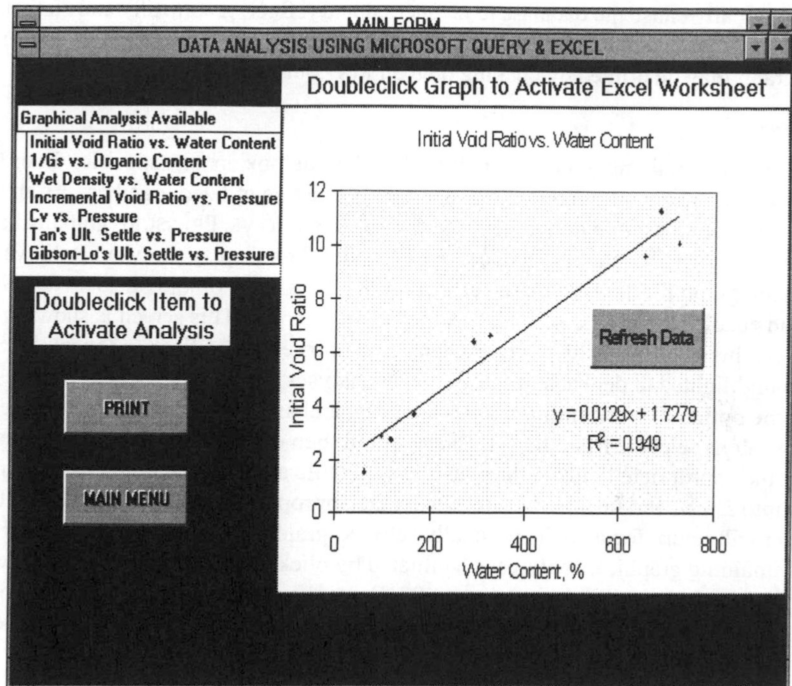

Figure 11 Consolidation Data Analysis using Microsoft Query and Excel.

without requiring any special input from the user other than a mouse click, *Excel* can directly query the database using *Query* and statistically analyze and plot the results. Moreover, as new data becomes available the relevant correlations can also be updated.

Figure 10 shows a list of the available test types and their relevant analyses after the DATA ANALYSIS button has been depressed on the MAIN MENU. In order to illustrate the role of DATA ANALYSIS in the DMSOS, once again the consolidation test will be used. Hence, a double-click on CONSOLIDATION ANALYSIS shown in Figure 10 reveals the form shown in Figure 11. A box in the upper left hand corner of this screen indicates the data plotting options that are available for the consolidation test. The first three options in this list represent cumulative test results based on data currently available in the database. Results of the statistical analysis, if appropriate, are included on the graph. Double-clicking on one of the graphical analyses listed initiates object linking and embedding (OLE) automation. OLE can be used to start applications, view and/or edit specific information, allow users to send and/or retrieve data from other applications and control the actions of other applications. Since, a complete description of OLE automation is beyond the scope of this paper, details outlining its use and application can be found in a related Microsoft publication (Microsoft Corp. 1994). Through OLE automation, the appropriate *Excel* spreadsheet is activated which then reveals the specific graph of interest. Double-clicking the REFRESH DATA button on the graph will then automatically cause the database to be queried and reflect, graphically and statistically, any new additional data that has been recently entered into the database. Additionally, the graph may be printed by clicking on the PRINT button.

The remaining graphical analyses listed in the box are site specific. In order to view a specific site of interest, *Excel* and *Query* must be utilized together. As an example, if one were to double-click on the VOID RATIO VS. PRESSURE graphical analysis, *Excel* and *Query* would be activated. *Excel* would then display the Void Ratio vs. Pressure graph for the last site selected (Figure 12). A click on the SELECT NEW SITE button enables the user to select a specific site of interest. This screen is shown in Figure 13. At this point, the site of interest denoted by project identification (example: SR 951 B) is highlighted under each of the graphical analyses list boxes. This form gives the user the option of performing any desired graphical analyses for any selected site. Once a new site is selected from the list, *Query* would then extract the pertinent information from the appropriate fields of the consolidation table associated with that site, import the data into *Excel*, and then plot the results on the appropriate graph. A click on the OK button will return focus to the originally selected graphical analysis. Movement between the remaining graphical analyses is facilitated by clicking the appropriate button on the graph. The graphical analyses for site S951B are shown in Figure 14. At this point, the graphs may be printed, the process repeated for a new site, or one may exit and return to the previous DATA ANALYSIS screen of the DMSOS (Figure 10).

PART II COMPUTER ADVISOR FOR ORGANIC SOIL PROJECTS (CAOSP)

Index Property Correlations

A comprehensive laboratory study was previously undertaken to determine the engineering properties of Florida organic soils (Stinnette 1992). This evaluation included the determination of index properties such as the moisture content, organic content, specific gravity and density. Correlations which have been determined consequent to the above study will be maintained and continuously updated in the Computer Advisor for Organic Soils Projects (CAOSP). The following linear correlations with the indicated coefficients of correlation (R^2) have been found to exist among index properties gathered from test sites in Central and South Florida (Stinnette 1992):

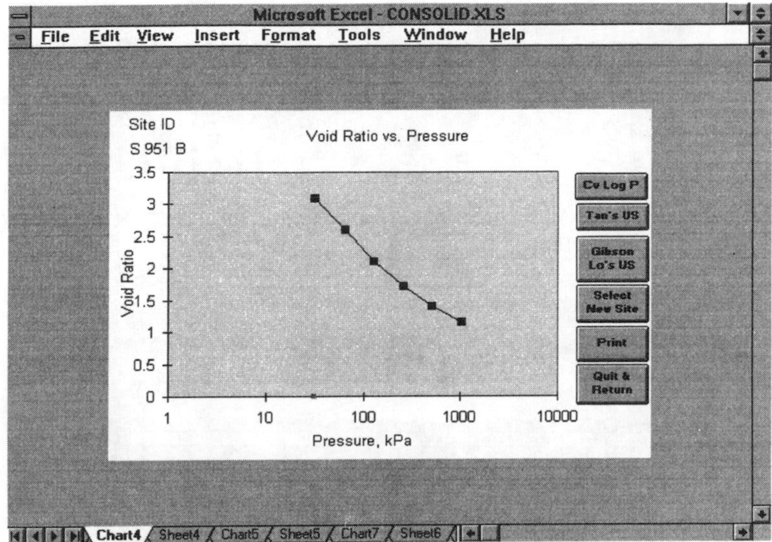

Figure 12 Void ratio vs. Pressure graphical analysis.

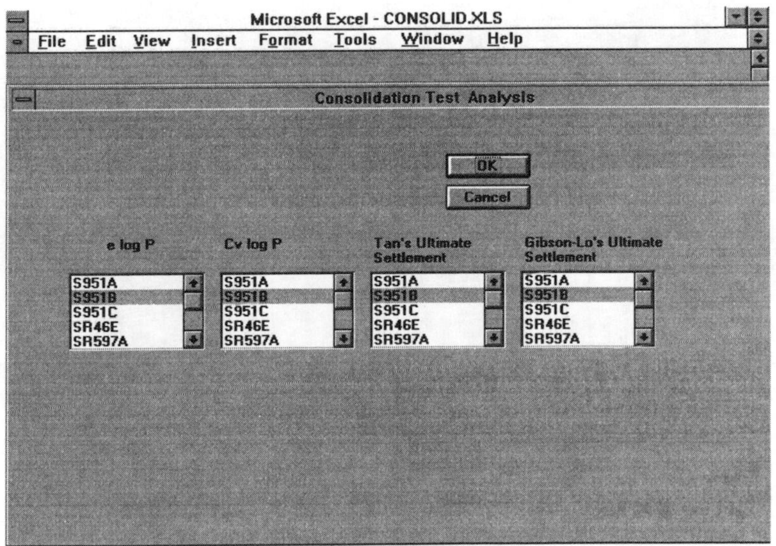

Figure 13 Site Selection Form.

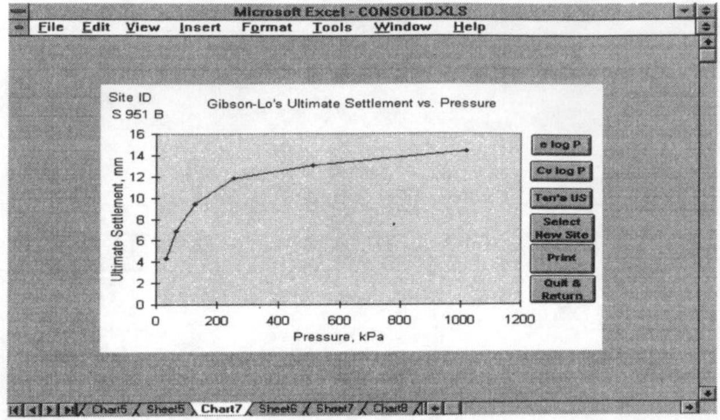

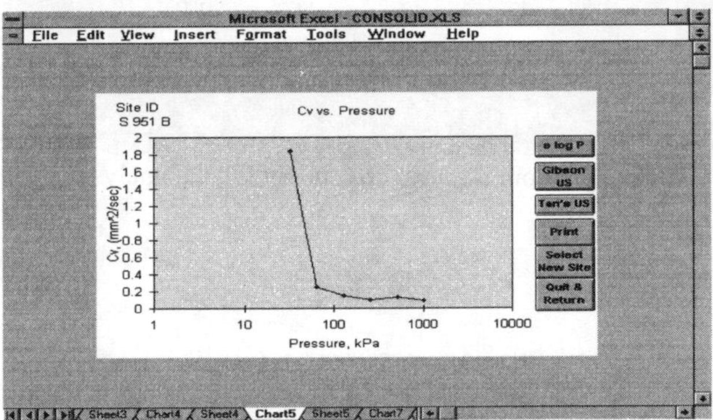

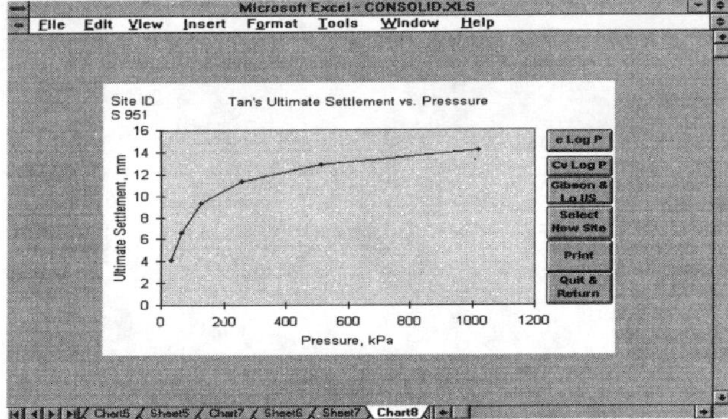

Figure 14 Graphical analysis for site S951B.

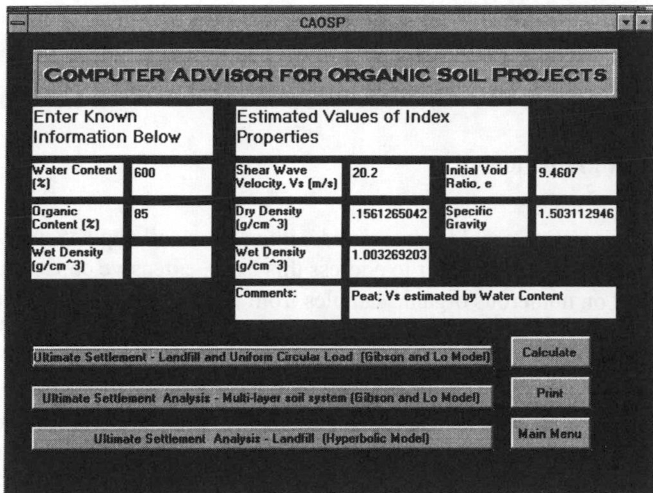

Figure 15 CAOSP.

Dry Density and Water Content;	$R^2 = 0.98$
Wet Density and Water Content;	$R^2 = 0.87$
Organic Content and Water Content;	$R^2 = 0.86$
Specific Gravity (1/Gs) and Organic Content;	$R^2 = 0.93$
Initial Void Ratio (e_o) and Water Content;	$R^2 = 0.95$
Initial Void Ratio (e_o) and Organic Content;	$R^2 = 0.96$

As more data becomes available from experimental research, efforts will be made to further verify these correlations. In addition to incorporating correlations determined by the USF researchers, organic soil correlations established by other researchers will also be included in CAOSP. Currently this includes investigations by Yaromko and Ses'kov (1979) in which the shear wave velocity of peat and organic mud are characterized by the moisture content. Clicking on the CAOSP button on the MAIN MENU reveals the screen shown in Figure 15. This screen shows the CAOSP user interface for estimating a number of index properties of any other organic soil site based on the moisture content and, if available, either the organic content or density. As an example, if the water content and organic content were determined to be 600% and 85%, respectively, then through the use of the CAOSP (Figure 15), values of the shear wave velocity (V_s), wet density (γ_w), dry density (γ_d), initial void ratio (e) and specific gravity (G_s) are estimated as:

V_s = 20.2 m/s

γ_w = 1.00 g/cm^3

γ_d = 0.156 g/cm^3
e = 9.46
G_s = 1.50

Ultimate Settlement Prediction

Rheological Model

A critical construction problem associated with organic soil deposits is the excessive long-term settlement. In order to address this issue, extensive consolidation tests were performed on numerous organic samples from across the state. The author's previous work (Stinnette 1992) also showed that the Gibson and Lo (1961) simplified expression for large time intervals accurately predicts the ultimate settlement for these samples. The ultimate settlement of each sample at all loading stages was then determined using appropriate rheological models and curve fitting techniques due to Gibson and Lo (1961) and Tan (1971), respectively. Since descriptions of these models can be found in previous work (Stinnette 1992), the basic equations are given in Appendix.

Based on the above laboratory data a new technique was then formulated to predict the ultimate settlement of normally consolidated (NC) organic soils based on the organic content and the consolidation pressure (Gunaratne et al. 1998). This technique enables quick and accurate estimation of the settlement of an organic deposit anticipated under a given loading condition and hence, it is invaluable for construction operations. It involves the development of separate expressions for primary and secondary compressibility parameters (*a* and *b*) of the Gibson and Lo (1961) model in terms of organic content and consolidation pressure. While a complete description of this rheological model is found in Gunaratne et al. (1998), the resulting expressions have been incorporated in the CAOSP.

However, the previously described model is valid in the field only for one-dimensional consolidation under extensively placed loads such as landfills. Stinnette (1996) extended the previous expressions to suit the field case of three-dimensional consolidation. This extension becomes important when, for instance, a circular surcharge is applied to preload the organic soil to induce settlement. The extended model predicts the ultimate field strain due to a pressure increase from σ'_1 to σ'_2 as (Stinnette, 1996):

$$\epsilon_{ult})_{field} = \int_{\sigma'_1}^{\sigma'_2} a_{lab}(\sigma')\, d\sigma' + \int_{\sigma'_1}^{\sigma'_2} b_{lab}(\sigma')\, d\sigma' + \int_{\sigma'_1}^{\sigma'_2} \frac{2\, \upsilon\, (K_o - K)}{E}\, d\sigma \qquad (1)$$

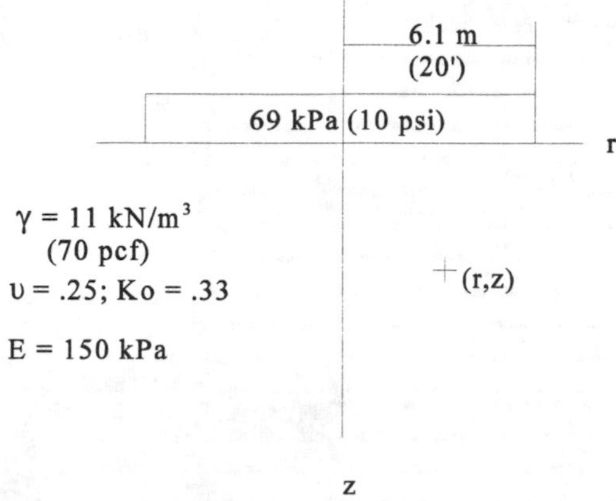

Figure 16 Example of imposed stress conditions in the field.

where

a and *b* are defined in Appendix A

υ	=	Poisson's Ratio
E	=	Elastic Modulus, kN/m^2
K_o	=	Coefficient of lateral earth pressure at rest
K	=	Coefficient of lateral earth pressure under loading

The following example illustrates the estimation of the anticipated ultimate settlement using the extended model (Eq. 1) incorporated in the CAOSP. The formulation of this model and the comparison of its prediction with field results is found in Stinnette (1996).

Figure 16 shows a circular load of radius of 6.1m (20') exerting a pressure of 69 kPa (10 psi) on the ground surface of a homogeneous organic soil. The soil has a density of 11 kN/m^3 (70 pcf), an organic content of 50%, a coefficient of at-rest lateral earth pressure (K_o) of 0.33 and an elastic modulus, E, of 150 kPa. It is desired to estimate the ultimate settlement due to the single stage loading in the upper 6 m of the deposit. A click on the ULTIMATE SETTLEMENT - LANDFILL AND UNIFORM CIRCULAR LOAD (GIBSON AND LO MODEL) button on the form shown in Figure 15 leads to the screen shown in Figure 17. The pertinent information required for the analysis is then entered in the appropriate boxes on the upper portion of the screen. If necessary, estimates of the engineering properties (E, K_o, OC, etc.) can be obtained by clicking on the TYPICAL PROPERTIES button. A mouse click on the CALCULATE button initiates the mathematical analysis. Once the analysis is performed utilizing *Visual Basic* programming coupled

Ultimate Settlement Analysis (Gibson & Lo model)

Input Parameters	
Deposit Height (m) - L	6
Organic Content (decimal) - L	.50
Effective unit weight (KN/m^3) - L	11
Height of New Surcharge (m) - L	4.39
Unit Weight of Surcharge Material (KN/m^3) - L	15.72
Depth of position to be analyzed* (m) - L	3
Height of Existing Surcharge (m)	
Radius of Surcharge (m)	6.1
Lateral Earth Pressure at Rest, Ko	.33
Elastic Modulus (kPa)	150

All input parameters are required for the circular surcharge centerline analysis- Only those marked with" L" (yellow box) are required for landfill analysis.
* Note: Below original ground surface

Typical Properties

Analysis of Input Data	
Stress Upper Limit, b, (kPa)	96.08
Stress Lower Limit, a, (kPa)	33
Strain (Gibson and Lo 1961) Landfill	0.208
Landfill Ultimate Settlement (m)	1.248
Extended Model's Strain for circular load	.2172
Circular Surcharge Ultimate Settlement (m)	1.3032

Calculate

Print

Main Menu

CAOSP

Figure 17 Ultimate settlement analysis - Landfill and uniform circular surcharge.

with *Mathcad*, the results are then displayed on the lower half of the screen (Figure 17). As seen from Figure 17, the ultimate settlement predicted from the extended model for the circular surcharge is 1.303 m. The 1-D "landfill" analysis prediction of 1.248 m is also given for comparison. The above analysis assumes that the soil is homogeneous. On the other hand, if the soil is layered, a multi-layer analysis should be conducted using an appropriate E_1/E_2 ratio (Figure 15).

Empirical Model

A simple curve-fitting technique was also found to give reasonable results for the estimation of the ultimate settlement of an organic soil. The method is also based on laboratory consolidation tests conducted on saturated samples of Florida organic soils with organic contents which varied from 25% to 90%. These soils were found to exhibit a hyperbolic strain versus time relationship and a hyperbolic ultimate strain versus pressure relationship. Utilizing these relationships, an empirical equation was developed to predict the ultimate strain of an organic soil based on the pressure increase and organic content. A complete formulation of this model and the comparison of its predictions to field results is found in Stinnette (1996). This model is also incorporated in the CAOSP and can be accessed by clicking on the ULTIMATE SETTLEMENT ANALYSIS- LANDFILL (HYPERBOLIC MODEL) button on the form shown in Figure 15.

PART III - DR QUALITY CONTROL SYSTEM

Dynamic replacement (DR) is a relatively new ground modification technique that has been used successfully to stabilize organic soil deposits by replacing the organic soil with sand columns. A quality control system that evaluates the degree of improvement of the organic soil layer at a given stage of treatment has been developed (Stinnette et al. 1997). In order to activate the quality control system, an accelerometer must be attached to the drop weight and monitored during impact. The quality control system evaluates the effects of treatment by progressively computing the dynamic settlement modulus (DSM) from the impact acceleration record during the entire treatment period. The role of DSM in evaluation of field DR is illustrated by using results from a field DR study on organic soils as follows.

The DR field pilot study site is located in Plant City, FL. A 0.2 ha area of the site was cleared and filled with approximately 1-1.5 m of sandy fill material (approx. 7% fines). This material acted as a work platform for the crane and Cone Penetration Test (CPT) rig and as initial replacement material for the dynamic replacement process. The thickness of the surficial organic deposit at the pilot study site was 1.7 m. The peat was characterized as primarily amorphous in nature having an organic content of 95%. A complete description of the field study and its results are given in Stinnette et al. (1996c).

During the pilot studies, the drop weight (3.64 tonne) was instrumented with a piezoelectric accelerometer and connected to the data acquisition system described previously. When the photoelectric cell beam (2.13 m above the ground surface) sensed the falling drop weight, data acquisition was initiated. Measurements are recorded during the impact for a duration of 1 second and were then stored in the computer (*.prn). The acceleration vs. time graph is automatically plotted on the *Visual Basic* data acquisition screen during testing (Figure 14). The file is then automatically analyzed by the DMSOS by clicking on DYNAMIC REPLACEMENT MACRO in the drop down menu on the DATA REDUCTION SCREEN (Figure 15). This action launches a specially written Excel spreadsheet which uses interactive macro programming. The values of acceleration, velocity and displacement vs. time are automatically calculated. Once the data has been reduced, both the continuous data such as the accelerometer record and discrete data such as the drop height, weight etc. are then stored in the *Access* database. Figure 18 shows the discrete data entry/retrieval form for the DR Pilot Study Results. In addition to pertinent test results being displayed, a double-click on the photo shown in the lower right portion of the screen plays the video of the field DR during that drop. A click on the GRAPHICAL ANALYSIS button reveals the DYNAMIC REPLACEMENT GRAPHICAL ANALYSIS SCREEN (Figure 19). As an example, a double-click on the ACCELERATION, VELOCITY AND DISPLACEMENT GRAPHICAL ANALYSIS initiates OLE automation and opens the appropriate *Excel* spreadsheet which then reveals the specific graph of interest (Figure 20). At this point, the DR data stored in the *Access* database may be retrieved for any impact number by clicking on the SELECT NEW TEST button which reveals the DR PILOT STUDY LOCATION FORM (Figure 21).

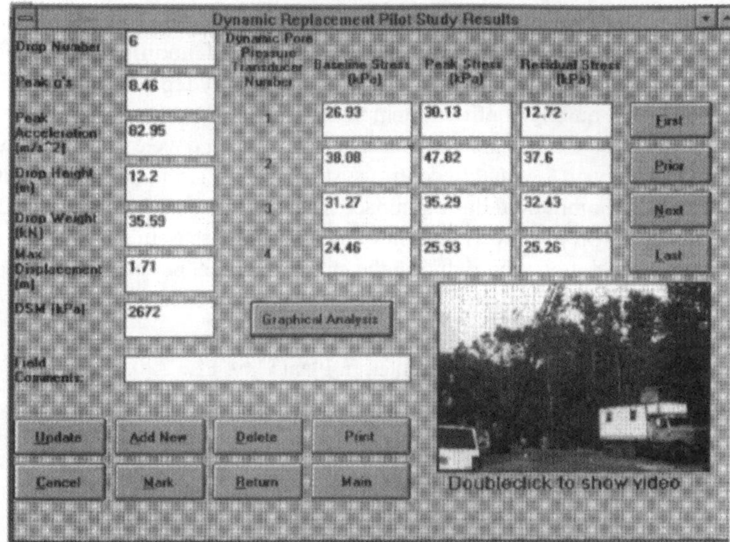

Figure 18 Dynamic replacement pilot study results form.

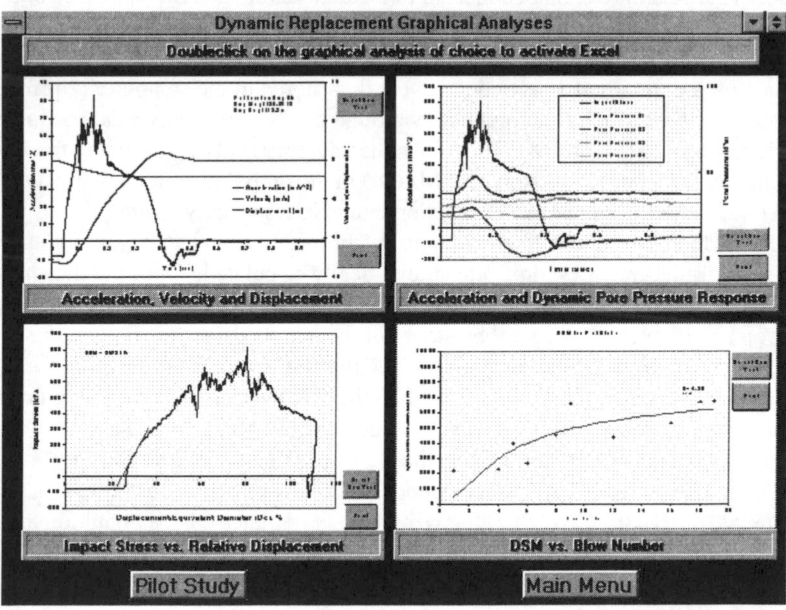

Figure 19 Dynamic replacement graphical analyses screen.

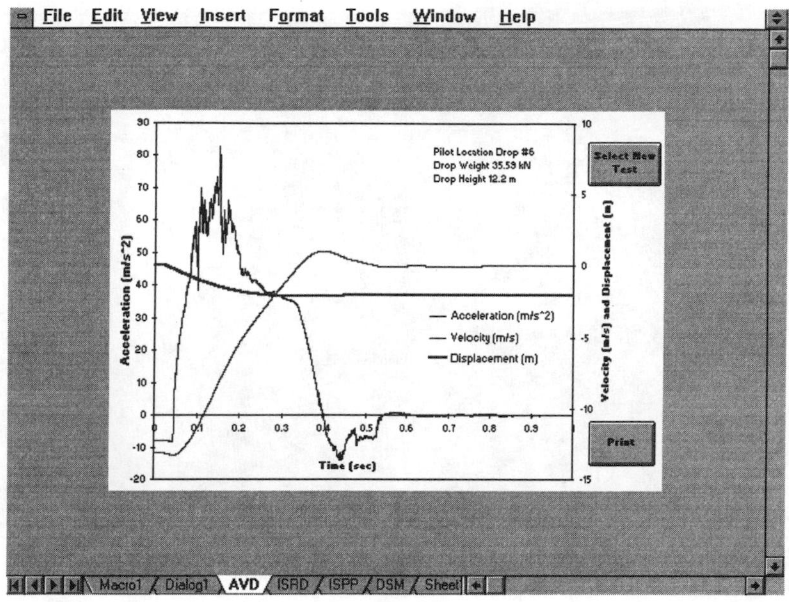

Figure 20 Acceleration, velocity and displacement graphical analyses screen.

The DSM is evaluated from the measured data as follows. First, the record of impact stress, σ, is determined from the acceleration record, a, by using Eq. (2)

$$\sigma = \frac{m\ a}{A} \tag{2}$$

where:

m = mass of the drop weight

A = base area of the drop weight

DR Pilot Study Location

Current data is from:

Pilot Study 6

Select new site from menu below and click on OK

Pilot Study 4
Pilot Study 5
Pilot Study 6

Graph of Acceleration, Velocity and Displacement — Show Graph

Impact Stress vs. Relative Displacement (DSM) — Show Graph

Graph of Acceleration and Dynamic Pore Pressure — Show Graph

Graph of DSM vs. Blow Number — Show Graph

OK

Cancel

Dialog1 / AVD / ISRD / ISPP / DSM / Sheet1 / Sheet2

Figure 21 DR pilot study location selection form.

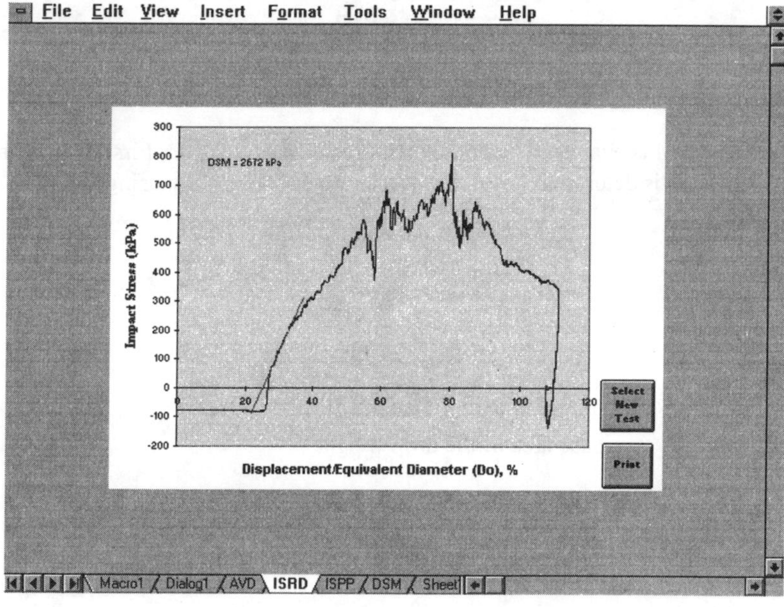

Figure 22 Impact stress vs. Relative displacement analysis.

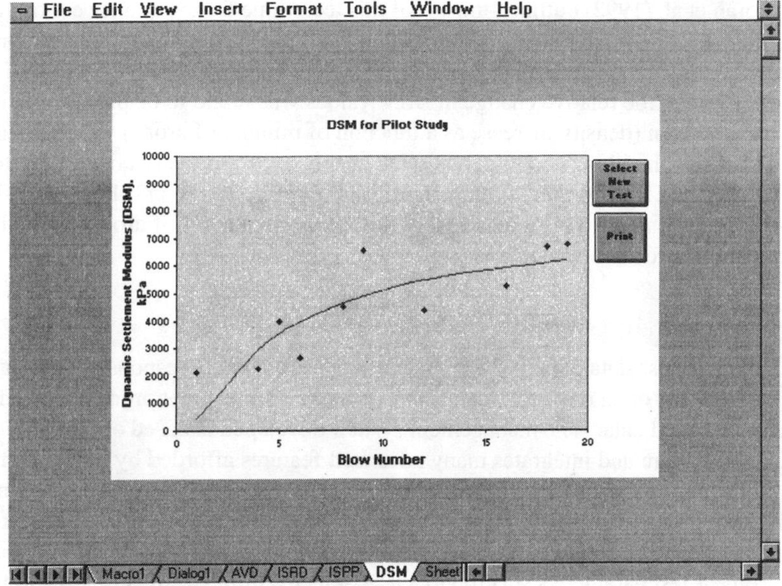

Figure 23 DSM vs. Blow Number analysis.

Then, the ratio of the instant displacement to the equivalent diameter, D_o (relative displacement), is then plotted against impact stress, σ. The dynamic settlement modulus, DSM, is then calculated as the slope of the tangent of the loading portion of the impact stress-relative displacement curve given by:

$$DSM = \frac{\Delta \sigma}{\Delta(\frac{d}{D_o})} \quad \textbf{(3)}$$

A click on the SHOW GRAPH button next to *Impact Stress vs. Relative Displacement* (Figure 21) reveals the graphical analysis of the DSM (Figure 22). Once the DSM is calculated for each blow, it is then stored in the *Access* database. The plot of *DSM vs. Blow Number* is displayed once the SHOW GRAPH button has been clicked (Figure 23).

Poran et al. (1992) outlined the use of this technique for the quality control of dynamic compaction on dry sand and report that the DSM values obtained based on experimental tests conducted in the laboratory have shown good correlation to soil densities. Further the relative change in DSM values was found to be proportional to the rate of densification (density increase as a function of number of drops). As seen in Figure 21, during the illustrated pilot study DR seems to be more effective in the first 15 drops, while after 15 drops there is an insignificant increase in the DSM, showing little additional improvement. This indicates that the optimum number of drops should be approximately 15 for this particular site.

CONCLUSION

An in-house data management system was developed in response to the data management needs of an ongoing organic soils research project that produces extensive laboratory and field data. The management system developed is based on the *Windows 3.1* operating system and integrates many advanced features afforded by this operating system. The main features of the DMSOS are: (1) automated data acquisition during consolidation, triaxial and field dynamic replacement tests, (2) analysis of automatically acquired or manually entered data, and (3) easy retrieval of general correlation results for organic soils or test parameters specific to a user selected organic soil site. In addition, a Computer Advisor for Organic Soils Projects (CAOSP) built into the DMSOS is capable of furnishing many useful engineering properties of an organic soil deposit once easily measurable parameters such as organic content or moisture content are input. Of these, the most critical properties from a construction perspective is the compressibility of the organic soil deposit. Also incorporated in the DMSOS is a quality control system developed for the evaluation of dynamic replacement (DR) of organic soils. The system facilitates rapid and efficient analysis and graphical presentation of the DR test results. These include the plots of acceleration, velocity and displacement of the drop weight for each impact. Additionally, the DSM parameter is computed for each blow and graphically presented such that its trend may be observed as treatment proceeds. From this plot, the optimum energy (number of blows) can be predicted when the change in the DSM becomes relatively insignificant. The system can easily be implemented in any laboratory with existing microcomputer facilities and it will accommodate any new *Windows*-based software developments in the foreseeable future.

ACKNOWLEDGMENTS

The author would like to thank research colleagues Dr. A. Mullins, Dr. S. Thilakasari, and Professor M. Gunaratne of the University of South Florida. Additionally, the author would like to thank Mr. Brian Jory and Ms. Terry Puckett of the Florida Department of Transportation, District 7 for their support during the study. Finally, the FDOT grant B-8452 WPI 0510665 is gratefully acknowledged.

APPENDIX ULTIMATE SETTLEMENT DETERMINATION OF ORGANIC SOIL

Gibson and Lo Rheological Model (1961)

The Gibson and Lo equation for large values of time is expressed as:

$$\varepsilon(t) = \Delta\sigma[\, a + b\, (1 - e^{(\frac{-\lambda\, t}{b})})] \tag{A1}$$

where

$\Delta\sigma$	=	effective stress increment
t	=	time
a	=	primary compressibility
b	=	secondary compressibility
λ/b	=	secondary compression rate factor

Hence, the ultimate settlement ($t \rightarrow \infty$) may then be expressed as:

$$\delta_{ult} = \Delta\sigma\; H\; [a + b] \tag{A2}$$

where

H	=	height of organic soil layer

Tan's Empirical Model (1971)

Tan's empirical relationship for large values of time is given as:

$$\frac{t}{\delta} = M\, t + C \tag{A3}$$

where

t	=	time
δ	=	total settlement at any time t
$M\&C$	=	empirical constants

It can be seen that as t becomes very large ($t \rightarrow \infty$) the ultimate settlement can be expressed as:

$$\delta_{ult} = \frac{1}{M} \qquad \text{(A4)}$$

REFERENCES

Adams, T.M., Tang, Agatha Y.S. and Wiegard, N. (1993). "Spatial data models for managing subsurface data." *J. of Computing in Civil Engineering*, 7(3), 260-277.

Benoit, J., P.A. de Alba and Sawyer, S.M. (1993). "National geotechnical experimentation sites central data repository." In *Geographic Information Systems and their Application in Geotechnical Earthquake Engineering,* 17-20.

Gibson, R.E. and Lo, K.Y. (1961). "A theory of consolidation of soils exhibiting secondary compression." *Acta Polytechnical Scandinavia*. Ci 10 296: Scandinavian Academy of Science.

Gunaratne, M., P. Stinnette, G. Mullins, C. Kuo and W.F Echelberger. (1998). "Compressibility relations for natural organic soil." *ASTM J. of Testing and Evaluation*, January.

Ishii, M., K. Ishimura and T. Nakayama. (1992). "Management and application of geotechnical data: the geotechnical data information system of the Tokyo metropolitan government." *Environ. Geol. Water Sci.*, 19 (3),169-178.

Lee, F.H., T.S. Tan, G.P. Karunaratne and Lee, S. (1990). "Geotechnical data management system." *J. of Computing in Civil Engineering*, 4 (3), 239-254.

Microsoft Corporation. (1994). *Microsoft Office Developer's Kit*, Microsoft Corporation.

Poran, C.J, Heh, K.S. and Rodriquez, J.A. (1992) "A new technique for quality control of dynamic compaction," in Grouting, Soil Improvement and Geosynthetics, *Proceedings of conference sponsored by ASCE, February 25-28, 1992,* Geotechnical Special Publication No. 30, ASCE, New York, pp. 915-926.

Rasdorf, W.J. and Spainhour, L.K. (1993). "Developing and implementing a conceptual computing materials design database." In *Proc. of ASME Computers in Engrg. Conf., San Diego, CA.*

Stinnette, P. (1992). *Engineering Properties of Organic Soils*. Masters Thesis, University of South Florida, Tampa, Florida.

Stinnette, P. (1996). *Geotechnical Data Management and Analysis System for Organic Soil.* Doctoral Thesis, University of South Florida, Tampa, Florida.

Stinnette, P, Gunaratne, M., Mullins, G. And Thilakasiri, S. (1997). "A quality control program for performance evaluation of dynamic replacement of organic soil deposits." *Geotechnical and Geological Engineering,*Vol. 15, No. 4, 283-302.

Sykora, D.W. and Koester, J.P. (1991). "Database of seismic body wave velocities and geotechnical properties." In *Geotechnical Engineering Congress 1991 ASCE Geotechnical Special Publication No. 27*, 690-700.

Tan, Swan Beng. (July 26 - August 1, 1971). "A method for estimating secondary and total settlement." In *Proc.of the Fourth Asian Regional Conf. on Soil Mechanics and Foundation Engrg.*, 2, Za-Chieh Moh, ed., Bangkok, Thailand.

Yaromko, V.N. and Ses'kov, V.E. (1979). "Elastic and dissipative properties of peats and organic muds." Translated from *Osnovaniya, Fundamenty I Mekhanika Gruntov*, 1, 11-13.

D.J. Bobrow [1] and S. Vaghar [2]

Review of GIS Developments in Geotechnical Instrumentation on the Central Artery/Tunnel Project, Boston, Massachusetts

---

REFERENCE: Bobrow, D.J., and Vaghar, S. , "**Recent GIS Developments in Geotechnical Instrumentation on the Central Artery/Tunnel Project, Boston, Massachusetts,**" *Field Instrumentation for Soil and Rock, ASTM STP 1358,* G.N. Durham and W.A. Marr, Eds., American Society for Testing and Materials, West Conshohocken, PA, 1999.

ABSTRACT: This paper discusses the Geographic Information System (GIS) application which has been developed for rapid analyses and reporting of instrumentation and survey data on the Central Artery/Tunnel Project, in Boston, Massachusetts. Recent developments of the GIS application have included the addition of modules for preparing groundwater and settlement contour plans, and integration of digital photographs into the projectwide mapping. A module for inserting geological cross sections is underway.

KEYWORDS: geographic information system (GIS), graphical interface, wide area network, contouring, digital photograph

---

The Central Artery/Tunnel (CA/T) Project is currently the largest government funded public works project in the United States. An existing 1950s elevated highway is being replaced by a tunnel alignment through downtown Boston. Figure 1 shows the project location; Contract C09A4 consists of jacking prefabricated concrete tunnel segments beneath an existing and operational railroad line.

The following sections provide an overview of the Project, the instrumentation program, and the GIS application which has been developed to store, process, and present the instrumentation information.

[1]Geotechnical Engineer, Bechtel/Parsons Brinckerhoff, Central Artery/Tunnel Project, 1 South Station, Boston, MA 02072.

[2]Loss Control Consultant, American International Group, 101 Federal Street, Boston, MA 02072.

A full account of this application can be found in a 1997 paper prepared by the authors (Vaghar et al , 1997).

All elevations mentioned in this paper are referred to the Central Artery/Tunnel Project Datum (PD), which is 100 feet below the National Geodetic Vertical Datum (Mean Sea Level).

Background

Boston is in the geographic center of the Boston Basin, an area with sedimentary, metamorphic and igneous rocks of late Precambrian and Cambrian age, overlain by glacial and post glacial soil deposits (Kaye 1982).

The Project site lies predominantly within an area that was once mostly covered by seawater, tidal marshes and estuaries. Most of these areas have been filled over the years, using clays dredged from the harbor, and sand and gravel from gravel pits and glacial drumlins, as well as building demolition rubble. The filling began in the late 1700s and generally ended in the 1960s (Ty 1987).

An idealized subsurface geologic profile of the Project is shown in Figure 2.

Instrumentation Program

The instrumentation program on the Project serves the following objectives (Vaghar et al. 1997):

1. Monitoring of ground and facilities deformations resulting from the construction, and forewarning of unforeseen conditions which may impact the safety of the personnel, the public, and the construction work.

2. Monitoring of changes in groundwater patterns and possible resulting ground movements, which may adversely impact the abutting facilities.

3. Providing pre-construction baseline condition data for comparison with construction and post-construction data, to aid in the evaluation of construction related impacts.

Figure 3 shows typical instrument installation details for selected instruments.

Overview of GIS Application

As mentioned above, a full account of the GIS application can be found in a 1997 paper prepared by the authors (Vaghar et al. 1997). The GIS application allows a user to retrieve, display and generate output from text and graphics databases for a specific area and then to relate this information to data collected over time for the selected

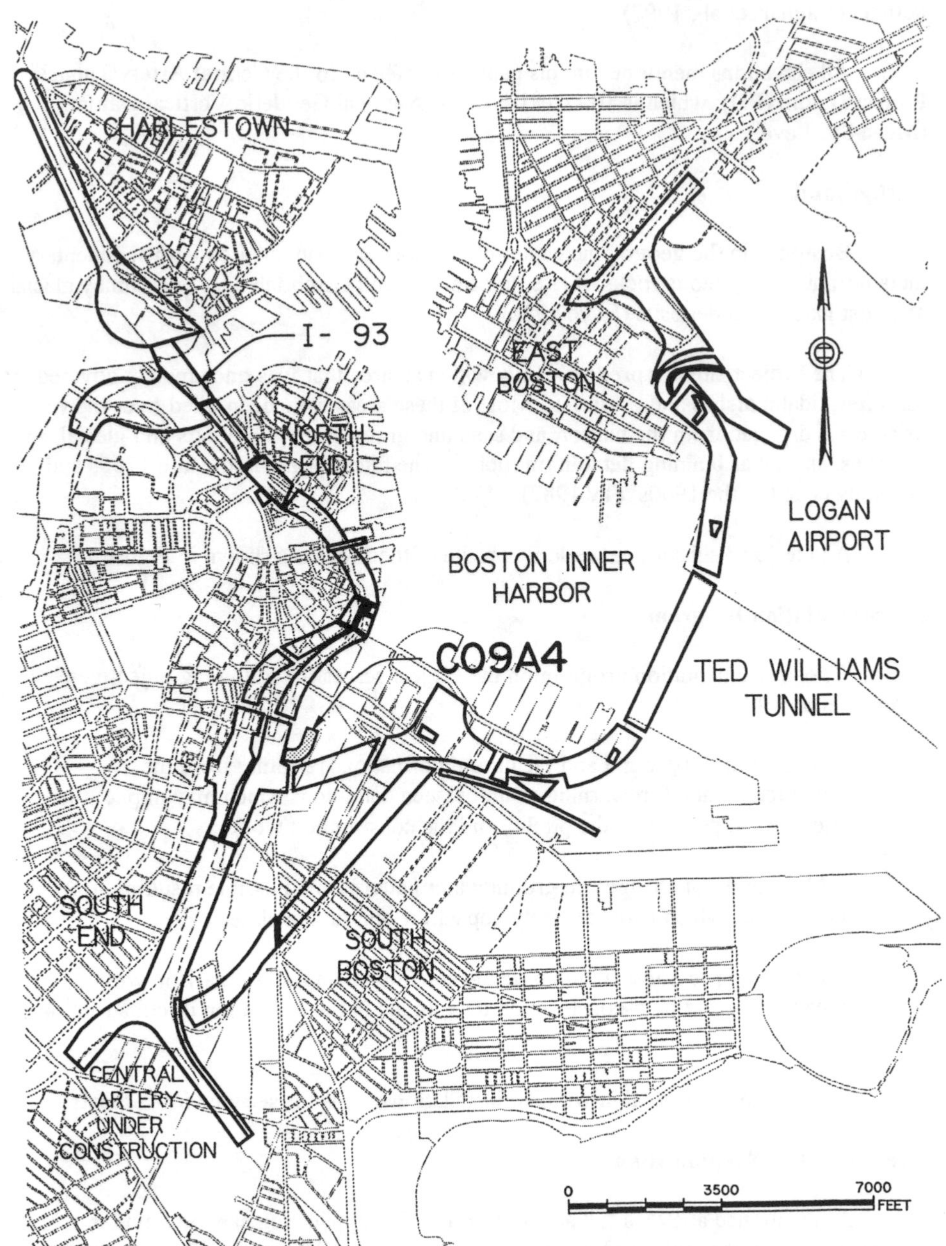

Figure 1 - *Location Plan*

instruments located in the specific contract. The text data are stored in Oracle, a relational database, while the graphics data such as mapping are stored in GDS, CA/T Project CADD system.

The Structured Query Language (SQL) used to communicate with the relational database allows the users to create the database tables and store and manipulate information in these tables, as well as maintain the database itself.

Once the data are entered and checked, they are available to users across a Wide Area Network (WAN). Users unfamiliar with the location of instruments can access a GDS graphical interface for location of the instruments.

Using GIS technology, spatial analysis capabilities allow the users to map tabular data from the relational database, interactively query graphics for associated tabular data, and restrict tabular data to a geographic area defined in GDS. Oracle and GDS are integrated using a Graphical User Interface. Using the GDS module SQL CADD, customized menus and basic programs written in GDS, the user is able to access and perform operations on both databases to generate customized maps, without an in-depth knowledge of either GDS or Oracle.

Use of the GDS graphical database helps with the evaluation of the data. When a change in the data is identified, the user can locate the area geographically, and review instrument readings. The user can zoom out to display an area as large as needed for the review. Once the desired instrument types are displayed, an attribute query can display specific categories of instruments within the group, such as instruments read within the past week, vibrating wire piezometers in clay, and instruments that exceed a predefined allowable value. The user may also perform the above mentioned queries, and color code any instruments that meet the criteria, by using pull-down menus.

Included in all reports sent to the Resident Engineer's office is a cover letter with an evaluation of the data. Any significant changes in readings are correlated with construction or natural activities and factors. When instruments exceed the predefined contract response values, a graph of the data accompanies the report. The Project uses plotting software to plot the instrumentation data. Oracle generates a text file containing all the data as well as default parameters used by the plotting program.

Recent Developments

The GIS application described above has been under concurrent development and use since 1994. New features are continuously developed and added, as new needs are identified. The following modules have recently been added.

- A module which generates groundwater and settlement contour plans.

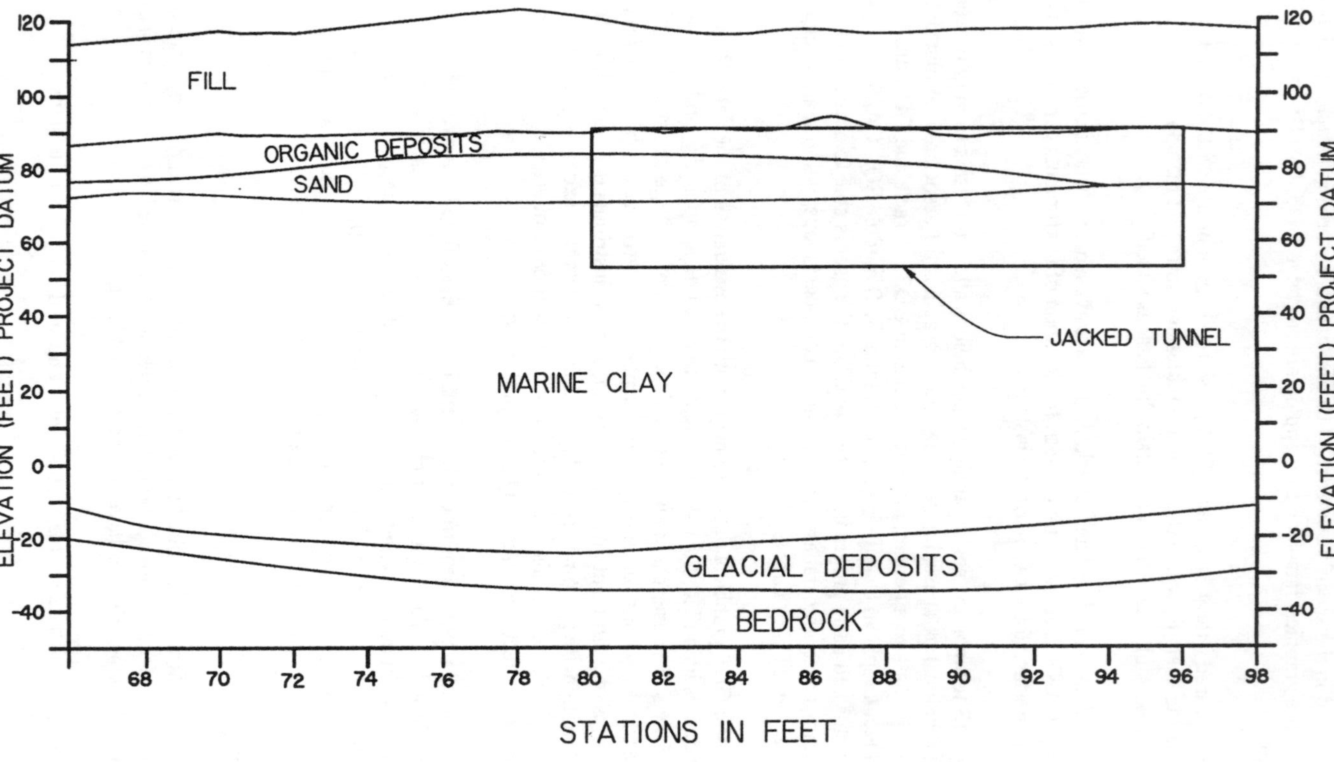

SCALE: VERTICAL = 10 x HORIZONTAL

Figure 2 - *Idealized C09A4 Subsurface Profile*

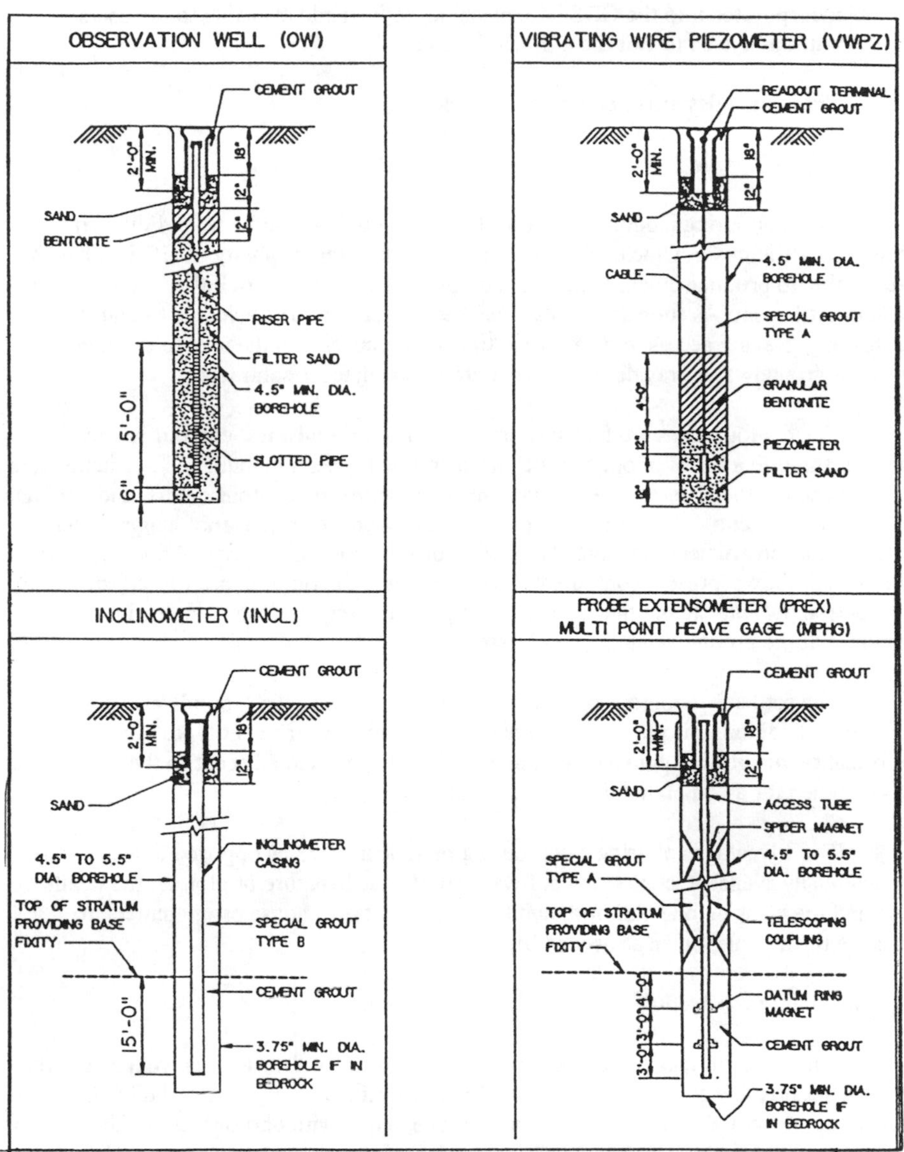

Figure 3 - *Typical Instrument Details*

• A module which has enabled site photographs taken with digital cameras to be incorporated into the GDS base mapping, to be made available to the users, without a need to exit the GIS application.

These modules are briefly described below.

Contour Plan Module

The contouring module integrates the speed, quality assurance, and the data crunching attributes of Oracle with the ease of use and graphic ability of GDS. Thus the user is able to produce quick, accurate, and up-to-date contours for either groundwater or ground settlement. As soon as new data has been entered into Oracle it is available for contouring. Having access to data for multiple contracts allows the user to evaluate the effect of groundwater drawdown in one contract on adjacent contracts.

The contour plans are fully customizable. The groundwater contour routine leads the user through a series of options such as instrument type, minimum and maximum date of readings, contour interval, spot elevations on instruments, contour colors, and geologic strata of instruments. The output is a contour plan with labels and spot height water levels at each instrument location. The option of selecting minimum and maximum dates of readings allows present contours to be compared with contours prior to construction. This also allows bracketing time so that during a pump test, for example, the immediate response of the groundwater can be assessed.

The settlement contour routine has similar features to the groundwater contour routine. In addition, it is possible to combine a number of instrument types, for example deformation monitoring points on buildings and on the ground adjacent to the building can be combined for a comprehensive plan coverage.

The advantages of using a contouring module in GDS as opposed to a commercially available contouring software are the facile nature of plotting the tabular data and the plans on one platform, and the ability to use in-house programmers to enhance the module, on an as needed basis.

Digital Photograph Module

The Project tracks or monitors over 250 buildings and structures, which abut the Project, for possible construction impact. Details of each building such as building date, ownership, structural ranking and foundation type, along with photographs of the exterior, are kept in a file for each building. Retrieving this information, in order to correlate the address with the instrumentation data has been a manual and rather time-consuming task, due in part to the nonsystematic nature of some of the building addresses.

Advantage was taken of the increasing commercial availability of digital cameras and associated software to add a module in the GIS application to locate and retrieve the building photographs electronically. The exterior of each building was photographed using a digital camera. The photographs were assigned an address and placed in a database. Through a link with GDS, these photographs can be viewed by clicking an icon on the building footprint on the project wide location plan. The photographs can also be printed or imported into other documents, in order to accompany instrumentation reports. Work is underway to add the building data retrieved in the form of a pop-up table. Using this module, buildings which the instrumentation data suggest may have been impacted, are readily identified, for possible further action such as owner notifications, and installation of additional instrumentation.

Future scheduled steps include the addition of interior photographs on selected buildings, particularly those which have been internally instrumented, with a link to a commercially available software which renders a "virtual reality" three dimensional picture of the interior of the building. This function is useful when evaluating buildings where the number of internal instrumentation exceeds several hundred, or where routine building access is difficult.

A follow up to the above module is to link the GIS application to the network of overhead video cameras which continuously monitor many areas of the Project and the downtown alignment corridor. This would enable the instrumentation data to be correlated with the causal construction activities on a real time basis.

Conclusion

Development of the GIS application has led to effective processing and evaluation of instrumentation data. The use of in-house programmers has allowed the development and enhancements to be directed by the user engineers, thus producing an application which is both user friendly and focused on the needs of the Project.

Recent additions such as groundwater and settlement plans and digital photograph modules go beyond the basic data processing, single instrument and tabular reports, and present the next logical steps in the evaluation of the instrumentation data.

Acknowledgment

Thanks are due to the Massachusetts Highway Department for permission to publish Project information.

References

Kaye, C.A., 1982, "Bedrock and the Quaternary Geology of the Boston Area", Geological Society of America, *Reviews in Engineering Geology*, Volume V.

Ty, R.K.S., 1987, "History and Characteristics of Man-made Fill in Boston and Cambridge", MSc. Thesis, Massachusetts Institute of Technology, pp. 132, 31 maps.

Vaghar, S., Bobrow, D.J., and Marcotte, T.A., 1997, "Instrumentation for Monitoring Ground Movements; Central Artery/Tunnel Project, Boston, Massachusetts", Rapid Excavation and Tunneling Conference, Las Vegas, Nevada.

Instrumentation for Measuring Physical Properties in the Field

Kanglin Li[1] and M. R. Reddy[2]

Use of Tensiometer for In Situ Measurement of Nitrate Leaching

---

REFERENCE: Li, K. and Reddy, M. R., "**Use of Tensiometer for In Situ Measurement of Nitrate Leaching,**" *Field Instrumentation for Soil and Rock, ASTM STP 1358*, G. N. Durham and W. A. Marr, Eds., American Society for Testing and Materials, West Conshohocken, PA, 1999.

Abstract: In order to monitor nitrate leaching from non-point source pollution, this study used tensiometers to measure insitu nitrate concentration and soil-moisture potential. Instead of filling the tensiometers with pure water, the study filled the tensiometers with nitrate ionic strength adjuster (ISA, 1 *M* $(NH_4)_2SO_4$). After the installation of the tensiometers at various depths along soil profiles, a portable pressure transducer was used to measure the soil moisture potential, and a nitrate electrode attached to an ion analyzer was used to measure the nitrate concentration insitu. The measurement was continuous and non-destructive. To test this method in the laboratory, eight bottles filled with pure sand were treated with known nitrate solutions, and a tensiometer was placed in each bottle. Measurements were taken every day for 30 days. Laboratory test showed a linear relationship between the known nitrate concentration and the tensiometer readings (R^2 = 0.9990). Then, a field test was conducted in a watermelon field with green manure mulch. Field data indicated a potential of nitrate leaching below the soil depth of 100 cm when crop uptake of nutrients was low.

Keywords: nitrate leaching, tensiometer, soil-moisture potential, ground water, porous media

---

To identify chemical transport in a soil system, frequent measurements are required to evaluate: 1) soil matric potential; 2) availability of plant nutrients; and 3) quality of ground water. Both moisture and nutrients in the topsoil support the growth of plants. However, nutrients found in deeper horizons of the soil are not desirable because they cause pollution of the ground water. Ground water is the source of drinking water to about 50% of the overall population and to more than 90% of the rural

---

[1,2]Assistant Professor and Professor, respectively, Department of Natural Resources and Environmental Design, North Carolina A&T State University, Greensboro, NC 27411.

population in the USA (Office of Technology Assessment 1990). Nitrate leaching is one of the most problematic and widespread of the vast number of potential ground water contaminants (Keeney 1986). Ground water with a nitrate nitrogen concentration > 10 mg L^{-1} is considered unsuitable for drinking in the USA (Canter 1997).

Tensiometers are widely used for irrigation management (Smajstrla and Locascio 1996, Peterson et al. 1993, Gaussoin et al. 1990) and for measurement of soil hydraulic potential (Hendrickx et al. 1990, Cassel and Klute 1986). Li et al. (1997) used a series of tensiometers to determine whether water movement and solute transport in intact saprolite materials were preferential. A tensiometer generally consists of a plastic tube closed at the bottom by a water-permeable porous ceramic cup. The tube is filled with pure water, then closed by a rubber septum stopper at the top. The permeable end is inserted into the soil with the tube usually in a vertical position. Water passes through the porous cup in either direction, depending on the water content of the soil. Water moving out of the tube into a dry soil creates a negative pressure head in the tensiometer. A pressure transducer is used to measure the pressure head. For a saturated soil, tensiometer readings are near zero kPa and could be as low as -80 to -100 kPa for dry soils. Therefore, the tensiometer is a device that contains a permanent internal volume of water capable of continuous ionic exchanges with soil solution. The solute concentration of the tensiometer solution is a function of the soil solution.

Lysimeters are used to collect solution from porous materials in order to measure solute concentrations in soil solution. The lysimeter is a sampler using a porous ceramic cup with a large diameter. With the cup in contact with the soil, soil solution is trapped in the lysimeter due to the negative pressure created inside the cup for a period of time. Some studies have summarized the advantage of this method for analyzing soil solution (Heinrichs et al. 1996, Moyer et al. 1996, Fraser et al. 1994). Ohte et al. (1997) used insitu lysimeters to study inorganic nitrogen discharged from forest soils. Hempel et al. (1995) studied mercury behavior in mercury contaminated sites by lysimeters. Vanclooster et al. (1995) monitored solute transport in a multi-layered sandy soil by a lysimeter. They found that all soil water was used to transport solute in the top of the soil. Studies also found some drawbacks of this method: 1) The accumulation of solution in the ceramic cup was irreversible (Bernhard and Schenck 1986). Ionic exchange between the lysimeter and the soil solution was limited; 2) it was difficult to obtain a volume of soil solution large enough for analytical purpose from a dry soil with a hydraulic potential value lower than -60 kPa (Saragoni et al. 1990); 3) the porous cups were clogged after a series of sampling cycles (Morisson and Lowery 1990, Grover and Lamborn 1970).

The ceramic cup of tensiometers can be used to capture solutes from soil systems as well as measure soil matric potential. Based on the ionic equilibrium between the soil solution and the tensiometer solution, Moutonnet et al. (1993) modified the tensiometer and used it to measure soil hydraulic potential, and then pumped solution from the tensiometer to measure nitrate. Finally, the solution was returned to the tensiometer for later measurement. They found that it required 8 days to establish an ionic equilibrium between the tensiometer and the soil solution in a silt loam soil. Their measurements with the solution from the tensiometers correlated with the amount of nitrate that had been consumed by the crop. During the growing season there was a minimal risk for contamination of water table by nitrate.

In this study, we filled a tensiometer with 1 *M* ammonium sulfate ($(NH_4)_2SO_4$) solution as an ionic strength adjuster (ISA) instead of pure water. After the establishment of a pressure and ionic equilibrium between the tensiometer and the soil solution, a pressure transducer was used to measure the soil hydraulic potential. Then, a nitrate electrode was directly inserted into the tensiometer to measure the nitrate concentration. The 1 *M* $(NH_4)_2SO_4$ solution maintains a desirable salt concentration in the tensiometer during the ionic exchange process (Mulvaney 1996, Greenberg et al. 1992). The objective of this study is to use tensiometers to monitor nitrate movement in soil profiles as well as to measure soil hydraulic potential as its original function. A field experiment was conducted in a soil with a seedless watermelon crop.

Materials and Methods

Tensiometers were assembled in the laboratory using a porous ceramic cup (O.D. = 2.223 cm), a piece of schedule 80 PVC tube (O.D. = 2.223 cm) and a piece of clear acrylic tube (O.D. = 1.59 cm). Various lengths of the PVC tube allowed the tensiometers to be installed at soil depths of interest. The unit was then filled with ISA solution (1 *M* $(NH_4)_2SO_4$). Finally, the open end of the acrylic tube was sealed by a rubber septum stopper.

Laboratory Calibration of the Tensiometers

Eight 1-L polypropylene bottles were filled with pure sand (diam. < 0.5 mm). A tensiometer with a length of 30 cm was inserted into the sand 2.5 cm above the bottom of each bottle. The bottles were then placed on a mechanical shaker to ensure a good contact between the porous cups and the sand. Each bottle was treated with 250 ml of solution with concentrations of 2, 10, 50, or 100 mg $L^{-1}$of NO_3^--N with a duplicate, respectively. The bottles were capped by lids with holes for the tensiometers to go through. Then, the holes were sealed with Teflon tape. Thus, the bottles were closed systems and able to maintain a stable hydraulic pressure for the calibration period.

Measurement was started from the first day of the installation. A portable pressure transducer (Marthaler et al. 1983) equipped with a hypodermic needle was used to measure the hydraulic potential in the bottles. By removing the septum stopper, a nitrate electrode was inserted into the solution through the open end of the acrylic tube. A portable Orion 710 pH/ISE meter was connected to the electrode. The millivolt readings of the concentrations of NO_3^--N in the tensiometers were read with the meter. Each reading was plotted against its known concentration of NO_3^--N to yield a regression coefficient and a R^2. A stable regression coefficient was found when an ionic equilibrium was established between the tensiometer solution and the sand solution. Measurements were taken for 30 days. The meter and the measurement were capable of being controlled by a computer.

Field Insitu Measurement of NO_3^--N and Hydraulic Potential

A field experiment was conducted at a site with a sustainable agricultural cropping system for seedless watermelon grown in rotation with green manure. The site is located at the Environmental Studies Center of North Carolina A&T State University. The soil was identified as an Enott-Like sandy loam (fine, mixed, thermic typic hapludalf) by the North Carolina Soil Survey Staff before the experiment was initiated. This survey also described the properties of the soil profile prior to the experiment (Table 1). The soil and landscape at this site represent many areas of the Piedmont region of southeastern USA.

Table 1-- *Soil Horizon, Depth, Texture, pH and Organic Carbon Content of the Experimental Site.*

HORIZON	DEPTH (cm)	TEXTURE	pH	ORGANIC CARBON (%)
AP	0 - 23	Sandy Loam	5.8	1.82
Bt	23-48	Clay	6.7	0.35
BC	48-64	S. Clay Loam	6.7	0.11
C	64-107	Sandy Loam	6.9	0.04
Cr	107-135	Sandy Loam	7.2	0.04

At the experimental site cover crops were planted in the fall of 1996 (Austrian winter peas, hairy vetch, and rye) then mowed by a flail mower on May 20, 1997 to leave the biomass as a uniform layer of mulch on the surface of the soil. The purpose of the mulch was to supply nutrients to the watermelon crop, conserve soil moisture, and control weeds and soil erosion. Tensiometers were installed in soil at the depths of 20, 40, 60, 80 and 100 cm with five replications. Seedlings of watermelon were transplanted on June 4, 1997. After a period of time for the tensiometer's equilibration, first measurements of hydraulic potential and nitrate concentration were taken in July 1997. A calibrated PH/ISE meter was used to measure NO_3^--N concentrations in the tensiometers at different depths. The measurements were taken each week.

Results and Discussion

Laboratory Calibration

The measured hydraulic potential remained almost a constant (-5.4 kPa) in all sand bottles during the period of the calibration (curve 4 in Figure 1). The regression analysis of each day resulted in a regression coefficient, an intercept and a R^2. In the initial stage (the first day) of the calibration (curve 1 of Figure 1), nitrate ions just started to migrate to the tensiometer through pores of the ceramic cup. The mV readings from the pH/ISE meter did not respond to the known concentrations of the

NO_3^--N in the sand solution. Curve 2 in Figure 1 shows the mV readings against the logarithm nitrate concentrations on the sixth day of the calibration. Although it was non-linear, a response was detected by the pH/ISE meter. The first linear response occurred on the 12th day with a regression coefficient of 39.7 mV (log(mg L^{-1}))$^{-1}$, an intercept of 265 mV (curve 3, Figure 1) and a R^2 = 0.9986; However, the regression parameters were not stable at this time.

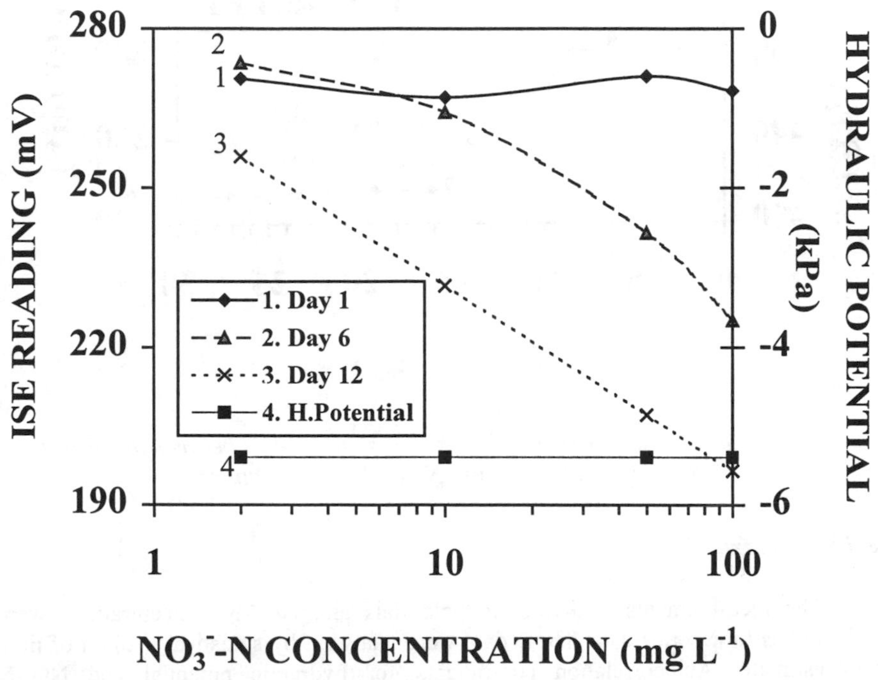

Figure 1 — *Plot of the pH/ISE Meter Readings as a Function of the Known NO_3^--N Concentrations Approaching the Ionic Equilibrium.*

The time for tensiometers to reach an ionic equilibrium between the ISA solution inside the tensiometers and the sand solution was determined by regression analysis between the pH/ISE readings and the known concentrations of NO_3^--N. As the calibration progressed with time, both the regression coefficient and the intercept approached their constants (Figure 2). Fifteen days were required for the ISA solution in the tensiometers to reach the nitrate equilibrium with the sand solution. The equilibrated regression coefficient was between -47 and -45 mV (log(mg L^{-1}))$^{-1}$ with a standard deviation of 0.53 mV (log(mg L^{-1}))$^{-1}$. The intercept was between 280 and 290 mV. Its standard deviation was 4.2 mV. The regression analysis had a R^2=0.9990.

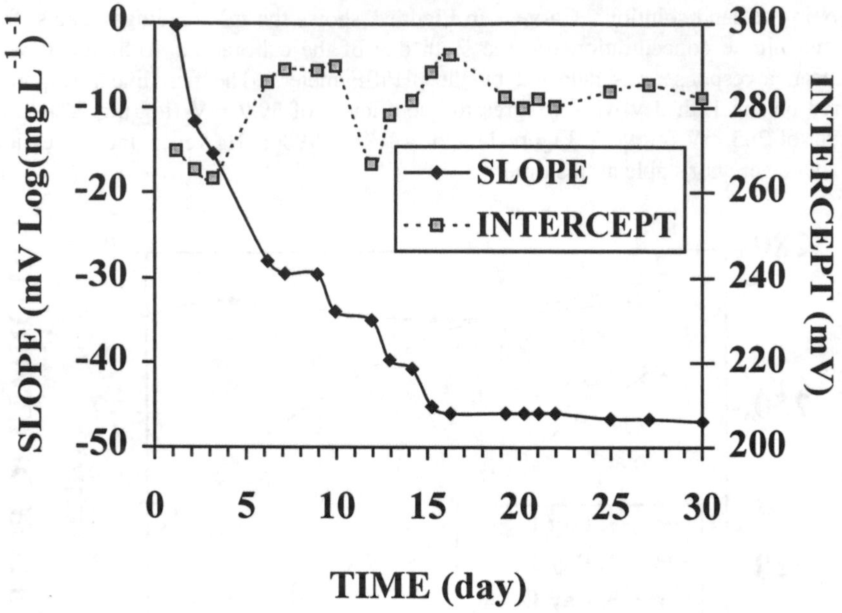

Figure 2 — *Slopes and Intercepts of the Regression Analysis between the pH/ISE Readings and the Known NO_3^--N Concentrations.*

Field Experiment

The measurements of hydraulic potentials and NO_3^--N concentrations were considered as time series data. These data were analyzed by statistical method of time series estimate. Autocorrelation coefficients for hydraulic potential and NO_3-N concentration were not significant and so general statistics and graphs are used in the following discussion. In July 1997 at the early stage of the watermelon growth, the surface soil (top 20 cm) maintained a high hydraulic potential. The moisture distribution over the entire profile was relatively uniform. As the demand by plant growth for water increased with growing season, hydraulic potential at the depths of 20 and 40 cm decreased rapidly because watermelon growth utilized the moisture mostly from the top 40 cm of the soil (Figure 3). The soil also supplied some moisture for the watermelon growth from as deep as 60 cm. Moisture status at depths of 80 and 100 cm fluctuated only slightly during this time period. When the hydraulic potential of the top 20 cm soil decreased to -60 kPa, the plots were irrigated on August 20, 1997. Also, occasional rain events replenished the top 20 cm soil with water. In general, insignificant amount of water from rainfall or irrigation percolated below a depth of 60 cm. Soil moisture below 60 cm depth remained stationary until the harvest of the watermelons.

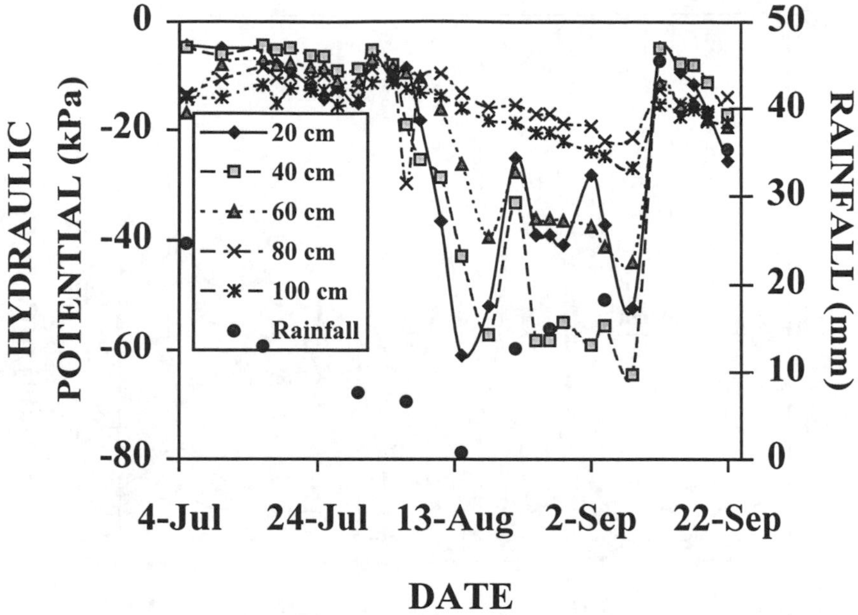

Figure 3 — *The Fluctuation of Soil Hydraulic Potential at Different Depths over Time and the Rainfall Data.*

Data in Figure 4 indicate that release of NO_3-N from the mulch composition was relatively low at the initial stage of watermelon growth (before July 29). The soil maintained a low level of nitrate concentration (< 10 mg L^{-1}). As the decomposition started to release NO_3^--N, a rapid accumulation of NO_3^--N occurred in the soil solution. This accumulation caused increases of NO_3^--N over the entire 100-cm profile until watermelon plants started their rapid consumption of nutrients. Most of NO_3^--N for watermelon growth was from top 20 cm soil, some from 20 to 40 cm depths. The watermelon rarely used nutrients from 60 cm depth and below. However, a decrease of NO_3^--N in soil solution below 60 cm depth happened after August 8. This implied a tendency of NO_3^--N leaching. Then a steady accumulation of NO_3^--N in the soil solution below 60 cm depth was observed. After the watermelons were ripe and harvested (September 10), decomposition of mulch was still in progress. NO_3^--N in the soil profile resumed its pace of accumulation and leaching to deeper soil. The content of NO_3^--N was 15 mg L^{-1} at 100 cm depth. It indicated a potential risk of leaching when crop consumption could not keep up with the release of NO_3^--N from the mulch decomposition. The results also show that each rainfall event caused sudden decreases of NO_3^--N in soil solution at depths of 20 to 60 cm due to dilution (Figure 4). But rain did not percolate to soil depths below 60 cm during the growing season.

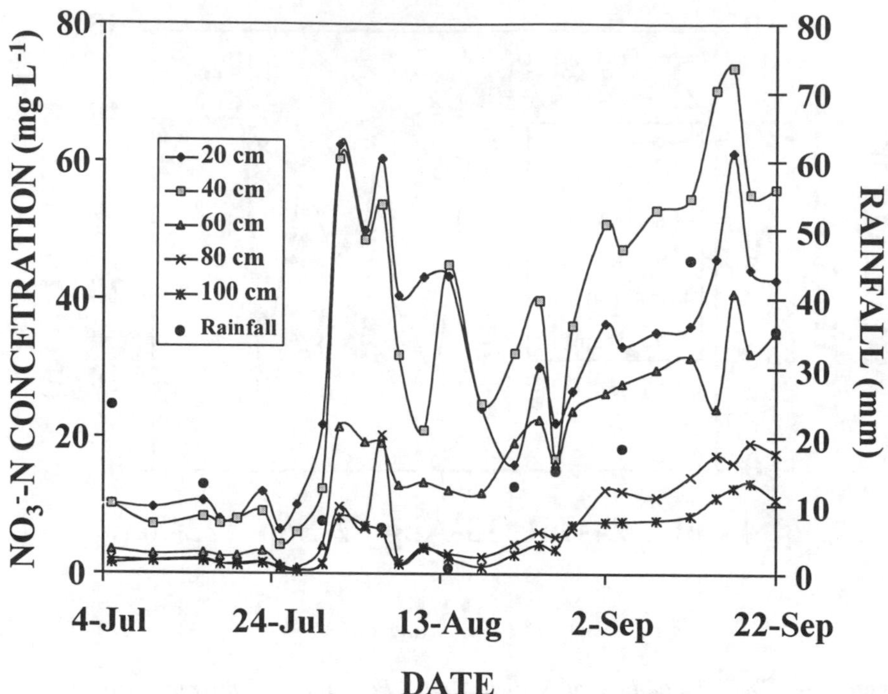

Figure 4 — *The Occurrences of NO_3^--N in Soil Solution during the Progress of Mulch Decomposition and the Nutrient Uptake by Watermelon Growth.*

The nitrate concentrations along the soil profile at four stages of the growing season were presented in Figure 5. When the mulch decomposition and the watermelon growth were at their initial stages (curve 1, Figure 5), the overall nitrate concentration in soil solution was low. As the depth increased, the nitrate concentration decreased. Then, mulch decomposition increased, but watermelon requirement for nutrients did not increase significantly. There was a high accumulation of NO_3^--N in the topsoil solution when mulch decomposition started to release NO_3^--N and watermelon growth did not have a great demand for nutrients (curve 2, Figure 5). When watermelon growth utilized NO_3^--N mostly from the top 20-cm soil, NO_3^--N concentration at the 40-cm depth was higher than that at the 20-cm depth (curve 3, Figure 5). After watermelons were ripe (September 10), NO_3^--N started to build up in the soil profile because the decomposition of the mulch was still in progress. Thus, nitrate leaching to deeper soil horizons increased the NO_3^--N concentrations to the soil depth of 100 cm, significantly (curve 4, Figure 5).

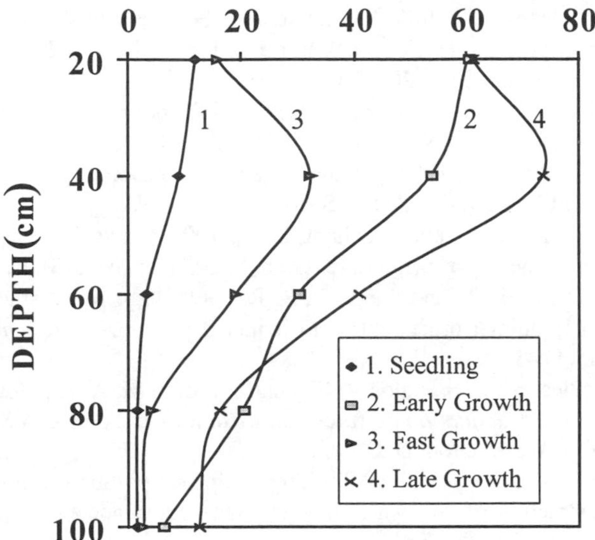

Figure 5 — *Typical NO_3^--N Profiles at Various Phases of Mulch Decomposition and Watermelon Growth.*

Conclusion

In this study, the tensiometers were filled with ISA solution instead of pure water in order to measure hydraulic potential and nitrate concentration in soil solution simultaneously and insitu. The tensiometer can be installed at any depth of interest. The time to reach an ionic equilibrium between the tensiometers and the test materials was 15 days. Results of this study showed the progressive drying of the soil due to deficit of irrigation and the consumption of the N nutrient by the watermelon crop. The nitrate measurement from 20 to 40 cm depths confirmed that watermelon growth consumed nutrients from these depths. Rainfall water only percolated the soil to a depth of about 60 cm. However, the later redistribution of nitrate with water can induce a potential risk of nitrate leaching when the nutrient requirement of the crop is low.

The advantage of this method of nitrate measurement is its rapidity and non-destructivity. With the electrode connected to a computer, this method is appropriate to obtain continuous readings of nitrate concentration in deeper soil horizons and water tables. When different ion selective electrodes are chosen, ISA solutions are available for other ions of interest. Thus, this method can also be used to measure different ionic contaminants in soil solution, subsurface soil or ground water.

REFERENCES

Bernhard, C., and Schenck, C., 1986, "Utilization des bougies poresues pour extraire la solution du sol dans le ried central de l'Ill en Alsace," *Bull. Groupe Fr. Humidimetrie Neutrinique*, 20:73-85.

Canter, L. W., 1997, "*Nitrate in groundwater,*" University of Oklahoma. Lewis Publishers. Boca Raton, New York, London, Tokyo.

Cassel, D. K., and Klute, A., 1986, "Water potential: Tensiometry," *Method of Soil Analysis*. Part 1. ASA, CSSA, and SSSA. Madison, WI., pp. 563-596.

Causoin, R. E., Murphy, J. A., and Branham, B. E., 1990, "A vertical installed, flush-mounted tensiometer for turfgrass research," *HortScience*, 25(8): 928-929.

Fraser, P. M., Cameron, K. C., and Sherlock, R. R., 1994, "Lysimeter study of the fate of nitrogen in animal urine returns to irrigated pasture," *European Journal of Soil Sci.*, 45(4):439.

Greenberg, A. E., Clesceri, L. S., Eaton, A. D., and Franson, M. A. H., 1992, "*Standard methods for the examination of water and wastewater*," APHA, AWWA, WDFF Publ. APHA, Washington, D. C.

Grover, B. L., and Lamborn, R. E., 1970, "Preparation of porous ceramic cups to be used for extraction of soil water having low solute concentrations," *Soil Sci. Soc. Am. Proc.*, 34:706-708.

Heinrichs, H., Bottcher, G., and Pohlmann, M., 1996, "Squeezed soil-pore solutes -- A comparison to lysimeter samples and percolation experiments." *Water, Air, and Soil Pollution*, 89(1/2): 189.

Hempel, M., Wilken, R. D., and Beyer, K., 1995, "Mercury contaminated sites - behavior of mercury and its species in lysimeter experiments," *Water, Air, and Soil Pollution*, 80(1/4): 1089.

Hendrickx, J. M. H., Wierenga, P. J., and Nash, M. S., 1990, "Variability of soil tension and soil water content," *Agric. Water Manage.*, 18:135-148.

Keeney, D., 1986, "Sources of nitrate to groundwater," *CRC critical reviews in environmental control*, Vol. 16, No. 3, pp. 257-304.

Li, K., Amoozegar, A., Robarge, W. P., and Buol, S. W., 1997, "Water movement and solute transport through saprolite," *Soil Sci. Soc. Am. J.*, 61:1738-1745.

Mackay, A. D., Sakadevan, K., and Hedley, M. J., 1994, "An insitu mini lysimeter with a removable ion exchange trap for measuring nutrient losses by leaching from grazed pastures," *Australian Journal of Soil Research,* 32(6): 1389.

Mathaler, H.P., Vogelsanger, W., Richard, F., and Wierenga, P.J., 1983, "A pressure transducer for field tensiometers," *Soil Sci. Soc. Am. J.*, 47:624-627.

Morrison, R. D., and Lowery, B., 1990, "Effect of cup properties, sampler geometry, and vacuum on the sampling rate of porous cup samplers," *Soil Sci.*, 140:308-316.

Moutnnet, P., Pagenel, J. F., and Fardeau, J. C., 1993, "Simultaneous field measurement of nitrate-nitrogen and matric pressure head," *Soil Sci. Soc. Am. J.*, 57:1458-1462.

Moyer, J., Saporito, W., Janke, L. S., L.S., and Rhonda, R., 1996, "Design, construction, and installation of an intact soil core lysimeter," *Agronomy Journal*, 88(2): 253.

Mulvaney, R. L., 1996, "Nitrogen - inorganic forms," *Method of Soil Analysis.* Part 3.. ASA, CSSA, and SSSA. Madison, WI., pp.1123-1184.

Office of Technology Assessment, 1990, "*Beneath the bottom line: Agriculture approaches to reduced agricultural contamination of groundwater*," OTA-F-418. U.S. Congress. U.S. Government Printing Office. Washington, D. C., pp. 3-20.

Ohte, N., Tokuchi, N., and Suzuki, M., 1997, "An insitu lysimeter experiment on soil moisture influence on inorganic nitrogen discharge from forest soil," *Journal of Hydrology*, 195: 78.

Peterson, D. L., Glenn, D. M., and Wolford, S. D., 1993, "Tensiometer-irrigation control valve," *Applied Engineering in Agriculture*, 9(3): 293-297.

Saragoni, H., Poss, R., and Oliver, R., 1990, "Dynamique et lixiviation des elements mineraux dans les terres de barre du sud du Togo," *Agron. Trop.* (Paris), 45:259-273.

Smajstrla, A.G., and Locascio, S. J., 1996, "Tensiometer-controlled, drip-irrigation scheduling of tomato," *Applied Engineering in Agriculture*, 12(3): 315-319.

Vanclooster, M., Mallants, D., Vanderborght, J., Diets, J., Van Orshoven, J., and Feyen, J., 1995, "Monitoring solute transport in a multi-layered sandy lysimeter using time domain reflectometry," *Soil Sci. Soc. Am. J.*, 59: 337-344.

R.A. Erchul[1]

Compaction Comparison Testing Using a Modified Impact Soil Tester And Nuclear Density Gauge

REFERENCE: Erchul, R. A., "**Compaction Comparison Testing Using a Modified Impact Soil Tester and Nuclear Density Gauge,**" *STP 1358, Field Instrumentation for Soil and Rock*, American Society for Testing and Materials, West Conshohocken, PA, 1999.

ABSTRACT: The purpose of this paper is to compare test results of a modified Impact Soil Tester (IST) on compacted soil with data obtained from the same soil using a nuclear density gauge at the U.S. Army Corp of Engineer's Buena Vista Flood Wall project in Buena Vista, Virginia. The tests were run during construction of the earth flood wall during the summer of 1996. This comparison testing demonstrated the credibility of the procedure developed for the IST as a compaction testing device. The comparison data was obtained on a variety of soils ranging from silty sands to clays. The Flood Wall comparison compaction data for 90% Standard Proctor shows that the results of the IST as modified are consistent with the nuclear density gauge 89 percent of the time for all types of soil tested. However, if the soils are more cohesive then the results are consistent with the nuclear density gauge 97 percent of the time. In addition these comparison tests are in general agreement with comparison compaction testing using the same testing techniques and methods on compacted backfill in utility trenches conducted earlier for the Public Works Department, Chesterfield County, Virginia.

KEYWORDS: compaction testing, impact soil tester, nuclear density gage, comparison testing

Introduction

The Impact Soil Tester (IST) was developed by Dr. Baden Clegg in Australia during the 1970s (Clegg 1978). The apparatus is commonly known as the Clegg Hammer of the Clegg Impact Soil Tester while the test is known as the Clegg Impact Test (CIT) or simply Impact Test. Although predominantly developed as an in situ soil strength test, the Impact Test has also been more recently used in the United Kingdom for the testing of backfill in utility trenches to determine if this material has been properly compacted

[1]Professor, Civil & Environmental Engineering Department, Virginia Military Institute, Lexington, Virginia, USA.

(Winter & Selby, 1991). The advantage of using the IST in compaction of utility trench backfill is its rapidity in taking a test and ease of operation.

The components of the IST are a guide tube, instrumented drop hammer and electronic readout (See Figure 1). Recent IST models have been available with the readout on the handles of the hammer, eliminating the meter box and cable as shown in Figure 1. The IST measures in units of Impact Value (IV) which is based on the peak deceleration of the 4.54kg (10 lbm) drop hammer measured in units of 10 gravities. The IV is measured by an internal accelerometer located in the tester and digitally displayed on the readout. The IST can be used on any type of soil and comes in four hammer sizes: 0.5 kg (1.1 lbm), 2.25 kg (5 lbm), 4.54 kg (10 lbm) and 20 kg (44 lbm). The choice of hammer size is a function of strength of soil, the desired depth of soil to be tested and, to a certain degree, the soil material types. In compaction testing the 4.54 kg (10 lbm) is most often used since it has an effective testing depth of approximately 150 mm (6 in) and its Impact Value scale of 0 to 100 covers the full range of materials encountered up to a CBR of about 70% (Clegg, 1983).

The operation procedure for using the IST is relatively straightforward, rapid and easily conducted. The operator places at least one foot on the edge of the IST guide tube flange to steady it during the test. The button on the readout box is depressed while the 4.54 kg (10 lbm) hammer is raised 457 mm (18 in.) and allowed to free fall. The readout button is held depressed while this step is repeated four consecutive times. With more recent models, the button is pressed and released and the four drops carried out while the circuit remains activated on a timer. In either case, the maximum value of the four blows remains displayed on the readout. Figure 1 shows an IST test in progress. The reading displayed after the fourth blow is commonly taken as the Impact Value and recorded.

The IST provides a strength parameter; it does not measure soil density directly. By understanding how soil strength relates to the compaction process, Impact Value may be used to indicate if compacted soils comply with the desired density specifications. The standard test method and additional information for determination of the IV of a soil can be found in ASTM D 5874-95. In addition, a modified testing procedure of measuring the accumulated depth of penetration of the hammer at the fourth blow was also employed during the comparison compaction testing. Testing with the IST using this modification is explained in this paper.

A Modified Testing Procedure

Research into the use of the IST has been ongoing at the Virginia Military Institute for several years. Both laboratory and field tests were conducted in conjunction with Chesterfield County, VA, Department of Utilities on utility trenches, focusing on the use of the IST in determining adequacy of trench backfill compaction. Specifically, the object of the work had been on the use of the IST in testing the backfill of utility trenches in lieu of other more expensive techniques, such as a nuclear density gauge.

Erchul and Meade (1990) and Erchul, et al. (1994) initially investigated the IST's effectiveness for compaction on various soils in the laboratory and field. They observed that an inverse non-linear relationship existed between the IV and the accumulated depth

Figure 1. Clegg Hammer

of the hammer penetration into the soil. It was thought that this relationship defined the energy curve developing from the impact of the hammer into the soil. They identified the relationship between the IV and the accumulated depth of penetration at the fourth blow (d) for the 4.54 kg (10 lbm) Impact Test Hammer as being:

$$IV = 60 / d \qquad (1)$$

where the depth (d) is measured in units of 2.54 mm (0.1 in).

Erchul and Meade (1990) developed a modified testing procedure for the IST that might permit a rapid prediction of compaction level or energy imparted into the soil in regard to attaining maximum density. For the modified testing procedure, a numerical scale of 0-20 in. depth units of 2.54 mm (0.1 in.) was added on the hammer shaft in order to measure depth of hammer penetration. Two acceptance zones were developed from previous field and laboratory testing data. These zones, Zone 90 Acceptance Criteria and Zone 95 Acceptance Criteria (See Figures 2 and 3) show what was considered acceptable combinations of IVdepth results for 90% and 95% Standard Proctor levels. In addition, statistical acceptance criterion developed for a set of eight tests on one lot of soil whereby 5 of 8 test results had to fall within the acceptance zone. This was based on a 10% risk of rejecting suitably compacted material.

Buena Vista Flood Wall Tests

Using ASTM D 5874-95 and the above mentioned methodology, six different areas on the Buena Vista flood wall project were tested multiple times for a total of 46 trials run. Each of the six areas contained a different type of soil ranging from CH (high plastic inorganic clays) to SM (silty sands). Initially a nuclear density gauge reading was taken at a specific location after compaction was conducted along a section of the flood wall. For comparison, immediately after this reading was taken the IST was used at eight positions at cardinal and inter-cardinal headings around the nuclear gauge location, approximately one foot away in each direction. The specifications called for 90% Standard Proctor for acceptance. The set of eight readings of IV and depth data were plotted on a Zone 90 Acceptance Criteria chart for each location. If more than three of the eight tests at each location fell outside the zone of acceptance, the soil was considered as failing to meet the specification. Conversely, if five or more results fell inside the zone, the compacted soil was considered passable at the 90% Standard Proctor compaction level.

The nuclear density gauge data were analyzed to determine water content and dry density at each location. The nuclear gauge results were compared with Standard Proctor data obtained in the laboratory on the same type of soil to determine whether the compacted soil passed or failed specifications according to the nuclear density gauge method.

The IST test results were then compared with the nuclear density gauge results to determine their agreement with the compaction of the test locations.

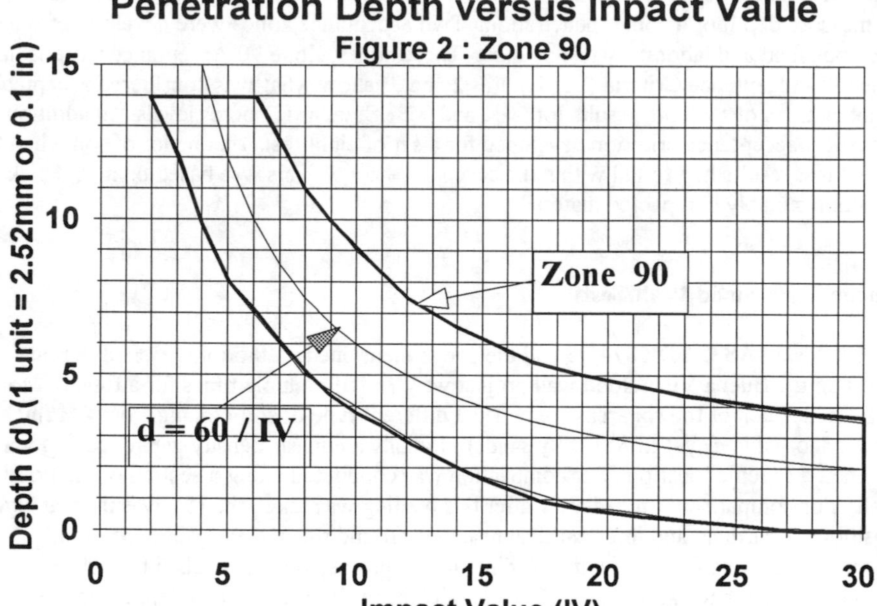
Penetration Depth versus Inpact Value
Figure 2 : Zone 90
Depth (d) (1 unit = 2.52mm or 0.1 in)
15
10
5
0
0
5
10
15
20
25
30
Impact Value (IV)
Zone 90
d = 60 / IV

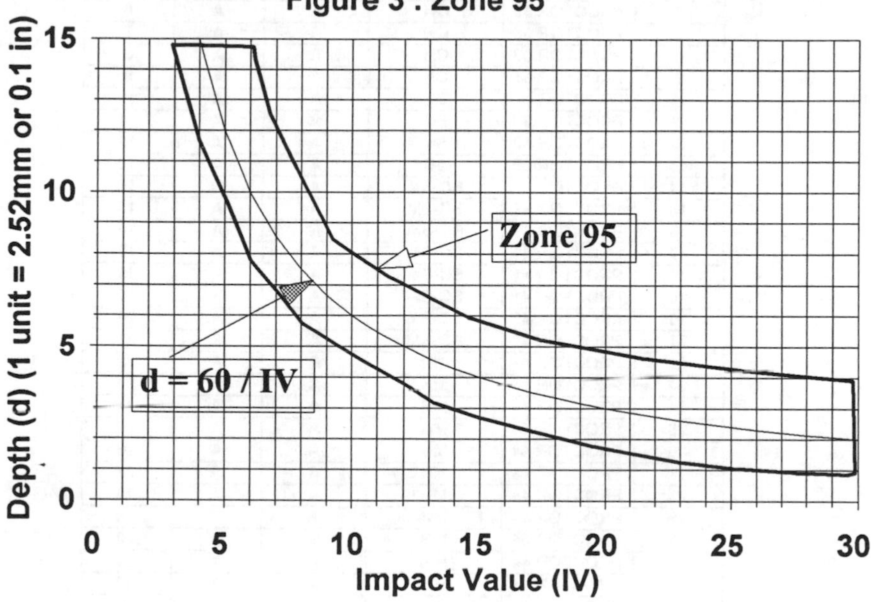
Penetration Depth versus Inpact Value
Figure 3 : Zone 95
Depth (d) (1 unit = 2.52mm or 0.1 in)
15
10
5
0
0
5
10
15
20
25
30
Impact Value (IV)
Zone 95
d = 60 / IV

Results

TABLE 1
RESULTS OF COMPARISON TESTS USING THE IMPACT SOIL TESTER (IST) AND THE NUCLEAR DENSITY GAUGE
A=ACCEPTED
R=REJECTED

TEST	SITE	SOIL TYPE	IST	GAUGE	TEST
1	PROCTOR 29	CH	A	A	1
2	PROCTOR 29	CH	A	A	2
3	PROCTOR 29	CH	A	A	3
4	PROCTOR 33	CH	A	A	4
5	PROCTOR 33	CH	A	A	5
6	PROCTOR 33	CH	A	A	6
7	PROCTOR 33	CH	A	A	7
8	PROCTOR 33	CH	A	A	8
9	PROCTOR 33	CH	A	A	9
10	PROCTOR 39	SC	A	A	10
11	PROCTOR 39	SC	A	A	11
12	PROCTOR 39	SC	A	A	12
13	PROCTOR 39	SC	A	A	13
14	PROCTOR 41	SC-SM	A	R	14
15	PROCTOR 41	SC-SM	A	A	15
16	PROCTOR 41	SC-SM	R	A	16
17	PROCTOR 41	SC-SM	A	A	17
18	PROCTOR 41	SC-SM	A	A	18
19	PROCTOR 41	SC-SM	A	R	19
20	PROCTOR 41	SC-SM	A	A	20
21	PROCTOR 41	SC-SM	A	R	21
22	PROCTOR 41	MH	A	A	22
23	PROCTOR 44	MH	A	A	23
24	PROCTOR 44	MH	A	A	24
25	PROCTOR 44	MH	A	A	25
26	PROCTOR 44	MH	A	A	26
27	PROCTOR 44	MH	A	A	27
28	PROCTOR 44	MH	A	A	28
29	PROCTOR 44	MH	A	A	29
30	PROCTOR 44	MH	R	A	30
31	PROCTOR 44	MH	A	A	31
32	PROCTOR 44	MH	A	A	32
33	PROCTOR 44	MH	A	A	33
34	PROCTOR 44	MH	A	A	34
35	PROCTOR 44	MH	A	A	35
36	PROCTOR 44	MH	A	A	36
37	PROCTOR 44	MH	A	A	37
38	PROCTOR 44	MH	A	A	38
39	PROCTOR 57	ML	A	A	39
40	PROCTOR 57	ML	A	A	40
41	PROCTOR 57	ML	A	A	41
42	PROCTOR 57	ML	A	A	42
43	PROCTOR 57	ML	A	A	43
44	PROCTOR 57	ML	A	A	44
45	PROCTOR 57	ML	A	A	45
46	PROCTOR 57	ML	A	A	46

All of the Zone 90 Acceptance Criteria and nuclear density gauge results are shown in Table 1 in terms of acceptance (A) or rejection (R) according to passing 90% Standard Proctor. Proctors 29, 33, 39, 41, 44 and 57 represent laboratory compaction tests for the six areas tested. Of the 46 trials run, five trials (11%) had Zone 90 Acceptance Criteria results that did not agree with the nuclear density gauge results. These data, though limited, imply that the Zone 90 Acceptance Criteria method has approximately an 89% consistency rate compared with the nuclear density gauge for all of the soil types tested. The following points need to be highlighted regarding the consistency rate:

- The nuclear density gauge was considered the standard (i.e. there was no other independent test to check the gauge),
- Only a limited amount of data were taken, and
- Retesting using the IST or nuclear density gauge at each site was not conducted due to contractor time constraints when test results were inconsistent.

The second purpose of this paper was to analyze the inconsistent data and propose some reasons for the differences. To do this, optimum moisture content, penetration depth, maximum dry density and Impact Values were studied for each trial and soil type. On examining these data only one trend was noted. Of these six areas, laboratory test Proctors 29, 33, 39 and 44 were obtained from areas consisting mostly of cohesive soils (CH, MH, SC). The other areas, Proctor 41 and 57, were mostly noncohesive soils (SC-SM, ML). Of the seventeen trials run on the mostly noncohesive soils, four results disagreed. This means that 76% of the trials run on the mostly noncohesive soils had Zone 90 Acceptance Criteria results that were consistent with the nuclear density gauge data. On the other hand, only one of twenty-nine trials run on the mostly cohesive soils disagreed with the nuclear density gauge making it consistent 97% of the time.

It is also important to comment on the data comparison results of the IST as modified and nuclear density gauge in earlier tests conducted in Chesterfield County, VA (Erchul *et al* 1994). During this testing, the Zone 95 Acceptance Criteria were used in 18 different areas in the county on a variety of soils. These data were compared to nuclear density gauge data from seven of the areas along with data obtained with another field testing procedure: the Standard One Point Proctor Test (ASTM D 698 - Method C). During this testing, the Zone 95 Acceptance Criteria results were consistent with the nuclear gauge 86% of the time, taking into account that the one marginal result returned by the Zone 95 Acceptance Criteria method would be considered as a rejection and be cause for retesting or recompacting and retesting . These results are consistent with the 85% agreement rate of the Zone 90 Acceptance Criteria data taken at the Buena Vista flood wall. The results using the Standard One Point Proctor Test were 71% in agreement with the nuclear density gauge in the Chesterfield County trials. In addition it must be commented that in these comparison tests, the nuclear density gauge was considered as the standard against which the other methods were compared.

Summary

The Impact Soil Tester (IST) in these tests has been shown to be a safe, quick, easy to operate, cost-effective way to test the acceptability of compacted soils. It is fairly light and has few parts, which makes it very mobile and practical for on the job use. The few parts of the IST also allow it to be set up and operated quickly and easily. With these advantages, the IST requires only minimal training and skill to operate. Compared to other soil compaction tests, the IST is cost effective because it involves little maintenance and provides easily obtainable data at a reasonable cost. Handling the instrument requires no special regulatory requirements or operator certificate. The IST gives direct and immediate results with no calculations required to determine the strength parameter. By using the Zone 90 Acceptance Criteria developed in earlier research as the basis for this comparison, this meant that a soil sample did not have to be taken back to the laboratory for further analysis when using the IST as modified as an acceptance tool regarding compacted soil passing or failing 90% Standard Proctor.

The comparison test procedures used for this paper show that the consistency of the Zone 90 Acceptance Criteria varies with the types of soils tested such as cohesive clays and highly plastic silts versus noncohesive soils. The Zone 90 Acceptance Criteria were more effective in testing cohesive soils when compared to a nuclear density gauge at 97% agreement but agreement decreased to 76% in noncohesive soils. Further testing using the Zone 90 Acceptance Criteria method and nuclear density gauge in noncohesive soils is recommended.

Using the modified IST procedure produced fairly reliable and compatible data in terms of Impact Values and penetration depths which were in agreement with the relationship developed in Equation 1 and other comparison tests conducted using the method on various types of soils. As a final comment, the eight readings required in the modified IST testing procedure allow for greater sampling which provides better results on a statistical basis regarding the testing of a lot after compaction.

References

Clegg, B., 1978, "An Impact Soil Tester for Low Cost Roads," *Proceedings*, 2nd Conference Road Engineering Association of Asia & Australia, pp. 58-65, Manila, Philippines.

Clegg, B., 1983, "Application of an Impact Test to Field Evaluation of Marginal Base Course Materials," *Proceedings*, Third International Conference Low Volume Roads, Arizona, Transport Research Record, 898, pp. 174–181, TRB, Washington, D.C., USA.

Erchul, R. A. and Meade, R. B., 1990, "Using the Modified Clegg Impact Hammer to Evaluate Adequacy of Compaction," prepared for the Department of Utilities, Chesterfield County, Virginia, USA.

Erchul, R. A., Decker, R. A., and Ackerman, P., 1994, "Comparison Results of Field Data Using 4.5 kg and 20 kg Clegg Impact Tester (CIT) with Other Standard Compaction Tests on Chesterfield County Soils," prepared for Department of Utilities, Chesterfield County, Virginia, USA.

Winter, M. G. and Selby, A. R., June 1991, "Clegg Meter Performance Assessment With Reference to the Reinstatement Environment," *Highway and Transportation*.

Wayne Saunders,[1] Richard Benson,[2] Frank Snelgrove,[3] and Susan Soloyanis[4]

Selecting Surface Geophysical Methods for Geological, Hydrological, Geotechnical, and Environmental Investigations: The Rationale for the ASTM Provisional Guide

---

REFERENCE: Saunders, W., Benson, R., Snelgrove, F., and Soloyanis, S., "**Selecting Surface Geophysical Methods for Geological, Hydrological, Geotechnical, and Environmental Investigations: The Rationale for the ASTM Provisional Guide,**" *Field Instrumentation for Soil and Rock, ASTM STP 1358*, G. N. Durham and W. A. Marr, Eds., American Society for Testing and Materials, West Conshohocken, PA, 1999.

ABSTRACT: The ASTM Provisional Guide (PS 78-97) for Selecting Surface Geophysical Methods was developed as a guide for project managers, contractors, geologists, and geophysicists to assist in selecting the most likely geophysical method or methods to conduct specific subsurface investigations. Numerous surface geophysical methods and techniques exist that can be used to determine subsurface soil and rock properties and their distribution. These same methods are also widely used to investigate and locate manmade structures such as buried objects and landfills. This paper discusses the general uses of surface geophysics and the use of the provisional guide. This paper is not intended to be used as the guide.

The ASTM Provisional Guide provides direction in selecting the most appropriate geophysical method or methods for a specific application under general site conditions. Secondary methods are also proposed that, under certain circumstances, should be evaluated before a final selection is made. Some typical conditions under which a primary or secondary method might or might not provide satisfactory results are given in the provisional guide. References for further information about selected methods and to method-specific ASTM guides are also provided. Secondary methods usually have less than desired performance, higher cost, or greater labor requirements as compared to the primary methods.

KEYWORDS: surface geophysics, site characterization

RATIONALE FOR THE ASTM GUIDE

The ASTM provisional guide (PS 78-97) for selecting surface geophysical methods was developed as a reference for project managers, contractors, geologists and geophysicists to assist in selecting the most likely geophysical method(s) for geological,

---

[1] Senior Geophysicist, SAIC, 11251 Roger Bacon Drive, Reston, VA 20190.

[2] President, Technos, Inc., 3333 Northwest 21st Street, Miami, FL 33142.

[3] Vice President, Geonics Limited, 1745 Meyerside Drive, Unit 8, Mississauga, Ontario, Canada L5T 1C6.

[4] Principal Scientist, Mitretek Systems, 4610 Fox Road, Cascade, CO 80809.

hydrological, geotechnical, and environmental subsurface investigations. It was developed by a small group of practicing geophysicists and approved by consensus ASTM vote. The purpose of the guide is to direct program managers to the generally acceptable surface geophysical methods for solving specific problems under simple or ideal site conditions. The benefits and limitations of specific surface geophysical methods are discussed in method-specific standards that have either been published or are under development. ASTM D 5777, Guide for Using the Seismic Refraction Method for Subsurface Investigations, is available as a published standard. Guides for ground penetrating radar, gravity, and resistivity methods are currently being balloted and should be available in late 1999. Guides for electromagnetic methods and seismic reflection are under development.

This paper will discuss the general uses of surface geophysics and the use of the provisional guide. Two examples, determining depth to bedrock and mapping a contaminant plume, that illustrate the use of the guide are provided.

SURFACE GEOPHYSICAL METHODS

Surface geophysical methods provide a means of mapping lateral and vertical variations in one or more physical properties of the subsurface. They can also be used to monitor temporal changes in these properties. Each geophysical method measures or responds to a specific physical, electrical, or chemical property of the soil, rock, pore fluids, or subsurface structure. Therefore, for a geophysical method to detect geologic conditions, buried structures, or contaminants there must be a sufficient variation in the property being measured for the condition or structure of interest to be detected. A contrast must be present in order for geophysical measurements to be meaningful. For example, the contact between fresh water and saltwater can be detected by electrical methods, which measure electrical properties. The contact between unconsolidated material (soil) and unweathered rock can be detected by seismic methods, which measure acoustical velocities. On the other hand, if a geophysical method is being using to detect a small change in water quality or to delineate the top of weathered rock whose properties are similar to the soil overlying it, it may not be successful. This is because the contrast in measured properties is likely to be small or gradational and therefore may not be detected by geophysical measurements.

Advantages of Surface Geophysical Methods

Surface geophysical measurements generally can be made quickly, are minimally intrusive, and enable interpolation between known points of control. Since a primary factor affecting the accuracy of geotechnical or environmental site characterization efforts is the number of sample points or borings, geophysical methods are commonly used to map large areas, locate specific targets, or to provide a laterally continuous image of the subsurface.

Types of Surface Geophysical Surveys

Geophysical measurements are commonly made by either profiling or sounding measurements. Profiling, either station by station or continuously, provides a means of assessing lateral changes in subsurface conditions at a given depth. Soundings provide a means of assessing depth and thicknesses of geologic layers or specific targets at a given location. Most surface geophysical sounding measurements can resolve three layers.

Products of Geophysical Surveys

Some surface geophysical methods provide data from which a preliminary interpretation can be made in the field, for example, ground penetrating radar, frequency domain electromagnetic (EM) profiling, DC resistivity profiling, magnetic profiling, and metal detector profiling. A map of radar anomalies or a contour map of the EM, resistivity, magnetics, or metal detector data can often be created in the field and used to locate anomalous conditions. Some methods (for example, time domain electromagnetics and DC resistivity soundings, seismic refraction, seismic reflection, and gravity measurements) require that the data be processed before any quantitative interpretation can be made.

Lateral and vertical boundaries determined by surface geophysical methods usually coincide with geological boundaries, and a cross-section produced from geophysical data may resemble a geological cross-section, although the two are not necessarily identical.

PROCESS FOR SELECTING SURFACE GEOPHYSICAL METHODS

The guide provides a table (Figure 1) for the selection of commonly used surface geophysical methods in four broad application areas:

- Natural geologic and hydrologic conditions - This includes assessing soil and unconsolidated layers, rock layers, depth to bedrock, depth to water table, fractures and fault zones, voids and sinkholes, soil and rock properties, and dam and lagoon leakage.

- Inorganic contaminants - This includes detecting and mapping inorganic plumes from landfills, saltwater intrusion, and soil salinity.

- Organic contaminants - This includes detecting and mapping light nonaqueous phase liquids, dissolved phase, and dense nonaqueous phase liquids.

- Man-made buried objects - This includes locating utilities, drums and USTs, UXO, abandoned wells, landfill and trench boundaries, forensic applications, and archeological features.

The surface geophysical methods considered in the table are those that are commonly used for geotechnical, hydrologic, or environmental applications and include: seismic refraction and reflection, DC resistivity, induced polarization (IP), spontaneous potential (SP), frequency and time domain electromagnetics, very low frequency (VLF) electromagnetics, pipe and cable locators, metal detectors, ground-penetrating radar, magnetics, and gravity. To use the table, identify the specific purpose of the investigation and examine the A- and B-rated methods for that purpose. A rating of "A" implies a primary choice of method. A rating of B implies a secondary or alternate choice of method. The methods must be further evaluated for use under site-specific conditions. The guide provides a brief discussion of each of the methods. References are also provided for further reading.

The process for selecting a surface geophysical method is illustrated in Figure 2 and is conducted as follows:

APPLICATIONS	GEOPHYSICAL METHODS												
	Seismic		Electrical			Electromagnetic							
	Refraction	Reflection	DC Resistivity	IP/ Complex Resistivity	SP	Frequency Domain	Time Domain	VLF	Pipe/Cable Locator	Metal Detectors	Ground Penetrating Radar	Magnetics	Gravity
Natural Geologic and Hydrologic Conditions													
Soil layers	A	B	A			B	A	B			A		
Rock layers	B	A									B		
Depth to bedrock	A	A	B			B	B	B			A		B
Depth to water table	A	A	B			B	B	B			A		
Fractures and fault zones	B	B	B			A	B	A			A	B	B
Voids and sinkholes	B		B			A					A		A
Soil and rock properties	A		A			A							
Dam and lagoon leakage			B		A	B					B		
Inorganic Contaminants													
Landfill leachate			A			A	A	B			B		
Saltwater intrusion			A			A	A	B			B		
Soil salinity			A			A							
Organic Contaminants													
Light, nonaqueous phase liquids			B			B	B				B		
Dissolved phase*													
Dense, nonaqueous phase liquids*													
Manmade Buried Objects													
Utilities						B			A	B	A		
Drums and USTs						A			A	A	A	A	
UXO										A	A	A	
Abandoned wells						B			B	B		A	
Landfill and trench boundaries	B		B			A	B				A		
Forensics			B			A			B	B	A	B	
Archaeological features	B	B	B			A					A	A	B

*See Natural Geologic and Hydrologic Conditions to characterize contaminant pathways.

A = primary choice of method
B = secondary choice of method

Figure 1. Table of surface geophysical methods and applications (from ASTM PS 78-97)

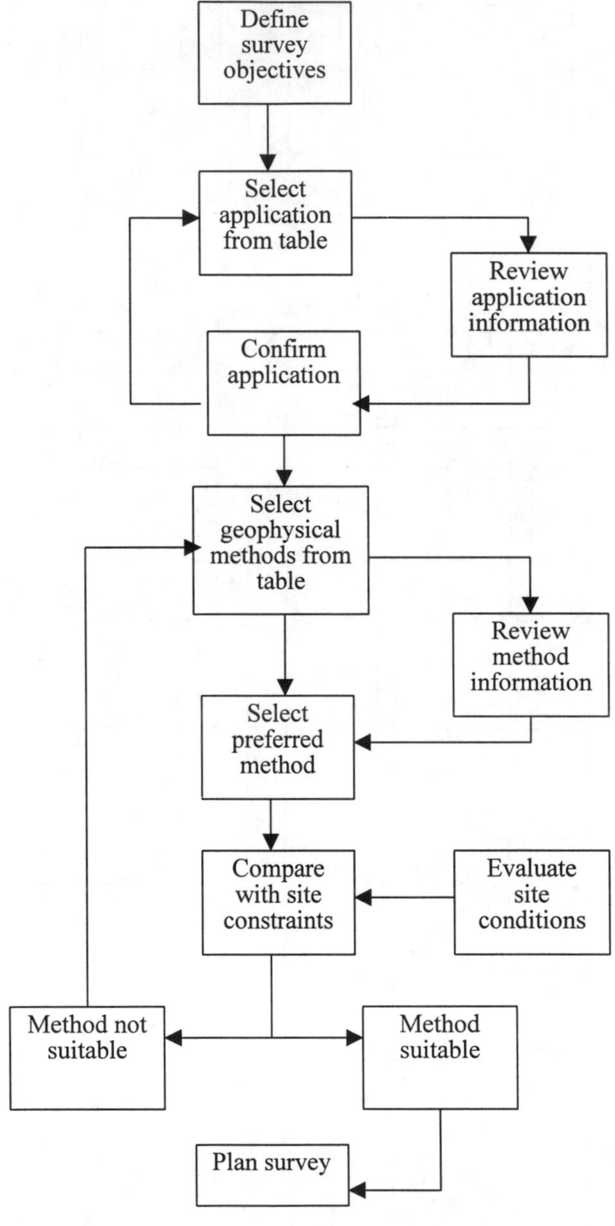

Figure 2. Process for selecting surface geophysical methods using ASTM PS 78-97

Select Application

The first step is to define the objectives of the survey and relate them to an application listed in the "Applications" column in the table. If the survey objective is not specified directly in the "Applications" list, several applications may have to be assessed by reviewing the section of the guide titled, "Discussion of Applications", before arriving at the "Application" most closely resembling the objective. In some instances there may be more than one objective that can be identified separately in the "Applications" list and carried along to the next step.

Select Applicable Geophysical Methods

For each "Application" there are usually several methods that can be used to conduct a geophysical survey, some of which are more generally used or preferred (A methods) than the others (B methods). Some of the unrated methods might under some circumstances be appropriate, although they are generally not used for the application.

In the cases where there is more than one "Application" required to cover the survey objectives, then the first set of methods reviewed would be A methods listed for both applications. For example, if the survey objective was to identify landfill boundaries and determine whether these are delineated by buried metal drums, an A-rated method or A in one category and B in the other might meet the survey requirements.

Confirm Selected Method(s)

At this stage, the selected method(s) are evaluated against the constraints that may be imposed by the survey site, topography, accessibility, cultural interferences, surface soils, and so on. If the method(s) selected fail at this stage, then different methods that will work within these constraints must be evaluated. Once the method(s) selected satisfy all the constraints then you are ready to plan the survey.

Review ASTM Guides and Referenced Literature

A review of the ASTM Guides for the selected method(s) and of the literature referenced in the guide will provide invaluable information for planning the survey.

EXAMPLES

The following examples illustrate the process by which appropriate surface geophysical methods to solve specific problems are chosen and implemented using the guide. Although the example investigations were not originally conceived and conducted using the ASTM guide, the case histories are useful to document the rationale for choosing surface geophysical methods.

Determining Depth to Bedrock

In this example, in which the problem is to determine depth to bedrock where there is a clay unit overlying bedrock, the A-rated method, radar, is eliminated because of the presence of clay; the A-rated method seismic reflection is eliminated because of cost and complexity; and the A-rated method, seismic refraction, is chosen.

Case History

The Farmington River valley in Simsbury, Connecticut, was deeply eroded by glacial ice during the Pleistocene. Stratified drift (delta and lake-bottom deposits), as much as 330 meters thick, accumulated in this overdeepened valley. The area between the towns of Terrys Plain and Avon is almost entirely underlain by fine to very fine sand, silt, and clay. Test holes and wells in the area penetrate thick deposits of very fine sand, silt, and clay. A study was conducted in this area by the U.S. Geological Survey (Melvin and Bingham, 1991) in order to assess the extent, lithology, and hydrogeologic characteristics of the stratified drift. Seismic refraction methods were used to map the depth to bedrock beneath the glacial deposits. The refraction method was chosen based on previous success of this technique in glacial valleys (Haeni, 1986 and 1988), the geology of the site, and amount and type of data needed. The results of one of the refraction profiles is shown in Figure 3. The depths to the water table and bedrock determined from seismic-refraction data agree with the data from well SI-363.

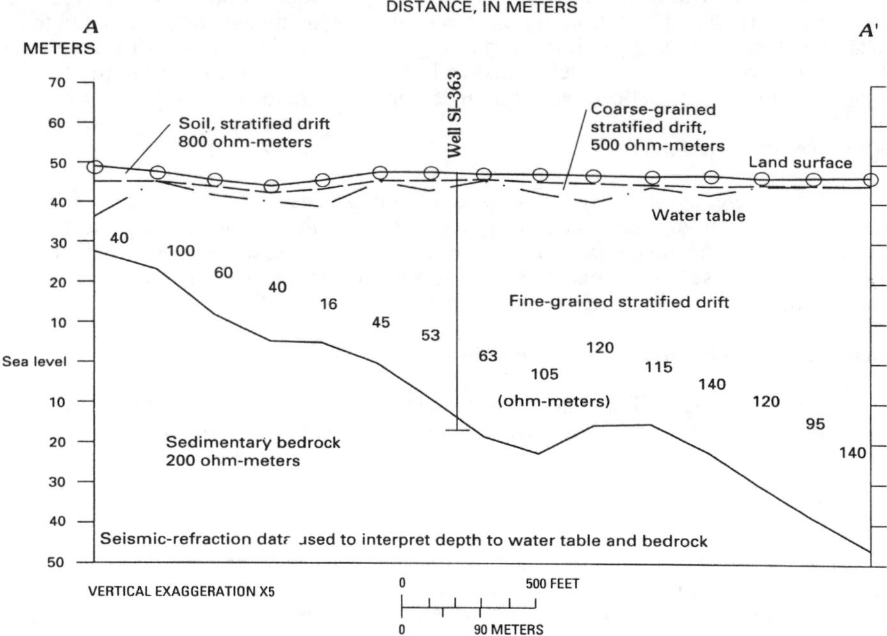

Figure 3. Cross Section A-A', Farmington River Valley (from Haeni, 1995)

Another study conducted by the U.S. Geological Survey, designed to evaluate surface geophysical methods in glaciated valleys (Haeni, 1995), used this site to test and compare several other geophysical methods. A DC resistivity sounding, a VLF terrain-resistivity profile, and an inductive terrain-conductivity profile were conducted along the same line as the previous refraction survey. Sources of cultural interference from railroad tracks and from a construction company workshop with power and telephone lines were present at the west end of the conductivity profile. The remainder of the area, which is used partly for agricultural activities, was a relatively flat flood plain free from cultural interference.

Although the two studies in this area did not use the ASTM guide, they illustrate the usefulness of it. In the guide, under the subheading of depth to bedrock, three geophysical methods are listed as primary methods: seismic refraction, seismic reflection, and ground penetrating radar. Ground penetrating radar would not have worked at this site since the stratified drift is fine-grained (i.e., electrically conductive) and very deep (up to 90 meters). Seismic reflection could have been used, but seismic refraction was a more economical choice because only the depth to rock information was needed and several miles of coverage were needed. Secondary methods in the ASTM Guide include DC resistivity and electromagnetics. A DC resistivity sounding was conducted at this site but was difficult to interpret due to the similar electrical properties of the silt and clay and the sedimentary bedrock. Electromagnetic methods are also limited by the lack of electrical contrast between the stratified drift and the bedrock.

In this example, the seismic refraction method is the best choice to find the depth to bedrock in an area with thick, fine-grained material overlying unweathered sedimentary bedrock.

Mapping a Contaminant Plume

In this example, in which the problem is to map leachate from an oil-field brine pit, the leachate is comparable to the plume created by saltwater intrusion. Two EM methods are used to define the plume.

Case History

An oil-field brine pit in Texas has been operational for forty years. The objective of the survey is to determine the lateral extent and vertical distribution of the brine pond leakage. There are three A-rated methods and two B-rated methods listed in the table in Figure 1. The resolution, ease of use and limitations of the methods described in the section of the guide titled "Discussion of the Geophysical Methods" should be evaluated and these criteria applied to the survey requirements. The requirements could be met, with some limitations, by the resistivity method but with such a large area to cover, the resulting survey could be quite slow and hence costly. The vertical resolution of the resistivity survey might also be somewhat limited.

Both lateral mapping and vertical distribution determination of the saline plume can be accomplished in a short time if the survey is carried out in two phases, using both EM methods. The frequency domain EM method should be used for a rapid reconnaissance survey to detect and map the areal extent of the plume. The time domain EM method should be used to define the vertical distribution of the plume and to derive a full cross-section using a series of soundings across the plume.

A frequency domain EM (ground conductivity) survey was conducted by making several transects around the brine pond. After detecting the plume, a few additional transects were made across the plume (Figure 4) to determine its full extent. Having defined the plume, a transect consisting of five time domain EM soundings was conducted across and extending beyond the plume to ensure full lateral coverage of the plume. A couple of additional soundings, one completely out of the plume for reference and one downgradient to determine its rate of dispersion, were also conducted. If a full 3D model of the plume had been required, several more time domain transects across the plume would have been necessary. Figure 5 shows the time domain sounding results (vertical resistivity profile) for a sounding in the middle of the plume compared to one completely out of the plume. The conductive saline plume is obvious; both upper and lower boundaries are clearly defined. Figure 6 shows the final cross-section of the plume.

In this example, two complementary EM methods are the best choice to delineate the plume.

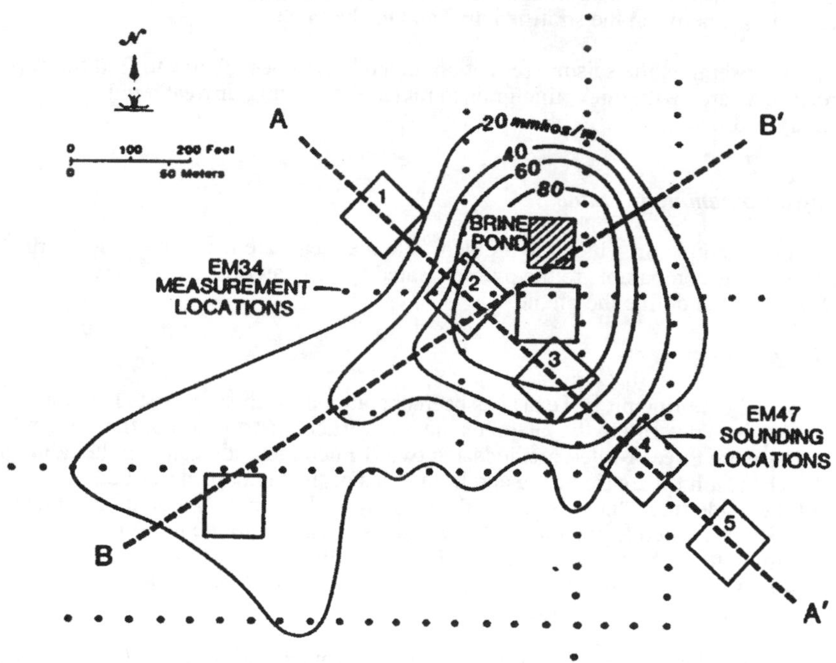

Figure 4. Location and apparent conductivity contour map of brine evaporation pit and plume derived from measurements with Geonics EM-34 at 20 meter coil separation and horizontal, co-planar magnetic dipoles (from Hoekstra, 1992).

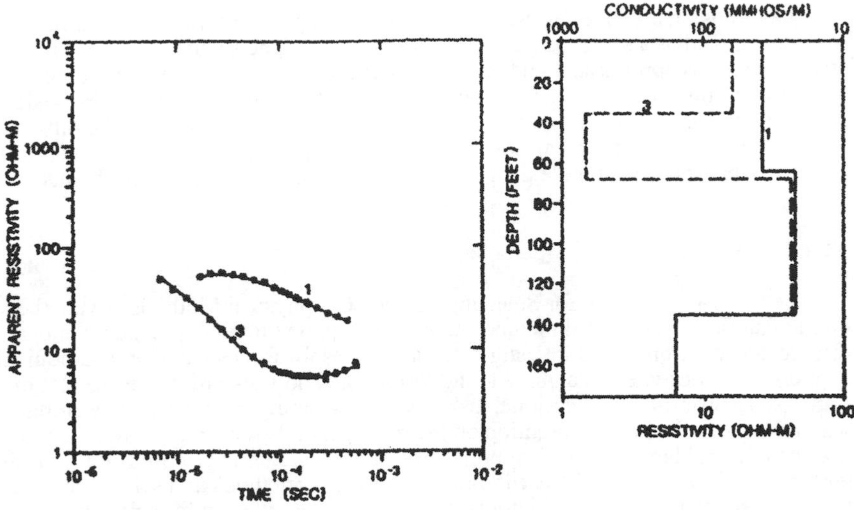

Figure 5. Apparent resistivity curves and their one-dimensional inversions for soundings 1 and 3 in Figure 4 (from Hoekstra, 1992).

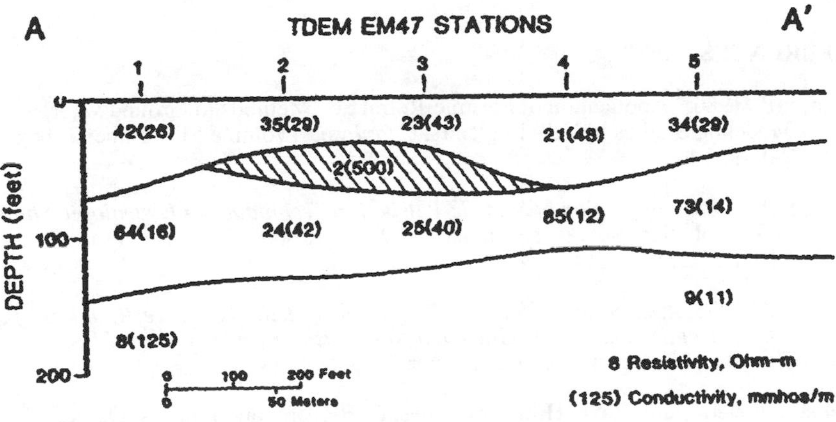

Figure 6. Geoelectric cross section derived from one-dimensional inversions of soundings along A-A' in Figure 4 (from Hoekstra, 1992).

STATUS OF GUIDE DEVELOPMENT

The Provisional Guide for Selecting Surface Geophysical Methods will be revised and balloted as a standard guide within two years. The standard guide is then reviewed, revised as appropriate, and reapproved every five years. Standard guides are being developed for each of the surface geophysical methods included in the Provisional Guide for Selecting Surface Geophysical Methods. The ASTM Guide for Planning and Conducting Borehole Geophysical Logging (D5753-95) provides similar information for the choice of appropriate borehole geophysical methods to solve specific problems.

CONCLUSIONS

The Provisional Guide for Selecting Surface Geophysical Methods covers the selection of surface geophysical methods commonly applied to geologic, geotechnical, hydrologic, and environmental investigations as well as for forensic and archaeological applications. It is designed as a guide to individual method standard guides so that the program manager or other professional tasked with evaluating proposals and writing scopes of work can focus their attention on the most generally acceptable methods to solve a specific problem. If physical properties of geologic units or pore fluids at a site are known, the guide can also be used to rule out primary methods and select secondary methods instead. The guide should not be used as the sole criterion for selecting geophysical methods and does not replace professional judgement or expertise.

ACKNOWLEDGMENTS

The authors thank the other Task Group and Subcommittee D18.01.02 members who write, edit, and revise the draft standard guides and all other ASTM Committee D 18 members who comment on and provide input to the guides at all stages of development.

REFERENCES

Haeni, F.P., 1986, "Application of Seismic-Refraction Methods in Groundwater Modeling Studies in New England", *Geophysics*, volume 51, number 2, page 236-249.

Haeni, F.P., 1988, *Application of Seismic-Refraction Techniques to Hydrologic Studies*, U.S. Geological Survey Techniques of Water-Resources Investigations Book 2, Chapter D2, 86 pages.

Haeni, F.P., 1995, *Application of Surface-Geophysical Methods to Investigations of Sand and Gravel Aquifers in the Glaciated Northeastern United States*, U.S. Geological Survey Professional Paper 1415-A.

Hoekstra, Pieter, Lahti, Raye, Hild, Jim, Bates, C. Richard, and Phillips, David, 1992, "Case Histories of Shallow Time Domain Electromagnetics in Environmental Site Assessment", *Ground Water Monitoring and Review*, volume 12, number 4, page 110-117.

Melvin, R.L., and Bingham, J.W., 1991, *Availability of Water from Stratified-Drift Aquifers in the Farmington River Valley, Simsbury, Connecticut*, U.S. Geological Survey Water-Resources Investigations Report 89-4140.

N. H. Osborne[1]

The Key Role of Monitoring in Controlling the Construction of the New Jubilee Line between Waterloo and Westminster, London

---

REFERENCE: Osborne, N. H., "**The Key Role of Monitoring in Controlling the Construction of the New Jubilee Line between Waterloo and Westminster, London,**" *Field Instrumentation for Soil and Rock, ASTM STP 1358*, G. N. Durham and W. A. Marr, Eds., American Society for Testing and Materials, West Conshohocken, PA, 1999.

Abstract:

Crucial to the success of Contract 102 of the Jubilee Line Extension Project (JLEP) was the detailed analysis of the impact of the construction process on the surrounding ground, structures and tunnels under construction. The construction works were to pass under or alongside some of London's most historic and prestigious buildings, consequently stringent limits were imposed upon ground and structure movements. To control these movements and remain within the limits set, a comprehensive monitoring system was required, comprising a wide variety of instruments linked to a geotechnical database. An extensive and integrated instrumentation system was designed to produce quality data and as no monitoring system existed that covered the unique requirements of the contract and of the Balfour Beatty Amec Joint Venture (BBA), the option was taken by BBA to design and develop a system in house.

The result was Geosys, a windows based monitoring system that could handle data from the various instruments on site, fulfilling a major role in the successful observational approach that was taken to control ground movement from both the tunnelling and the grouting. Geosys enabled electronic data to be reviewed in "real time" and the surveyed precise levelling results could be reviewed within half an hour of survey, if required; feedback loops into the design resulted in a major impact upon the construction decision making processes. The data handling and processing developed within Geosys are described. The subsequent geotechnical benefits to the contract, through the application of high quality processed monitoring data to the control of the ground movements and the construction works, are discussed.

Keywords:

Jubilee, monitoring system, instrumentation, tunnelling, ground movement, data, grouting, observation

---

[1]Geotechnical Engineer, Land Transport Authority of Singapore, 207 River Valley Road, # 02-61 U.E. Square, 238275, Singapore.

Introduction

This paper is concerned with construction of Contract 102 of the JLEP and focuses on the monitoring of ground movements as a consequence of the construction processes undertaken. These include over 2.5 km of twin running tunnel from Green Park to Waterloo, connecting adits, concourses and platform tunnels; a 35 metre deep diaphragm wall box for a new station at Westminster, a new ticket hall at Waterloo and associated compensation and permeation grouting.

All this construction had a major impact upon the structures along the route and tight control upon ground and building movements needed to be maintained to ensure that damage did not occur to structures such as the British Rail Viaduct out of Charing Cross, the RAC Club and Big Ben. The control of ground movements had the potential to be a very complex task due to the limited clearances between structures, grouting shafts, grouting arrays and tunnels. This control was achieved through numerous different types of instruments, that were targeted to monitor specific construction events and scenarios with information transferred to Geosys, a specifically designed monitoring system with a large database storage capacity, multi-user access and a rapid processing time for data. It had the facility to present the processed data in numerous flexible and simple formats; the database also needed to be flexible with the ability to automatically accept and order vast amounts of data from a wide variety of instruments in a wide variety of different formats.

Prior to the major construction works, a robust, reliable and integrated instrumentation system was designed with specific monitoring objectives and installed. This comprised of a wide range of instruments with in excess of 6000 precise levelling points, used as primary control, backed up by over 2000 electrolevel beams, subsurface inclinometers, extensometers, piezometers, strain gauges and assorted instrumentation. As construction events impacted upon the instrumentation, readings were taken either manually, by a surveyor, or remotely by datalogger and modem, and presented to Geosys for processing and storage. A great number of different construction events were occurring at different locations throughout the site. At one stage there were sixteen active tunnel faces, ten separate locations of compensation grouting and the ongoing construction of the two stations. The monitoring staff required to cope numbered 70, including surveyors, monitoring engineers and technicians. Consequently the amount of monitoring data to be processed was enormous, with a maximum of 7500 observations taken during a 24 hour period and 102,950 electronic readings also collated during a 24 hour period. The efficient and rapid capturing, processing and outputting of data by Geosys was of paramount importance to allow fast response to ground movements .

Data Acquisition, Processing, Storage, Archive and Output

A sophisticated communications interface was required to enable survey results from precise levelling, electronic data from instruments on site and records of grout injected to be accepted into Geosys. The survey results could be sent directly from the field via modem and mobile phone or downloaded and presented to Geosys at the office in a Comma Separated Variable (CSV) format, these results could be accepted every three minutes and processed immediately. Data from the electronic instruments was gained through "real time" and hourly communications with 80 dataloggers throughout the site.

The grouting records were downloaded from the grouting sub-contractor's (Amec-Geocisa JV) database and imported into Geosys at whichever frequency was required. Once the data had been acquired by Geosys they were processed, this included the validation of the data, the application of calibration factors and assignment to the correct database in the correct chronological order. The data were stored in both raw downloaded format and processed format, with an error log maintained of any corrupted data and any failures in communications, for future scrutiny if the integrity of the data be questioned or for future back analysis purposes.

Having processed the data, Geosys could produce plots in several flexible formats or output data files for further analysis almost instantaneously. The selection of plots available include displacement against time, displacement against distance over time, with the worst case angle of distortion calculated and illustrated, or a combination of ground movement and actual grout injected over a defined area produced as a contour plot. It is these plots that are reviewed and analysed to control ground movements by making adjustments to the ongoing construction activities.

Control of Heave Induced Through Grouting

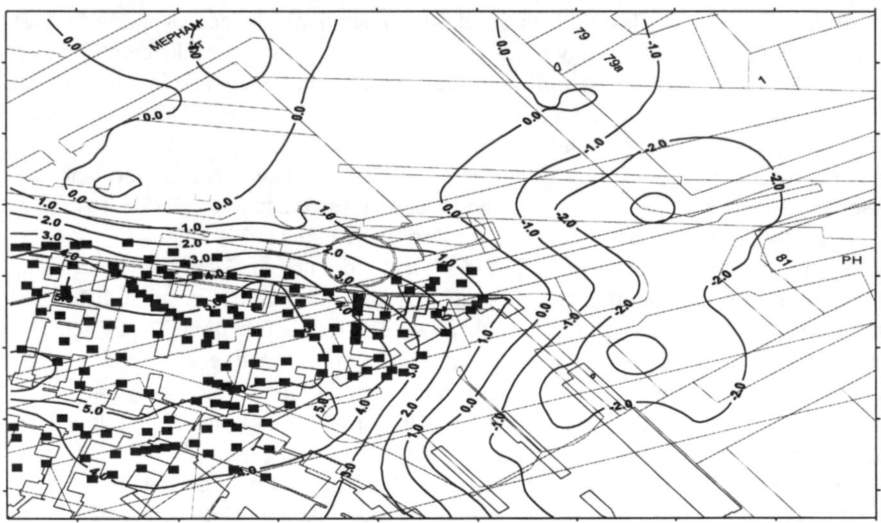

■ **- Grout Locations**

Figure 1 - Ground Heave (mm) Overlain by Grout Injected (l) at Waterloo, 04 – 09/08/1996

Two grouting techniques were utilised that created heave of the ground. Permeation grouting which was carried out where the cover of London Clay to the crown was less than 6m, to reduce permeability and to increase strength of the overlying Thames Gravels. This resulted in heaving of the ground as the grouting began to achieve its objectives and the ground tightened. Observational compensation grouting was also carried out in targeted areas to induce heave compensating for the post tunnelling immediate and consolidation settlements. Both types of grouting needed to be controlled closely by monitoring as the heave could induce stresses in structures within the proximity leading to the risk of damage. Control was achieved by observational methods through regular monitoring and the review of set control levels and targets. Grouting was controlled by adjusting grout volumes, pumping pressures, grouting locations, or changing grout mix formulations.

Survey information was presented as contour plots with the option to choose the base date, either to show movement from contract datum or to reflect a particular sequence of construction activities. To assist with this review the actual injected grout volumes and locations could be overlaid (Figure 1). Note the small localised heave created in the area targeted by the injection of 33 cubic litres of grout.

Concurrent Grouting and Tunnelling

To constrain settlement to a minimum, a concurrent grouting strategy was developed by Amec-Geocisa JV such that compensation grout was injected ahead, behind and to the sides of the advancing tunnel face in an attempt to compensate for as much immediate settlement as was possible, without impacting on the safety of the tunnel below or creating excessive heave ahead of the tunnel face. Again, close control was required of the ground, structures and tunnel under construction with constant review of monitoring observations, grout volumes and locations, to make any adjustments necessary to optimise tunnel speed and for grout operations to achieve their objectives.

Figure 2 illustrates the control achieved whilst progressing under the RAC Club. The RAC Club was particularly sensitive to heave and settlement due to the limited cover between the tunnel crown and the building foundations, a minimum cover of 7m. As tunnelling progressed, intense monitoring with a rapid output of relevant data for analysis was required to enable adjustments to the grouting configuration and volumes to be made, if necessary. Over the 120 hour period as the tunnel progressed below the building, 43 level runs were undertaken on the 50 precise levelling points within the building. All this data was processed and plotted within an hour of the reading being taken, giving great control over the movement of the building, maintaining its level to within 3mm of its undisturbed level. The heave and subsequent settlement can be clearly identified beyond the +/- 1mm tolerance of the surveying which materialises after the grouting.

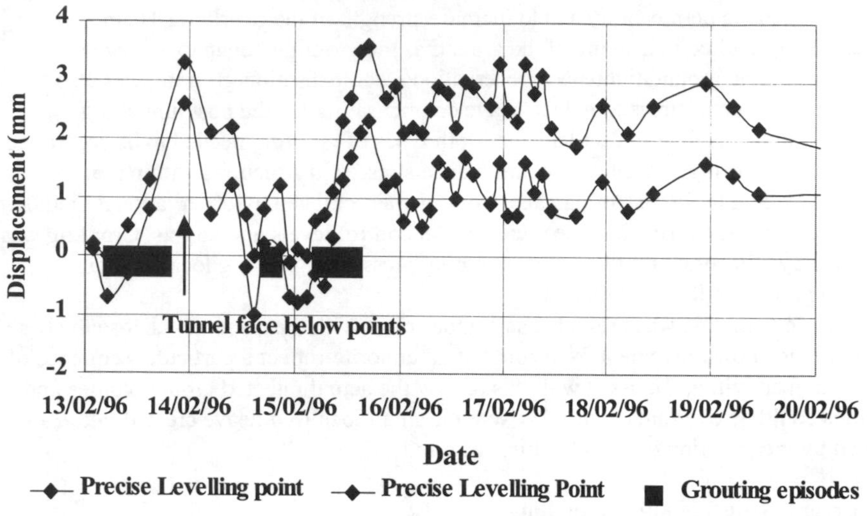

Figure 2 – Control of Ground Movement Through Grouting

Control of Big Ben

A great deal of construction activity had an impact upon Big Ben including two tunnel drives, two platform enlargements, the construction of a jacking chamber, shafts, a pipejack and the construction of Westminster station box. All had the potential to create settlement with various predictions made including K.G.Higgins et al. (1996) who through numerical analysis predicted a maximum settlement of 45mm to the North face of Big Ben and 24.4mm to the South giving an increase in tilt of 1:728. A separate settlement analysis by Mott MacDonald produced settlement of 20.4mm of the North face and 8.5mm at the South face resulting in an increase in tilt of 1:1261. Close control was required, with to date 1380 level runs taken on each of the 35 precise levelling points surrounding the structure. Other instrumentation used to monitor the structure were electrolevels, electrolevel inclinometers for subsurface lateral movements, a Gedometer to monitor the tilt at the top of the tower, an optical plum; and a total station for verticality monitoring.

The control and correlation between different instruments achieved is depicted in the following figures. In Figure 3, two independent methods of monitoring the changing tilt of Big Ben at both the top and bottom of the tower are illustrated. The various excavation events increase the tilt with grouting applied to compensate. By using this monitoring, decisions were made as to the measures for controlling the movement of this historic structure, changing grout volumes or locations, for example.

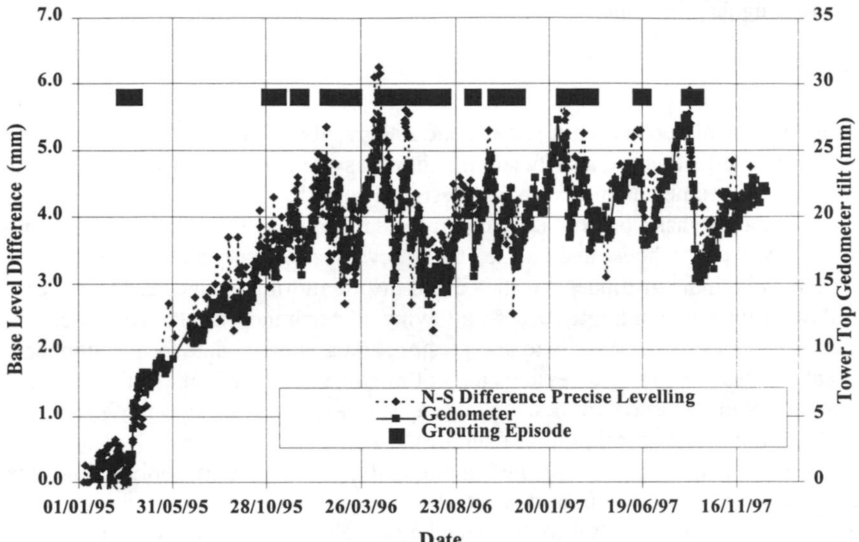

Figure 3 - Big Ben Movement, Gedometer Versus Level Difference

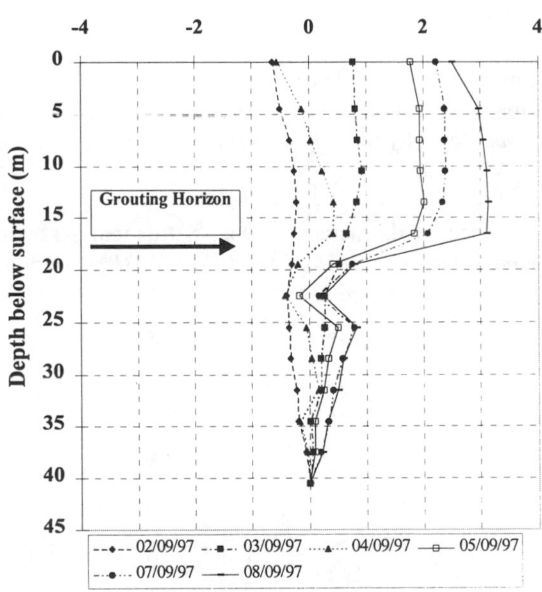

Figure 4 – Subsurface Displacement at Big Ben During Observational grouting

Further effects of the grouting can be seen from the sub-surface electrolevel inclinometers. At the grouting horizon during one particular grouting episode, a lateral movement of 3 mm towards Big Ben can be observed (figure 4) as the grout displaces the ground in both horizontal and vertical directions. Observation of this lateral movement played an important role in refinning the volumes of grout required to adjust the tilt of Big

Ben and ensuring that the impact on adjacent structures was minimal.

Conclusion

The importance of both a robust, reliable and integrated instrumentation system that combines with an effective and an efficient monitoring system was identified early within the project. The instrumentation system was designed such that it targeted construction activities with the emphasis on producing quality and meaningful data. As the monitoring system was developed in house and on site, it was developed to meet the specific requirements of both the instruments installed and the monitoring engineers. The system was windows driven with unlimited access, allowing all personnel simultaneous use. Three key features were: its capacity to accept, store and assimilate data form a number of different sources; the rapid processing and plotting time of vast amounts of information from these varied sources; and the capacity to present simple and flexible outputs in the form of graphical plots or data files.

The Observational Method was applied to much of the construction works and for this to function effectively, tight monitoring control was required to allow engineering decisions to be made. This control was provided by a well planned and integrated instrumentation system, providing quality data, and combining with a rapid processing system providing relevant detailed data for analysis almost instantaneously.

References

Higgins, K.G., et al. April 1996, "Numerical Modelling of the Influence of Westminster Station Excavation and Tunnelling on the Big Ben Clock Tower," *Proc. Sym. Geotechnical Aspects ofUunderground Excavations in Soft Ground,* London 1996, pp. 447-453.

Mott Macdonald, January 1996, "Settlement Assessment of Big Ben Clock Tower," *Commissioned Report No. 24634/F&G/103B*, Croydon, United Kingdom, 1996.

Author Index

R

S

T

V

W

Y

Z

Subject Index

RECEIVED
JAN - 3 2000